# Lecture Notes in Mathematics





For further volumes:
http://www.springer.com/series/304

András Némethi • Ágnes Szilárd

# Milnor Fiber Boundary of a Non-isolated Surface Singularity

Springer

András Némethi
lfréd Rényi Institute of Mathematics
Hungarian Academy of Sciences
Reáltanoda utca 13-15
1053 Budapest
Hungary
nemethi@renyi.hu

Ágnes Szilárd
lfréd Rényi Institute of Mathematics
Hungarian Academy of Sciences
Reáltanoda utca 13-15
1053 Budapest
Hungary
szilard@renyi.hu

ISBN 978-3-642-23646-4 e-ISBN 978-3-642-23647-1
DOI 10.1007/978-3-642-23647-1
Springer Heidelberg Dordrecht London New York

Lecture Notes in Mathematics ISSN print edition: 0075-8434
ISSN electronic edition: 1617-9692

Library of Congress Control Number: 2011940287

Mathematics Subject Classification (2010): 32Sxx, 14J17, 14B05, 14P15, 57M27



Printed on acid-free paper

Springer is part of Springer Science+Business Media (www.springer.com)

*to*

*Miksa, Marcella, Balázs, Jankó*

# Acknowledgements

Both authors, in the period of writing this book, were partially supported by several grants: the EU Marie Curie TOK "BUDALGGEO" Grant; joint Hungarian OTKA grants T049449, K61116; the first author by OTKA grant K67928; and both authors by NKTH-TÉT Austro-Hungarian, Spanish-Hungarian and French-Hungarian bilateral grants.

The authors express their gratitude for the support and assistance of their place of work, the Rényi Institute of Mathematics of the Hungarian Academy of Sciences, Budapest, and for the excellent work atmosphere.

They are also grateful to A. Parusinski, R. Randell, C. Sabbah, J. Seade and D. Siersma for their advices during the preparation of the manuscript, and bringing the authors attention to certain articles and results form the literature.

The authors also wish to thank the devoted work of the referees: their detailed suggestions definitely improved the presentation of the material of the present book.

In some of our computations we used the computer program *Mathematica.*

# Contents

# Chapter 1
# Introduction

## 1.1 Motivations, Goals and Results

**1.1.1.** The origins of the present work go back to some milestones marking the birth of singularity theory of complex dimension $\geq 2$. They include the Thesis of Hirzebruch (1950) containing, among others, the modern theory of cyclic quotient singularities; Milnor's construction of the exotic 7-spheres as plumbed manifolds associated with "plumbing graphs"; Mumford's article about normal surface singularities [79] stressing for the first time the close relationship of the topology with the algebra; the treatment and classification of links of singularities by Hirzebruch and his students in the 1960s (especially Brieskorn and Jänich, and later their students) based on famous results on classification of manifolds by Smale, Thom, Pontrjagin, Adams, Kervaire and Milnor, and the signature theorem of Hirzebruch. Since then, and since the appearance of the very influential book [77] of Milnor in 1968, the theory of normal surface singularities and isolated hypersurface singularities produced an enormous amount of significant results. In all of them, the *link* of an isolated singularity plays a central role.

In the presence of a smoothing, like in the case of hypersurfaces, the link appears also as the boundary of the smoothing, the Milnor fiber. This interplay has enormous consequences.

First, the link of a singularity is the boundary of an arbitrary small neighbourhood of the singular point, hence one can *localize* the link in *any arbitrarily small* representative. Hence, by resolving the singularity, the link appears as the boundary of a small tubular neighbourhood of the exceptional locus, that is, as a plumbed 3-manifold. In this way, e.g. for isolated surface singularities, a bridge is created between the link and resolution: the resolution at topological level is codified in the resolution graph, which also serves as a plumbing graph for the link.

On the other hand, the same manifold is the boundary of the Milnor fiber, that "nearby fiber" whose degeneration and monodromy measures the complexity of the singularity. This interplay between the two holomorphic fillings of the link, the resolution and the Milnor fiber, produces (perhaps) the nicest index-theoretical

A. Némethi and Á. Szilárd, *Milnor Fiber Boundary of a Non-isolated Surface Singularity*,
Lecture Notes in Mathematics 2037, DOI 10.1007/978-3-642-23647-1_1,

relations of hypersurface singularity theory: relations of Durfee [29] and Laufer [58] valid for hypersurfaces, and generalizations of Looijenga, Wahl [68, 134], Seade [113] for more general smoothings. It culminates in deformation theory describing the miniversal deformation spaces in some cases.

The links of normal surface singularities, as special oriented 3-manifolds represented by negative definite plumbing graphs, started to have recently a significant role in low-dimensional topology as well: they not only provide crucial testing manifolds for the Seiberg–Witten, or Heegaard Floer theory of 3-manifolds, and provide ground for surprising connections between these topological invariants and the algebraic/analytic invariants of the singularity germs (see for example [87] and [88] and the references therein), but they appear also as the outputs of natural topological classification results, as solutions of some universal topological properties – for example, the rational singularities appear as $L$-spaces [88], that is, 3-manifolds with vanishing reduced Heegaard Floer homology. Similarly, the classification of symplectic fillings – or more particularly, the classification of rational ball fillings – of some 3-manifolds find their natural foreground in some singularity links, see [90] and [127] and references therein.

**1.1.2.** Although the literature of isolated singularities is huge, surprisingly, the literature of non-isolated singularities, even of the non-isolated hypersurface singularities, is rather restricted. One of the main difficulties is generated by the fact that in this case the link is not smooth.

On the other hand, the boundary of the Milnor fiber is smooth, but (till the present work) there was no construction which would guarantee that it also appears as the boundary of any arbitrary small representative of an isolated singular germ. Lacking such a result, it is hard to prove in a conceptual way that this manifold is a plumbed manifold, even in the case of surfaces.

The present work aims to fill in this gap: we provide a general procedure that may be used to attack and treat non-isolated hypersurface singularities. First, the localization property guarantees the existence of a plumbing representation. But the strategy and the presentation is not limited to the plumbing representation of the boundary (the proof of this fact is just the short Proposition 11.3.3), we target a uniform conceptual treatment of the involved invariants of the singular germs including its connection with the normalization, transversal type singularities and different monodromy operators. Although the whole presentation is for surface singularities, some results can definitely be extended to arbitrary dimensions.

More precisely, some of the conceptual results of the present work are the following. Below, the germ of the holomorphic function $f : (\mathbb{C}^3, 0) \to (\mathbb{C}, 0)$ defines a complex analytic hypersurface singularity with 1-dimensional singular locus $\Sigma$. Its zero set $\{f = 0\}$ is denoted by $V_f$, and its Milnor fiber by $F$. Its oriented boundary $\partial F$ is a connected oriented 3-manifold.

1. The oriented 3-manifold has a plumbing representation. In fact, we not only prove this result, but provide a concrete algorithm for the construction of the plumbing graph: given *any* germ $f$, anyone, with some experience in blowing

ups and handling equations of resolutions, is able to determine the graph after some work. (For the algorithm, see Chap. 10.) The output graphs, in general, are not irreducible, are not negative definite, hence in their discussion the usual calculus – blowing up and down $(-1)$-rational curves – is not sufficient. In our graph-manipulations we will use the *calculus of oriented 3-manifolds* as it is described in Neumann's foundational article [94] (in fact, we will restrict ourself to a "reduced class of graph-operations"). The corresponding background material and preliminary discussions are presented in Chap. 4. For some interesting examples and peculiarities see Sect. 1.2.

In this direction some particular results were known in the literature. The fact that the boundary of the Milnor fiber is plumbed was announced by F. Michel and A. Pichon in [73,74], cf. Remark 10.2.12(c), and simple examples were provided in [19,73,76] obtaining certain lens spaces and Seifert manifolds. Randell [108] and Siersma [118,119] determined the homology of the boundary $\partial F$ for several cases. Moreover, they characterized via different criteria those situations when $\partial F$ is homology sphere; see also Remark 2.3.5.

However, the present work uses a different and novel strategy compared with the existing literature of non-isolated singularities. Moreover, it provides the plumbing graph for arbitrary germs (even for which ad hoc methods are not available), and it points out for the first time in the literature that for these boundary manifolds one needs to use "extended", general plumbing graphs – where the edges might have negative weights as well. Such graphs were not used at all in complex algebraic geometry before. The sign decorations of edges are irrelevant for trees, but are crucial in the presence of cycles in the plumbing graph, as it is shown in many examples in this work.

The reason for the appearance of the negative edges is the following: certain parts of the graph behave as "usual graphs of complex algebraic geometry" but with opposite orientation. This fact is the outcome of the *real analytic* origin of the plumbing representation, see the next Sect. 1.1.3.

2. Recall that for any germ $g : (\mathbb{C}^3, 0) \to (\mathbb{C}, 0)$ such that the pair $(f, g)$ forms an ICIS (isolated complete intersection singularity), by a result of Caubel [19], $g$ determines an open book decomposition on $\partial F$ (similarly as the classical Milnor fibration is cut out by the argument of $g$). For any such $g$, our method determines, as an additional decoration in the plumbing graph, the "multiplicity system" of this open book decomposition too (for definitions, see 4.1.8).
3. The boundary $\partial F$ consists of two parts (a fact already proved by D. Siersma [118, 119] and used by F. Michel, A. Pichon and C. Weber in [73, 74, 76] too): the first, $\partial_1 F$, is the complement of an open tubular neighbourhood of the strict transforms of $\Sigma$ in the link of the normalization of $V_f$. The second one, $\partial_2 F$, can be recovered from the transversal plane curve singularity types of $\Sigma$ together with the corresponding vertical monodromy actions. Already determining these two independent pieces can be a non-trivial task, but the identification of their gluing can be incomparably harder. The present work clarifies this gluing completely (in fact, resolves it so automatically, that if one does not look for the phenomenon deliberately, one will not even see it).

4. The monodromy action on $\partial F$, in general, is not trivial. In fact, its restriction to the piece $\partial_1 F$ is trivial, but the monodromy on $\partial_2 F$ can be rather complicated. In order to understand the monodromy action on $\partial_2 F$, we need the *vertical* monodromies of the transversal singularity types, and, in fact, via the Wang exact sequence, we need their Jordan block structure corresponding to eigenvalue one; on the other hand, the monodromy action on $\partial_2 F$ is induced by the *horizontal* monodromy of the transversal singularity types. In the body of the paper we determine this commuting pair of actions, and as a by-product, the homology of $\partial F$ and the characteristic polynomial of the algebraic monodromy action on $H_1(\partial F)$ too (under certain assumptions).

The discussion includes the study of some monodromy operators of the ICIS $(f, g)$ as well, as detailed below.

**1.1.3.** The "Main Algorithm" is based on a special construction: we take an arbitrary germ $g : (\mathbb{C}^3, 0) \to (\mathbb{C}, 0)$ such that $\Phi = (f, g)$ forms an ICIS. The topology of such a map is described in Looijenga's book [67], and it will be used intensively. We recall the necessary material in Chap. 3.

In general, $\Phi$ provides a powerful tool to analyse the germ $f$ or its $g$-polar properties. Usually, for an arbitrary germ $f$ with 1-dimensional singular locus, one takes a plane curve singularity $P : (\mathbb{C}^2, 0) \to (\mathbb{C}, 0)$ and considers the composed function $P \circ \Phi$. For certain germs $P$, this can be thought of as an approximation of $f$ by isolated singularities. For example, if $P(c, d) = c + d^k$, then $P \circ \Phi = f + g^k$ is one of the most studied test series, the Iomdin series of $f$ associated with $g$ [49, 62].

If we wish to understand the geometry of $P \circ \Phi$, for example its Milnor fiber, then we need to analyse all the intersections of $\{P = \delta\}$ with the discriminant $\Delta$ of $\Phi$, and we have to understand the whole monodromy representation of $\Phi$ over the complement of $\Delta$ – a very difficult task, in general. On the other hand, if we "only" wish to determine some "correction terms" – for example, $i(f) - i(f + g^k)$ for an invariant $i$ –, it is enough to study $\Phi$ only above a neighbourhood of the link of the distinguished discriminant component $\Delta_1 := \Phi(V_f)$.

This fact has been exploited at many different levels, and for several invariants, see e.g. the articles of Lê and Teissier initiating and developing the "theory of discriminants" [60, 62, 131]; the article of Siersma [117] about the zeta function of the Iomdin series, or its generalizations by the first author, cf. [81] and the references therein.

Using this principle, in [92] we determined the *links* of members of the Iomdin series, that is, the links of isolated singularities $f + g^k$ for $k \gg 0$. In that work the key new ingredient was the construction of a special graph $\Gamma_{\mathscr{C}}$, dual to a curve configuration $\mathscr{C}$ in an embedded resolution of $V_{fg} \subset \mathbb{C}^3$ localized above a "wedge neighbourhood" of $\Delta_1$ (for the terminology and more comments see Sect. 7.1).

The point is that the very same graph $\Gamma_{\mathscr{C}}$ not only contains all the information necessary to determine the links of the Iomdin series (and the correction terms $i(f) - i(f + g^k)$ for several invariants $i$), but it is the right object to determine $\partial F$ as well.

The bridge that connects $\partial F$ with the previous discussion (about series and discriminant of the ICIS $(f, g)$) is realized by the following fact. Let $k$ be a sufficiently large even integer. Consider the local *real* analytic germ

$$\{f = |g|^k\} \subset (\mathbb{C}^3, 0).$$

Then, for large $k$, its link (intersection with an arbitrary small sphere) is a smooth oriented 3-manifold, independent of the choice of $k$, which, most importantly, *is diffeomorphic with* $\partial F$, cf. Proposition 11.3.3. In particular, it means that $\partial F$ can be 'localized': it appears as the boundary of an arbitrary small neighbourhood of an analytic germ! But, in this (non-isolated) case, the corresponding space-germ is not complex, but real analytic.

As a consequence, after resolving this real analytic singularity, the tubular neighbourhood of the exceptional set provides a plumbing representation $G$ of $\partial F$. The point is that the graph $\Gamma_{\mathscr{C}}$ codifies all the necessary information to recover the topology of the resolution and of the plumbing: the plumbing graph $G$ of $\partial F$ appears as a "graph covering" of $\Gamma_{\mathscr{C}}$. The Main Algorithm (cf. Chap. 10) provides a pure combinatorial description of $G$ derived from $\Gamma_{\mathscr{C}}$. (The necessary abstract theory of "coverings of graphs", developed in [86], is reviewed in Chap. 5.)

The method emphasizes the importance of real analytic germs, and their necessity even in the study of complex geometry.

In Chaps. 13–18 we determine several related homological invariants from $\Gamma_{\mathscr{C}}$ (characteristic polynomials of horizontal/vertical monodromies, Jordan block structure, cf. below). In fact, we believe that $\Gamma_{\mathscr{C}}$ contains even more information than what was exploited in [92] or here, which can be the subject of future research.

On the other hand, finding the graphs $\Gamma_{\mathscr{C}}$ can sometimes be a serious job. Therefore, we decided to provide examples for $\Gamma_{\mathscr{C}}$ in abundance in order to help the reader understand the present work better, and support possible future research as well.

As different embedded resolutions might produce different graphs $\Gamma_{\mathscr{C}}$, readers with more experience in resolutions might find even simpler graphs in some cases. Each $\Gamma_{\mathscr{C}}$ can equally be used for the theory worked out.

**1.1.4.** The above presentation already suggests that the geometry of $\Phi$ near the discriminant component $\Delta_1$ is reflected in the topology of $\partial F$ as well. Technically, this is described in the commuting actions of the horizontal and vertical monodromies on the fiber of $\Phi$. Similarly, one can consider the horizontal/vertical monodromies of the local transversal types of $\Sigma$, associated with $\Delta_1$. The determination of these two pairs of representations is an important task (independently of the identification of $\partial F$), and it is crucial in many constructions about non-isolated singularities.

We wish to stress that our method provides (besides the result regarding $\partial F$) a uniform discussion of these monodromy representations, and gives a clear procedure to determine the corresponding characters of the $\mathbb{Z}^2$-representations and the characteristic polynomials with precise closed formulae. This can be considered as the *generalization* to isolated complete intersection singularities *of A'Campo's formula* [2–4], valid for hypersurface singularities. Moreover, we determine even the Jordan block structures in those characters which are needed for the homology of $\partial F$ and its algebraic monodromy. This is a generalization of results of Eisenbud–Neumann [33] regarding Jordan blocks of algebraic monodromies associated with 3-dimensional graph manifolds.

**1.1.5.** The material of the present work can be grouped into several parts.

The first part contains results regarding the germ $f$, the ICIS $\Phi$ and the graph $\Gamma_{\mathscr{C}}$ read from a resolution. This is an introductory part, where we list all the background material needed from the literature.

The second part contains the description of $\partial F$, its invariants, and establishes the main connections with other geometrical objects.

Finally, the third part (but basically everywhere in the chapters), we present many examples, among them treatments of specific classes of singularities as homogeneous, suspensions, cylinders, etc.

Any example is given in two steps: first, the graph $\Gamma_{\mathscr{C}}$ has to be determined, a more or less independent task. This can be done in many different ways using one's preferred resolution tricks. Nevertheless, in most of the cases, it is not a trivial procedure, except for special cases such as cylinders or homogeneous germs. Then, in the second step, we run the Main Algorithm to get the plumbing graph of $\partial F$ or its monodromy, or the open book decomposition of $g$ living on $\partial F$.

Our examples test and illustrate the theory, emphasizing the new aspects: the obtained graphs are not the usual negative definite graphs provided by resolutions of normal surface singularities, not even the dual graphs of complex curve configurations of a complex surface (where the intersections of curves are always positive). They contain pieces with "opposite" orientation, hence vertices may have to be connected by negative-edges too. Indeed, in the resolution of the real singularity $\{f = |g|^k\}$, some singularities are *orientation reversing* equivalent with complex Hirzebruch–Jung singularities; hence the corresponding Hirzebruch–Jung strings should be sewed in the final graph with opposite orientation.

In Chaps. 20–23 for certain families (basically, for "composed singularities") we also provide alternative, topological constructions of $\partial F$.

**1.1.6.** The titles of the chapters and sections already listed in the Contents were chosen so that they would guide the reader easily through the sections.

**1.1.7.** Most of the theoretical results of the present work were obtained in 2004–2005; the Main Algorithm was presented at the Singularity Conference at Leuven, 2005. Since that time we added several examples and completed the theoretical part. A completed version [93] was posted on the Algebraic Geometry preprint server on 2 September 2009.

## 1.2 List of Examples with Special Properties

In order to arouse the curiosity of the reader, and to exemplify the variety of 3-manifolds obtained as $\partial F$, we list some peculiar examples. They are extracted from the body of the work where a lot more examples are found.

In the sequel we use the symbol $\approx$ for *orientation preserving diffeomorphism*. If $M$ is a 3-manifold with fixed orientation, then $-M$ is $M$ with opposite orientation.

For certain choices of $f$, the boundary $\partial F$ might have one of the following peculiar properties:

**1.2.1.** $\partial F$ cannot be represented by a negative definite plumbing, but $-\partial F$ admits such a representation, even with all edge decorations positive, see Example 10.4.5.

**1.2.2.** Neither $\partial F$, nor $-\partial F$ can be represented by a negative definite plumbing, see Sect. 19.7(4a).

**1.2.3.** $\partial F$ can be represented by a negative definite graph, but it is impossible to arrange all the edge decorations positive. Hence, such a graph cannot be the graph of a normal surface singularity. See 10.4.4.

**1.2.4.** $\partial F$ can be represented by a negative definite graph with all edge decorations positive. If $\partial F$ is a lens space then this property is automatically true. On the other hand, examples with this property and which are not lens spaces are rather rare. The examples 19.8(e-f) are Seifert manifolds with three special orbits.

**1.2.5.** In all the examples, $\partial F$ is not orientation preserving diffeomorphic with the link of the normalization of $V_f$ (evidently, provided that the singular locus of $V_f$ is 1-dimensional); a fact already noticed in [73].

**1.2.6.** There are examples when $\partial F$ is orientation *reversing* diffeomorphic with the link of the normalization of $V_f$, see 19.8(d).

**1.2.7.** There are examples when $\partial F \approx -\partial F$. In the world of negative definite plumbed manifolds, if both $M$ and $-M$ can be represented by negative definite graphs then the graph is either a string or it is cyclic [94]. Here in 21.1.3 we provide examples for $\partial F \approx -\partial F$ with plumbing graphs containing an arbitrary number of cycles. See also Sect. 19.7(3b).

**1.2.8.** It may happen that $\partial F$ fibers over $S^1$ with rank $H_1(\partial F)$ arbitrary large, cf. 21.1.

**1.2.9.** $\partial F$ might be non-irreducible, cf. 19.8.2(c) and 20.1.7.

**1.2.10.** Any $S^1$-bundles with non-negative Euler number over any oriented surface can be realized as $\partial F$, cf. 23.1 and 19.10.7.

**1.2.11.** The monodromy on $\partial F$, in general, is not trivial, the algebraic monodromy might even have Jordan blocks of size two, cf. 20.2.

**1.2.12.** There exist pairs of singularity germs with diffeomorphic $\partial F$ but different characteristic polynomials of $H_1(\partial F)$, and/or different mixed Hodge weight filtrations on $H_1(\partial F)$, and/or different multiplicities, cf. 19.9.3.

# Part I
# Preliminaries

# Chapter 2
# The Topology of a Hypersurface Germ $f$ in Three Variables

## 2.1 The Link and the Milnor Fiber $F$ of Hypersurface Singularities

Let $f : (\mathbb{C}^n, 0) \to (\mathbb{C}, 0)$ be the germ of a complex analytic function and set $(V_f, 0) = (f^{-1}(0), 0)$. Its singular locus $(Sing(V_f), 0)$ consists of points $\Sigma := \{x : \partial f(x) = 0\}$.

Our primary interest is the *local structure of* $f$, namely a collection of invariants and properties containing information about local ambient topological type of $(V_f, 0)$ – sometimes called the "local Milnor package" of the germ. To start with, we fix some notations: $B_\epsilon$ is the closed ball in $\mathbb{C}^n$ of radius $\epsilon$ and centered at the origin; $S_\epsilon = S_\epsilon^{2n-1}$ is its boundary $\partial B_\epsilon$; $D_r$ denotes a complex disc of radius $r$, while $D_r^2$ is a bidisc. Usually, $S^k$ denotes the $k$-sphere with its natural orientation, and $T^\circ$ the interior of the closed ball or tubular neighbourhood $T$.

The next theorem characterizes the homeomorphism type of the triple $(B_\epsilon, B_\epsilon \cap V_f, 0)$ showing its local *conic structure*. In the general case of semi-analytic sets it was proved by Lojasiewicz [66], the case of germs of complex algebraic/analytic hypersurfaces with isolated singularities was established by Milnor [77], while the generalization to non-isolated hypersurface singularities was done by Burghelea and Verona [18]:

**Theorem 2.1.1.** *[18, 66, 77] There exists $\epsilon_0 > 0$ with the property that for any $0 < \epsilon \le \epsilon_0$ the homeomorphism type of $(B_\epsilon, B_\epsilon \cap V_f)$ is independent of $\epsilon$, and is the same as the homeomorphism type of the real cone on the pair $(S_\epsilon, S_\epsilon \cap V_f)$, where 0 corresponds to the vertex of the cone.*

The intersection $K := V_f \cap S_\epsilon$ is called the *link* of $(V_f, 0)$ $(0 < \epsilon \le \epsilon_0)$. By Milnor [77],

$$K \text{ is } (n-3)\text{-connected.} \tag{2.1.2}$$

Moreover, $K$ is an oriented manifold provided that $f$ has an isolated singularity. By the above theorem, the local structure is completely determined by $K$ and its

A. Némethi and Á. Szilárd, *Milnor Fiber Boundary of a Non-isolated Surface Singularity*, Lecture Notes in Mathematics 2037, DOI 10.1007/978-3-642-23647-1_2,

embedding into the $(2n-1)$-sphere. A partial information about this embedding is provided by the complement $S_\epsilon \setminus K$. The fundamental result of Milnor in [77] states that it is a locally trivial fiber bundle over the circle.

**Theorem 2.1.3.** *[77] There exists $\epsilon_0$ with the property that for any $0 < \epsilon \leq \epsilon_0$ the map $f/|f| : S_\epsilon \setminus K \to S^1 = \{z \in \mathbb{C} : |z| = 1\}$ is a smooth locally trivial fibration. Moreover, for any such $\epsilon$, there exists $\delta_\epsilon$ with the property that for any $0 < \delta \leq \delta_\epsilon$, the restriction $f : B_\epsilon^\circ \cap f^{-1}(\partial D_\delta) \to \partial D_\delta$ is a smooth locally trivial fibration. Its diffeomorphism type is independent of the choices of $\epsilon$ and $\delta$. Furthermore, these two fibrations are diffeomorphic.*

Either of the fibrations above is referred to as the *local Milnor fibration* of the germ $f$ at the origin; its fiber is called the *Milnor fiber*. In this book we will deal mainly with the second fibration. Let $F_{\epsilon,\delta} := B_\epsilon \cap f^{-1}(\delta)$ be the Milnor fiber of $f$ in a small *closed* Milnor ball $B_\epsilon$ (where $0 < \delta \ll \epsilon$). Sometimes, when $\epsilon$ and $\delta$ are irrelevant, it will be simply denoted by $F$. It is a smooth oriented $(2n-2)$-manifold with boundary $\partial F_{\epsilon,\delta}$. If $f$ has an isolated singularity then $\partial F_{\epsilon,\delta}$ and $K$ are diffeomorphic.

The geometric monodromy (well defined up to an isotopy) of the Milnor fibration $\{F_{\epsilon,e^{i\alpha}\delta}\}_{\alpha\in[0,2\pi]}$ is called the *Milnor geometric monodromy* of $F_{\epsilon,\delta}$. It induces a Milnor geometric monodromy action on the boundary $\partial F_{\epsilon,\delta}$. This restriction can be chosen the identity if $\Sigma$ is empty or a point, otherwise this action can be non-trivial.

If $f$ has an isolated singularity then the fibration and its fiber $F$ have some very pleasant properties.

**Theorem 2.1.4.** *If $\Sigma = \{0\}$ then the following facts hold.*

*(a) [77] The homotopy type of $F$ is a bouquet (wedge) of $(n-1)$-spheres; their number, $\mu(f)$, is called the "Milnor number" of $f$.*
*(b) [77] The Milnor fibration on the sphere provides an open book decomposition of $S_\epsilon$ with binding $K$. In particular, the closure of any fiber $(f/|f|)^{-1}(e^{i\alpha})$ $(\alpha \in [0, 2\pi])$ is the link $K$.*
*(c) (Monodromy Theorem, see [15, 22, 59, 67] for different versions, or [41, 56] for a comprehensive discussions) Let $M : H_{n-1}(F, \mathbb{Z}) \to H_{n-1}(F, \mathbb{Z})$ be the algebraic monodormy operator induced by the geometric monodormy, and let $P(t)$ be its characteristic polynomial. Then all the roots of $P$ are roots of unity. Moreover, the size of the Jordan blocks of $M$ for eigenvalue $\lambda \neq 1$ (respectively $\lambda = 1$) is bounded by $n$ (respectively by $n-1$).*

**Example 2.1.5.** If $n = 2$, then $(V_f, 0) \subset (\mathbb{C}^2, 0)$ is called plane curve singularity. Even if $f$ is isolated, it might have several local irreducible components, let their number be $\#(f)$. Then the Milnor number $\mu(f)$ satisfies Milnor's identity [77]:

$$\mu(f) = 2\delta(f) - \#(f) + 1, \tag{2.1.6}$$

where $\delta(f)$ is the *Serre-invariant*, or *delta-invariant*, or the number of double points concentrated at the singularity [116].

$\delta(f) = 0$ if and only if $\mu(f) = 0$, if and only if $f$ is smooth.

The link $K$ is diffeomorphic to $\#(f)$ disjoint copies of $S^1$. The pair $(S_\epsilon, K)$ (that is, the embedding $K \subset S_\epsilon$, where $S_\epsilon = S^3$) is called the *embedded link* of $f$ and creates the connection with the classical knot theory.

For more details about invariants of plane curve singularities, see [16, 136]; about the topology of isolated hypersurface singularities in general, see [5, 77, 114].

For the convenience of the reader we recall the definition of the *open book decomposition* as well.

**Definition 2.1.7.** *An open book decomposition of a smooth manifold $M$ consists of a codimension 2 submanifold $L$, embedded in $M$ with trivial normal bundle, together with a smooth fiber bundle decomposition of its complement $p : M \setminus L \to S^1$. One also requires a trivialization of the a tubular neighborhood of $L$ into the form $L \times D$ such that the restriction of $p$ to $L \times (D \setminus 0)$ is the map $(x, y) \mapsto y/|y|$.*

*The submanifold $L$ is called the binding of the open book, while the fibers of $p$ are the pages.*

**2.1.8.** The above Theorem 2.1.4 about isolated hypersurface singularities became a model for the investigation of non-isolated hypersurface germs as well. For these germs similar statements are still valid in some weakened forms. For example, $F$ is a parallelizable manifold of real dimension $2n - 2$, it has the homotopy type of a finite CW-complex of dimension $n - 1$ [77]. Moreover, by a result of Kato and Matsumoto [53],

$$F \text{ is } (n - 2 - \dim \Sigma)\text{-connected.} \tag{2.1.9}$$

For example, if $n = 3$ and $\dim \Sigma = 1$ then $F$ is a connected finite CW-complex of dimension 2, and $K$ is connected of real dimension 3. This is the best one can say using the general theory. Since all the spaces $K$, $F$, and $\partial F$ might have non-trivial fundamental groups, these spaces are extremely good sources for codifying important information, but their study and complete characterization is much harder than the study of simply connected spaces.

## 2.2 Germs with 1-Dimensional Singular Locus: Transversal Type

In this section we restrict ourself to the case of a complex analytic germ $f : (\mathbb{C}^3, 0) \to (\mathbb{C}, 0)$ whose singular locus $(\Sigma, 0)$ is 1-dimensional. Denote by $\cup_{j=1}^s \Sigma_j$ the decomposition of $\Sigma$ into irreducible components. As we have already mentioned, in this case $K$ is singular too: its singular part is $L = \cup_j L_j$, where $L_j := K \cap \Sigma_j$.

An important ingredient of the topological description of the germ $(V_f, 0)$ and of the local Milnor fibration is the collection of *transversal type singularities*, $T\Sigma_j$, associated with the components $\Sigma_j$ $(j \in \{1,\dots,s\})$ [49, 62]; their definition follows.

$T\Sigma_j$ is the equisingularity type (that is, the embedded topological type) of the local plane curve singularity $f|_{Sl_q} : (Sl_q, q) \to (\mathbb{C}, 0)$, where $q \in \Sigma_j \setminus \{0\}$, and $(Sl_q, q)$ is a transversal smooth complex 2-dimensional slice-germ of $\Sigma_j$ at $q$. The topological type of $f|_{Sl_q}$ is independent of the choice of $q$ and $Sl_q$. Similarly, its Milnor fiber $(f|_{Sl_q})^{-1}(\delta) \subset Sl_q$ is independent of $q$ and $\delta$ (for $\delta$ small), and it will be denoted by $F'_j$. We write $\mu'_j$ for the Milnor number, $\#T\Sigma_j$ for the number of irreducible components, and $\delta'_j$ for the Serre-invariant.

The monodromy diffeomorphism of $F'_j$, induced by the family $[0, 2\pi] \ni \alpha \mapsto (f|_{Sl_q})^{-1}(\delta e^{i\alpha})$, is called the *horizontal monodromy of* $F'_j$, and is denoted by $m'_{j,hor}$. The diffeomorphism induced by the family $s \mapsto (f|_{Sl_{q(s)}})^{-1}(\delta)$, above an oriented simple loop $s \mapsto q(s) \in \Sigma_j \setminus \{0\}$, which generates $\pi_1(\Sigma_j \setminus \{0\}) = \mathbb{Z}$ (and with $\delta$ small and fixed), is called the *vertical monodromy*. It is denoted by $m'_{j,ver}$. Both monodromies are well-defined up to isotopy, and they commute up to an isotopy.

For a possible explanation of the names "horizontal/vertical", see 3.1.10.

The primary goal of the present work is the study of the *boundary* $\partial F$ *of the Milnor fiber* $F$, although sometimes we will provide results regarding the Milnor fiber itself or about the ambient topological type as well (but these will be mostly immediate consequences of results regarding the boundary of $F$).

## 2.3 The Decomposition of the Boundary of the Milnor Fiber

As in 2.2, let $\partial F$ denote the boundary of the Milnor fiber $F$ of $f$, that is $\partial F = \partial F_{\epsilon,\delta} = S_\epsilon \cap f^{-1}(\delta)$. By the above discussion it is a smooth oriented 3-manifold.

Siersma in [118] provides a natural decomposition

$$\partial F = \partial_1 F \cup \partial_2 F,$$

which will be described next.

Let $T(L_j)$ be a small closed tubular neighbourhood of $L_j$ in $S_\epsilon$, and denote by $T^\circ(L_j)$ its interior. Then $\partial_2 F = \cup_j \partial_{2,j} F$ with $\partial_{2,j} F = \partial F \cap T(L_j)$, and $\partial_1 F = \partial F \setminus \cup_j T^\circ(L_j)$. The parts $\partial_1 F$ and $\partial_2 F$ are glued together along their boundaries, which is a union of tori.

**Theorem 2.3.1.** *[118]*

1. *For each $j$, the natural projection $T(L_j) \to L_j$ induces a locally trivial fibration of $\partial_{2,j} F$ over $L_j$ with fiber $F'_j$ (the Milnor fiber of $T\Sigma_j$) and monodromy $m'_{j,ver}$ of $F'_j$. This induces a fibration of $\partial(\partial_{2,j} F)$ over $L_j$ with fiber $\partial F'_j$.*

2. *The Milnor monodromy of $\partial F$ can be chosen in such a way that it preserves both $\partial_1 F$ and $\partial_2 F$. Moreover, its restriction on $\partial_1 F$ is trivial, and also on the gluing tori $\partial(\partial_1 F) = -\cup_j \partial(\partial_{2,j} F)$.*
3. *The Milnor monodromy on $\partial_2 F$ might be nontrivial. This monodromy action on each $\partial_{2,j} F$ is induced by the horizontal monodromy $m'_{j,hor}$ acting on $F'_j$. (Since it commutes with $m'_{j,ver}$, it induces an action on the total space $\partial_{2,j} F$ of the bundle described in part 1.)*

Notice that by Theorem 2.3.1(1), since $F'_j$ is connected, we also obtain that

$$\partial_{2,j} F \text{ is connected.} \tag{2.3.2}$$

**Remark 2.3.3.** One has the following relationship connecting the boundary $\partial F$ and the link $K$. Definitely, $K$ has a similar decomposition $K = K_1 \cup K_2$, where $K_2 = \cup_j K_{2,j}$, $K_{2,j} := T(L_j) \cap K$, and $K_1 := K \setminus \cup_j T^\circ(L_j)$. Then $K_1 \approx \partial_1 F$, hence $\partial K_1 \approx \partial(\partial_1 F)$ as well. On the other hand, each $K_{2,j}$ has the homotopy type of $L_j$. More precisely, the homeomorphism type of $K$ can be obtained from $\partial F$ by the following "surgery": one replaces each $\partial_{2,j} F$ – considered as the total space of a fibration with base space $L_j$ and fiber $F'_j$ –, by a total space of a fibration with base space $L_j$ and whose fiber is the real cone over $\partial F'_j$.

In particular, if each transversal type singularity $T\Sigma_j$ is locally irreducible, then $K$ is an oriented topological 3-manifold (since the real cone over $\partial F'_j$ is a topological disc).

For another construction/characterization of $K$ see 7.5.10.

**Corollary 2.3.4.** *$\partial F$ is connected.*

*Proof.* Use Remark 2.3.3 and the fact that $K$ is connected, cf. (2.1.2). □

**Remark 2.3.5.** Consider a germ $f$ as above with 1-dimensional singular locus and let $M_q : H_q(F,\mathbb{Z}) \to H_q(F,\mathbb{Z})$ be the monodromy operators acting on the homology of the Milnor fiber. Furthermore, let $M'_{j,ver}$ be the algebraic vertical transversal monodromy induced by $m'_{j,ver}$.

Randell and Siersma in the articles [108, 118, 119] determined the homology of the link $K$ and of the boundary $\partial F$ for several cases. Moreover, they characterized via different criteria those situations when the link $K$ and $\partial F$ are homology spheres:

1. [108, (3.6)] $K$ is a homology sphere if and only if $\det(M_q - I) = \pm 1$ for $q = 1, 2$.
2. [118, 119] $\partial F$ is a homology sphere if and only if $\det(M_2 - I) = \pm 1$ and $\det(M'_{j,ver} - I) = \pm 1$ for any $j$.
3. [118] (in [118] attributed to Randell too) $\partial F$ is a homology sphere if and only if $K$ is homology sphere and $\det(M'_{j,ver} - I) = \pm 1$ for any $j$.

In fact, these statements were proved for arbitrary dimensions. For more comments on these properties see 24.4.3.

# Chapter 3
# The Topology of a Pair $(f, g)$

## 3.1 Basics of ICIS: Good Representatives

In many cases it is convenient to add to the germ $f$ another germ, say $g$, such that the pair $(f, g)$ forms an *isolated complete intersection singularity* (ICIS in short). Traditionally, one studies the *g-polar geometry* of $f$ in this way, generalizing the classical polar geometry, when $g$ is a generic linear form. This method, suggested by Thom and developed by Lê Dũng Tráng [60–62] and Teissier [130, 131], computes certain invariants of $f$ by *induction on the dimension*. This lead to the polar invariants of Teissier, the carrousel description of the monodromy by Lê, and later to the study of certain invariants of series and "composed singularities" by Siersma [117] and the first author [81–83], or of Lê cycles and the numbers of Massey [70–72]; see [132] as well. These techniques have generalizations in the theory of one-parameter and equisingular deformations, initiated by Zariski and Teissier, producing great results such as the Lê Ramanujam Theorem [63] and recent work of Fernández de Bobadilla, see [10, 11] and references therein.

However, as an independent strategy, the germ $g$ might also serve as an auxiliary object to determine abstract *g-independent invariants* of $f$. For example, this book contains the description of $\partial F$ in terms of a pair $(f, g)$.

In all the above methods the key ingredient is the fiber structure of the ICIS $(f, g)$.

First, we provide some basic definitions and properties of isolated complete intersection singularities. Although, they are defined generally in the context of germs $(\mathbb{C}^n, 0) \to (\mathbb{C}^k, 0)$, we will keep our specific dimensions $n = 3$ and $k = 2$; in this way we also fix the basic notations we will need.

For more details regarding this section see the book of Looijenga [67].

**3.1.1.** Consider an analytic germ $\Phi = (f, g) : (\mathbb{C}^3, 0) \to (\mathbb{C}^2, 0)$. The map $\Phi$ defines an ICIS if the following property holds: if $I \subset \mathcal{O}_{\mathbb{C}^3,0}$ denotes the ideal generated by $f, g$ and the $2 \times 2$ minors of the Jacobian matrix $(d\Phi)$ in the local algebra $\mathcal{O}_{\mathbb{C}^3,0}$ of convergent analytic germs $(\mathbb{C}^3, 0) \to (\mathbb{C}, 0)$, then $\dim \mathcal{O}_{\mathbb{C}^3,0}/I < \infty$.

A. Némethi and Á. Szilárd, *Milnor Fiber Boundary of a Non-isolated Surface Singularity*, Lecture Notes in Mathematics 2037, DOI 10.1007/978-3-642-23647-1_3,

In other words, the scheme-theoretical intersection $\Phi^{-1}(0) = \{f = 0\} \cap \{g = 0\}$ has only an isolated singularity at the origin. In particular, $(Sing(V_f)) \cap V_g = \{0\}$ and $V_f$ intersects $V_g$ in the complement of the origin transversally in a smooth punctured curve.

**Example 3.1.2.** Let $f : (\mathbb{C}^3, 0) \to (\mathbb{C}, 0)$ be an analytic germ with 1-dimensional critical locus. Then any generic *linear* form $g : (\mathbb{C}^3, 0) \to (\mathbb{C}, 0)$ has the property that the pair $\Phi = (f, g)$ forms an ICIS.

In particular, any such $f$ can be completed to an ICIS $\Phi = (f, g)$.

The *critical locus* $(C_\Phi, 0)$ of $\Phi$ is the set of points of $(\mathbb{C}^3, 0)$, where $\Phi$ is not a local submersion. Its image $(\Delta_\Phi, 0) := \Phi(C_\Phi, 0) \subset (\mathbb{C}^2, 0)$ is called the *discriminant locus* of $\Phi$.

**Lemma 3.1.3.** *[67, 2.B] Fix a germ $\Phi$ as above. Then there exist a sufficiently small closed ball $B_\epsilon \subset \mathbb{C}^3$ of radius $\epsilon$, and a bidisc $D^2_\eta \subset \mathbb{C}^2$ with radius $0 < \eta \ll \epsilon$ such that:*

1. *the set $(\Phi^{-1}(0) \setminus \{0\}) \cap B_\epsilon$ is non-singular;*
2. *$\partial B_{\epsilon'}$ intersects $\Phi^{-1}(0)$ transversally for all $0 < \epsilon' \leq \epsilon$;*
3. *$C_\Phi \cap \Phi^{-1}(D^2_\eta) \cap \partial B_\epsilon = \emptyset$; and the restriction of $\Phi$ to $\Phi^{-1}(D^2_\eta) \cap \partial B_\epsilon$ is a submersion.*

**Definition 3.1.4.** *The map $\Phi : \Phi^{-1}(D^2_\eta) \cap B_\epsilon \to D^2_\eta$ with the above properties is called a "good" representative of the ICIS $\Phi$. In the sequel, in the presence of such a good representative, $\Sigma_\Phi$ will denote the intersection of the critical locus with $\Phi^{-1}(D^2_\eta)$ and $\Delta_\Phi$ its image in $D^2_\eta$.*

Also, we prefer to denote the local coordinates of $(\mathbb{C}^2, 0)$ by $(c, d)$.

With these notations one has the following fibration theorem, cf. 2.8 in [67]:

**Theorem 3.1.5.** *(i) $\Phi : B_\epsilon \cap \Phi^{-1}(D^2_\eta) \longrightarrow D^2_\eta$ is proper. The analytic sets $\Sigma_\Phi$ and $\Delta_\Phi$ are 1-dimensional, and the restriction $\Phi|_{\Sigma_\Phi} : \Sigma_\Phi \to \Delta_\Phi$ is proper with finite fibers.*

*(ii) $\Phi : (\Phi^{-1}(D^2_\eta - \Delta_\Phi) \cap B_\epsilon, \Phi^{-1}(D^2_\eta - \Delta_\Phi) \cap \partial B_\epsilon) \to D^2_\eta - \Delta_\Phi$ is a smooth locally trivial fibration of a pair of spaces.*

**Definition 3.1.6.** *A fiber $F_{c,d} = \Phi^{-1}(c, d) \cap B_\epsilon$, for $(c, d) \in D^2_\eta - \Delta_\Phi$, is called a Milnor fiber of $\Phi$, while the fibration itself is referred to as the Milnor fibration of $\Phi$. The fiber sometimes is denoted by $F_\Phi$ too.*

*For any fixed base point $b_0 = (c_0, d_0) \subset D^2_\eta - \Delta_\Phi$, one has the natural* **geometric monodromy representation***:*

$$m_{geom,\Phi} : \pi_1(D^2_\eta - \Delta_\Phi, b_0) \longrightarrow Diff^\infty(F_{b_0})/isotopy.$$

*It induces the* **algebraic monodromy representation**

$$M_\Phi : \pi_1(D^2_\eta - \Delta_\Phi, b_0) \to Aut H_*(F_{b_0}, \mathbf{Z}).$$

**Proposition 3.1.7.** *[67] The Milnor fiber $F_{c,d}$ of $\Phi$ is connected.*

**3.1.8.** If $f$ has a 1-dimensional singular locus, then the singular locus $\Sigma = \Sigma_f$ is a subset of $\Sigma_\Phi$ and $\Phi(\Sigma) = \{c = 0\}$ is an irreducible component of the discriminant $\Delta_\Phi$. By convention, we denote this component by $\Delta_1$. Then $\Phi^{-1}(\Delta_1) \cap \Sigma_\Phi$ is exactly $\Sigma$. Recall that the irreducible components of $\Sigma$ are denoted by $\{\Sigma_j\}_{j=1}^s$. Part (i) of Theorem 3.1.5 guarantees that the restriction $\Phi : \Sigma_j \to \Delta_1$ is a branched covering for any $1 \leq j \leq s$. Let $d_j$ denote its degree. Note that this agrees with the degree of the restriction of the map $g$ to $\Sigma_j$.

In general, it is extremely difficult to determine either the geometric monodromy $m_{geom,\Phi}$, or the algebraic monodromy representation $M_\Phi$, hence it is hard to recover information about the global Milnor fibration. This is mainly due to the fact that the fundamental group $\pi_1(D^2_\eta - \Delta_\Phi, b_0) = \pi_1(\partial D^2_\eta \setminus \Delta_\Phi, b_0)$ is non-abelian, in general. Nevertheless, the fundamental group of a small tubular neighbourhood of $\partial D^2_\eta \cap \Delta_1$ in $\partial D^2_\eta$ is abelian, hence the fiber structure above it can be understood more easily. The representation restricted to the fundamental group of this tubular neighbourhood still contains key information about the geometry of the fibration "near $\Delta_1$", hence about the singular locus of $f$.

The next definition targets this restriction of the representation.

For any fixed $c_0$, set $D_{c_0} := \{c = c_0\} \cap D^2_\eta$. Then, if $|c_0| \ll \eta$, the circle $\partial D_{c_0}$ is disjoint from $\Delta_\Phi$. Consider the torus $T_\delta := \cup_{c_0} \partial D_{c_0}$, where the union is over $c_0$ with $|c_0| = \delta > 0$. Hence, for $0 < \delta \ll \eta$, the restriction of $\Phi$ on $\Phi^{-1}(T_\delta)$ is a fiber bundle with fiber $F_\Phi$.

**Definition 3.1.9.** *The monodromy above a circle in $T_\delta$, consisting of points with fixed $d$-coordinates, is called the horizontal monodromy of $\Phi$ near $\Delta_1$, and it is denoted by $m_{\Phi,hor}$. Similarly, the monodromy above a circle in $T_\delta$, consisting of points with fixed $c$-coordinates, (e.g., above $\partial D_\delta$) is the vertical monodromy of $\Phi$ near $\Delta_1$; it is denoted by $m_{\Phi,ver}$.*

They are defined up to an isotopy, and they commute up to an isotopy.

**Remark 3.1.10.** Usually, in our figures, in $D^2_\eta$ we take the $c$-coordinate as the horizontal, while the $d$-coordinate as the vertical axis. Hence, the circle in $T_\delta$ with $d$ constant is a "horizontal" circle, while $c$ a circle with $c$ constant is "vertical".

**Remark 3.1.11.** Set $(V_g, 0) := g^{-1}(0)$ in $(\mathbb{C}^3, 0)$. Clearly, $F_{\epsilon,\delta}$ and $\Phi^{-1}(D_\delta)$ can be identified, where the second space is considered in $B_\epsilon \cap \Phi^{-1}(D^2_\eta)$, and its "corners" are smoothed. Under this identification, $F_{\epsilon,\delta} \cap V_g$ corresponds to $\Phi^{-1}(\delta, 0)$. Hence $\partial F_{\epsilon,\delta}$ and $\partial\Phi^{-1}(D_\delta)$ can also be identified in such a way that $\partial F_{\epsilon,\delta} \cap V_g$ corresponds to $\partial\Phi^{-1}(\delta, 0)$. Notice that $\partial\Phi^{-1}(D_\delta)$ consists of two parts, one of them being $\Phi^{-1}(\partial D_\delta)$, the other the complement of the interior of $\Phi^{-1}(\partial D_\delta)$ defined as

$$\partial'\Phi^{-1}(D_\delta) := \cup_{(\delta,d)\in D_\delta} \partial(\Phi^{-1}(\delta, d)).$$

By triviality over $D_\delta$ of the family $\cup_{(\delta,d)\in D_\delta} \partial(\Phi^{-1}(\delta, d))$, the part $\partial'\Phi^{-1}(D_\delta)$ is diffeomorphic to the product $D_\delta \times \partial\Phi^{-1}(\delta, 0)$.

## 3.2 The Milnor Open Book Decompositions of $\partial F$

Besides the geometry of the ICIS, the germ $g$ provides a different package as well. These are invariants determined by the *generalized Milnor fibration*, or open book decomposition, induced by $\arg(g) = g/|g|$ on $\partial F$.

Before we describe it, we recall an immediate natural generalization of Milnor's result 2.1.4(b) valid for isolated complete intersections, which was established by Hamm. Although, again, the result is valid for any map $(\mathbb{C}^n, 0) \to (\mathbb{C}^k, 0)$; we state it only for $n = 3$ and $k = 2$.

**Theorem 3.2.1.** *[43] Assume that $\Phi = (f, g) : (\mathbb{C}^3, 0) \to (\mathbb{C}^2, 0)$ is an ICIS and $f$ has an isolated singularity at the origin. Let $K_f$ be the link of $f$ in a sufficiently small sphere $S_\epsilon$. Then $g$ defines an open book decomposition in $K_f$ with binding $K_f \cap V_g$ and fibration $g/|g| : K_f \setminus V_g \to S^1$. The pages are diffeomorphic to the fibers of $\Phi$.*

At first glance it is not immediate what the right generalization of Hamm's result would be for the case when $f$ has 1-dimensional singular locus, since the link is singular.

The generalization was established by Caubel. The next results are either proved or follow from the statements proved in [19]:

**Theorem 3.2.2.** *1. The argument of the restriction of $g$ on $\partial F_{\epsilon,\delta} \setminus V_g$ defines an open book decomposition on $\partial F_{\epsilon,\delta}$ with binding $\partial F_{\epsilon,\delta} \cap V_g$, and fibration $g/|g| : \partial F_{\epsilon,\delta} \setminus V_g \to S^1$.*
*2. The fibration $g/|g| : \partial F_{\epsilon,\delta} \setminus V_g \to S^1$ is equivalent to the fibration $\Phi : \Phi^{-1}(\partial D_\delta) \to \partial D_\delta$ with monodromy $m_{\Phi,ver}$ $(0 < \delta \ll \eta)$.*
*3. Moreover, this structure is compatible with the action of the Milnor monodromy on $\partial F_{\epsilon,\delta}$ in the following sense. The restriction of the Milnor monodromy of $\partial F_{\epsilon,\delta}$ on a tubular neighbourhood of $\partial F_{\epsilon,\delta} \cap V_g$ is trivial, and its restriction on $\partial F_{\epsilon,\delta} \setminus V_g$ is equivalent to the horizontal monodromy of $\Phi^{-1}(\partial D_\delta)$ over the oriented circle $\{|c| = \delta\}$ (induced by the local trivial family $\{\Phi^{-1}(\partial D_c)\}_{|c|=\delta}$).*

*Proof.* The first part is stated and proved in Proposition 3.4 of [19]. Although the second part is not stated in [loc.cit.], it follows from the proof of Proposition 3.4. and by similar arguments as the proof of Theorem 5.11 in Milnor's book [77]. The last monodromy statement can be proved the same way. □

## 3.3 The Decomposition of $\partial F$ Revisited

Next, we present how one can recover the decomposition $\partial F_{\epsilon,\delta} = \partial_1 F \cup \partial_2 F$, cf. 2.3, from the structure of $\Phi$ via the identification of $F_{\epsilon,\delta}$ with $\Phi^{-1}(D_\delta)$, cf. Remark 3.1.11.

For any $j \in \{1, \ldots, s\}$, let $T_j^\phi$ be a small closed tubular neighbourhood in $\mathbb{C}^3$ of $\Phi^{-1}(\partial D_0) \cap \Sigma_j$. Then, for $\delta$ sufficiently small, and for any $(\delta, d) \in \partial D_\delta$, the fiber $\Phi^{-1}(\delta, d)$ intersects $\partial T_j^\phi$ transversally. In particular, $\Phi^{-1}(\partial D_\delta) \cap T_j^\phi$ and $\Phi^{-1}(\partial D_\delta) \setminus (\cup_j T_j^\phi)$ are fiber bundles over $\partial D_\delta$.

**Proposition 3.3.1.** *One has the following facts:*

1. *There is an orientation preserving homeomorphism*

$$\partial F_{\epsilon,\delta} \longrightarrow \Phi^{-1}(\partial D_\delta) \cup \partial' \Phi^{-1}(D_\delta)$$

*which sends a tubular neighbourhood $T(V_g)$ of $V_g$ onto $\partial' \Phi^{-1}(D_\delta)$ and $\partial F_{\epsilon,\delta} \setminus T^\circ(V_g)$ onto $\Phi^{-1}(\partial D_\delta)$ (identifying even their fiber structures, cf. 3.2.2). Under this identification, $T(V_g) \subset \partial_1 F_{\epsilon,\delta}$, and the fibration*

$$g/|g| : \partial_1 F_{\epsilon,\delta} \setminus T^\circ(V_g) \to S^1$$

*corresponds to*

$$\Phi : \Phi^{-1}(\partial D_\delta) \setminus (\cup_j T_j^{\phi,\circ}) \to \partial D_\delta,$$

*while*

$$g/|g| : \partial_{2,j} F_{\epsilon,\delta} \to S^1 \quad \text{to} \quad \Phi : \Phi^{-1}(\partial D_\delta) \cap T_j^\phi \to \partial D_\delta.$$

*The identifications are compatible with the action of the Milnor/horizontal monodromies (over the circle $|c| = \delta$).*

2. *For each $j \in \{1, \ldots, s\}$, the fibration $g/|g| : \partial_{2,j} F_{\epsilon,\delta} \to S^1$ can be identified with the pullback of the fibration $\partial_{2,j} F_{\epsilon,\delta} \to L_j$ (cf. 2.3.1) under the map $\arg(g|L_j) : L_j \to S^1$, which is a regular cyclic covering of $S^1$ of degree $d_j$. Therefore, the fiber of $g/|g| : \partial_{2,j} F_{\epsilon,\delta} \to S^1$ is a disjoint union of $d_j$ copies of $F'_j$ and the monodromy of this fibration is*

$$m^\Phi_{j,ver}(x_1, \ldots, x_{d_j}) = (m'_{j,ver}(x_{d_j}), x_1, \ldots, x_{d_j-1}).$$

*The action of the Milnor monodromy on $\partial_{2,j} F_{\epsilon,\delta}$ restricted to the fiber of $g/|g|$ is the "diagonal" action:*

$$m^\Phi_{j,hor}(x_1, \ldots, x_{d_j}) = (m'_{j,hor}(x_1), \ldots, m'_{j,hor}(x_{d_j})).$$

*Proof.* The first part follows by combining arguments of [77] and [19], as in the second part of 3.2.2. The point is that when we "push out" $\Phi^{-1}(\partial D_\delta)$ along the level sets of $\arg(g)$ into $\partial F_{\epsilon,\delta}$, this can be done by a vector field which preserves a tubular neighbourhood of each $\Sigma_j$. The second part is standard, it reflects the fiber structure of $\Phi$, see e.g. [81] or [117]. □

Above, by Theorem 3.2.2 and by the structure of open books, the fibrations $\partial F_{\epsilon,\delta} \setminus V_g$ and $\partial F_{\epsilon,\delta} \setminus T^\circ(V_g)$ over $S^1$, both induced by $g/|g|$, are equivalent. A similar fact is true for the fibrations $\partial_1 F_{\epsilon,\delta} \setminus V_g$ and $\partial_1 F_{\epsilon,\delta} \setminus T^\circ(V_g)$.

## 3.4 Relation with the Normalization of the Zero Locus of $f$

The space $\partial_1 F$ and the fibration $g/|g| : \partial_1 F_{\epsilon,\delta} \setminus V_g \to S^1$ have another "incarnation" as well.

In order to see this, let $n : (V_f^{norm}, n^{-1}(0)) \to (V_f, 0)$ be the normalization of $(V_f, 0)$. For the definition and general properties of the normalization of 2-dimensional analytic spaces, see the book of Laufer [57] or the monograph of L. Kaup and B. Kaup [54] . Note that each local irreducible component of $(V_f, 0)$ lifts to a connected component of the normalization, hence $(V_f^{norm}, n^{-1}(0))$ stands here for a multi-germ of normal surface singularities. Moreover, any normal surface singularity has at most an isolated singularity, but usually this germ is not a hypersurface germ, its embedded dimension can be arbitrarily large.

If $(X, 0)$ is an irreducible normal surface singularity, represented in some affine space, say $(X, 0) \subset (\mathbb{C}^N, 0)$, then similarly as for hypersurface singularities one defines its link $K_X$ as $X \cap S_\epsilon \subset S_\epsilon \subset \mathbb{C}^N$, for $\epsilon$ sufficiently small [57,67]. Moreover, $K_X$ is connected (see, for example, [57, 4.1]).

Furthermore, if $g : (X, 0) \to (\mathbb{C}, 0)$ is an analytic non-constant germ on $(X, 0)$, then similarly as in the cases of Milnor 2.1.4(b) and Hamm 3.2.1 one gets an open book decomposition of $K_X$ with binding $K_X \cap V_g$ and projection $g/|g| : K_X \setminus V_g \to S^1$ [21,43,61].

Let us return to our situation. We denote the link of the multi-germ $(V_f^{norm}, n^{-1}(0))$ by $K^{norm}$. It is the disjoint union of all the links of the components of $(V_f^{norm}, n^{-1}(0))$. Consider as well the lifting $g \circ n : (V_f^{norm}, n^{-1}(0)) \to (\mathbb{C}, 0)$ of $g$, which determines an open book decomposition on $K^{norm}$ with binding $K^{norm} \cap V_{g\circ n}$ and Milnor fibration

$$\arg(g \circ n) : K^{norm} \setminus V_{g\circ n} \to S^1.$$

Furthermore, for any $j \in \{1, \dots, s\}$ let us denote by $St(\Sigma_j) \subset V_f^{norm}$ the strict inverse image of $\Sigma_j$, that is the closure of $n^{-1}(\Sigma_j \setminus 0)$. Set $St(\Sigma) := \cup_j St(\Sigma_j)$. Then, $K^{norm} \cap V_{g\circ n} \cap St(\Sigma) = \emptyset$, and

$$\arg(g \circ n) : (K^{norm} \setminus V_{g\circ n}, St(\Sigma)) \to S^1$$

is a locally trivial fibration of a pair of spaces.

Usually $St(\Sigma_j)$ is *not* irreducible. An upper bound for the number of its irreducible components is the number of components of $\partial F'_j$, or equivalently, the number of irreducible branches $\#T\Sigma_j$ of the local transversal type $T\Sigma_j$. Nevertheless, $|St(\Sigma_j)|$ can sometimes be strictly smaller, see the discussion and examples of Sects. 7.5 or 10.3. Compare also with 10.3.6.

**Proposition 3.4.1.** *Let $T_{St(\Sigma)}$ be a small closed tubular neighbourhood of $St(\Sigma)$ in $\partial V_f^{norm}$. Then the following facts hold:*

*(a) $\partial_1 F$ is orientation preserving diffeomorphic to $K^{norm} \setminus T^{\circ}_{St(\Sigma)}$. In particular, the number of connected components of $\partial_1 F$ is the number of irreducible components of $f$.*

*(b) The fibrations of the pairs of spaces*

$$\arg(g \circ n) : (K^{norm} \setminus (V_{g\circ n} \cup T^{\circ}_{St(\Sigma)}), \partial T_{St(\Sigma)}) \to S^1$$

*and*

$$\arg(g) : (\partial_1 F_{\epsilon,\delta} \setminus V_g, \partial(\partial_1 F_{\epsilon,\delta})) \to S^1$$

*are equivalent.*

*In particular, for any $j$, the number of tori along which the connected space $\partial_{2,j} F$ is glued to $\partial_1 F$ agrees with the number of irreducible components of $St(\Sigma_j)$.*

*Proof.* The normalization map is an isomorphism above the regular part of $V_f$. □

The above facts show clearly that $\partial_1 F$ is guided by the link of the normalization, while $\partial_2 F$ by the local behaviour near $\Sigma$.

**3.4.2.** By the results of the above subsections, the fiber $F_{g,\partial F}$ of the fibration $\arg(g) : \partial F \setminus V_g \to S^1$ provided by Theorem 3.2.2 can be compared with the fiber $F_{g,K^{norm}}$ of the fibration $\arg(g \circ n) : K^{norm} \setminus V_{g\circ n} \to S^1$.

Indeed, by 3.3.1 and the above discussion one obtains that the fiber $F_{g,K^{norm}}$ intersects $St(\Sigma)$ in $N := \sum_j \#T\Sigma_j \cdot d_j$ points.

**Corollary 3.4.3.** *1. The fiber $F_{g,\partial F}$ can be obtained as follows: take the fiber $F_{g,K^{norm}}$ and the $N$ intersection points of it with $St(\Sigma)$, delete some small disc neighbourhoods of these points, and then, for each $j \in \{1,\dots,s\}$, glue to the resulting surface with boundary $d_j$ copies of $F'_j$ along their boundaries.*

*2. In particular, at the level of Euler characteristics, one has*

$$\chi(F_{g,\partial F}) = \chi(F_{g,K^{norm}}) + \sum_j d_j(1-\mu'_j - \#T\Sigma_j) = \chi(F_{g,K^{norm}}) - 2\cdot \sum_j d_j\delta'_j.$$

*3. Assume that* $\dim \Sigma_f = 1$. *Then, for any germ $g$ such that $(f,g)$ is an ICIS, the Euler characteristics of the pages of the two open book decompositions induced by the argument of $g$ on $\partial F$ and $K^{norm}$ are not equal. More precisely, one has the strict inequality:*

$$\chi(F_{g,\partial F}) < \chi(F_{g,K^{norm}}).$$

*Proof.* (1) follows from the above discussions, (2) rewrites (1) at the Euler characteristic level and uses the Milnor identity (2.1.6), while (3) follows from the fact that the Serre invariant is strictly positive for a non-smooth plane curve singularity. □

# Chapter 4
# Plumbing Graphs and Oriented Plumbed 3-Manifolds

## 4.1 Oriented Plumbed Manifolds

The first goal of the present work is to provide a plumbing representation of the 3-manifold $\partial F$, where $F$ is the Milnor fiber of a hypersurface singularity $f : (\mathbb{C}^3, 0) \to (\mathbb{C}, 0)$ with 1-dimensional singular locus. The construction will be compatible with the decomposition of $\partial F$ into $\partial_1 F$ and $\partial_2 F$, hence it also provides plumbing representations for these oriented 3-manifolds with boundary.

Even more, for any $g$ such that the pair $(f, g)$ forms an ICIS, as in Sect. 3.1, we will also provide a plumbing representation of the pair $(\partial F, \partial F \cap V_g)$ and of the multiplicity system of the generalized Milnor fibrations $\partial F \setminus V_g$, $\partial_1 F \setminus V_g$ and $\partial_2 F$ over $S^1$ induced by $g/|g|$.

In this Chapter we recall the necessary definitions and relevant constructions. Regarding plumbed 3-manifolds and plumbing calculus we follow Neumann's seminal article [94] with small modifications, which will be explained below.

**4.1.1. The plumbing graph.** For any graph $\Gamma$, we denote the set of vertices by $\mathcal{V}(\Gamma)$ and the set of edges by $\mathcal{E}(\Gamma)$. If there is no danger of confusion, we denote them simply by $\mathcal{V}$ and $\mathcal{E}$.

In the case of plumbing graphs of *closed* 3-manifolds, any vertex has two decorations, both integers: one of them is the Euler obstruction, or *'Euler number'*, while the other one is the *'genus'*, written as $[g_v]$ and omitted if it is zero. Furthermore, the edges also have two possible decorations: $+$ or $-$. In most of the cases we omit the decoration $+$, nevertheless we prefer to emphasize the sign $-$ with the symbol $\ominus$.

Although, for plumbing representations of links of normal surface singularities we need only the sign $+$, and for such graphs the intersection matrix associated with the graph is always negative definite, in the present situation both restrictions should be relieved. Nevertheless, all our 3-manifolds are *oriented*, hence we will restrict ourselves to *'orientable plumbing graphs'*, cf. [94, (3.2)(i)]. These are characterized by $g_v \geq 0$ for any vertex $v$.

A. Némethi and Á. Szilárd, *Milnor Fiber Boundary of a Non-isolated Surface Singularity*, Lecture Notes in Mathematics 2037, DOI 10.1007/978-3-642-23647-1_4,

**4.1.2. The plumbing construction.** Fix a *connected* plumbing graph $\Gamma$.

The oriented plumbed 3-manifold $M(\Gamma)$ associated with $\Gamma$ is constructed using a set of $S^1$-bundles $\{\pi_v : B_v \to S_v\}_{v\in\mathscr{V}}$, whose total space $B_v$ has a fixed orientation. They are indexed by the set of vertices $\mathscr{V}$ of the plumbing graph, so that the base-space $S_v$ of $B_v$ is a closed orientable real surface of genus $g_v$, and the Euler number of the bundle $B_v$ is the Euler number decoration of the vertex $v$ on the graph. Then one glues these bundles corresponding to the edges of $\Gamma$ as follows. First, one chooses an orientation of $S_v$ and of the fibers compatible with the orientation of $B_v$. Then, for each edge adjacent to $v$ one fixes a point $p \in S_v$, an orientation preserving trivialization $D_p \times S^1 \to \pi_v^{-1}(D_p)$ above a small closed disc $D_p \ni p$, and one deletes its interior $D_p^\circ \times S^1$. Here, similarly as above, $S^1$ is the unit circle with its natural orientation. Then, any edge $(v, w)$ of $\Gamma$ determines $\partial D_p \times S^1$ in $B_v$ and $\partial D_q \times S^1$ in $B_w$, both diffeomorphic to $S^1 \times S^1$. They are glued by an identification map $\epsilon\left(\begin{smallmatrix} 0 & 1 \\ 1 & 0 \end{smallmatrix}\right)$, where $\epsilon = \pm$ is the decoration of the edge.

If we allow disconnected plumbing graphs, for example $\Gamma$ is the disjoint union of the plumbing graphs $\Gamma_1$ and $\Gamma_2$, written as $\Gamma = \Gamma_1 + \Gamma_2$, then by convention $M(\Gamma)$ is the *oriented connected sum* $M(\Gamma) = M(\Gamma_1)\#M(\Gamma_2)$. Furthermore, sometimes it is convenient to allow the empty graph too. It corresponds to $M(\emptyset) = S^3$.

For more details, see [94], page 303.

In order to codify certain additional geometric information (e.g., information on links in the 3-manifold, or boundary components), we will be working with plumbing graphs that have **extra decorations**, as made precise in the next subsections.

**4.1.3. Oriented links in oriented closed plumbed 3-manifolds** are represented on the graph by *arrowhead vertices*; the other "usual" vertices will be called *non-arrowheads*. Arrowhead vertices have no "Euler number" or "genus" decorations. Each arrowhead is connected by an edge to some non-arrowhead vertex $v$, and this edge has a sign-decoration $+$ or $\ominus$, similarly to the edges connecting two non-arrowhead vertices (whose significance was explained in the preceding subsection). Any arrowhead supported by a non-arrowhead $v$ codifies a generic $S^1$-fiber of $B_v$, while the sign-decoration of the supporting edge determines an orientation on it. The correspondence is realized as follows. For each non-arrowhead vertex $v$ choose an orientation of $S_v$ and of the fibers as in the previous subsection. This is used in the gluing of the bundles too, but it also identifies the orientation of link-components: if an arrowhead is supported by a $+$-edge then the link component inherits the fiber-orientation, otherwise the orientation is reversed. The disjoint union of these oriented $S^1$-fibers, indexed by the set of all arrowheads, constitutes an oriented link $K$ in the oriented plumbed 3-manifold $M = M(\Gamma)$.

There is an exception to the above description when the graph consists of a double arrow. In this case the 3-manifold is $S^3$, and the arrows represent two Hopf link-components. If the sign-decoration of the double-arrow is $+$ then both Hoph link-components inherit the orientation of the oriented Hopf $S^1$-fibration, otherwise the orientation of one of them is reversed.

Usually we write $\mathscr{A}$ for the set of arrowheads, and $\mathscr{W}$ for the set of non-arrowheads, that is $\mathscr{V} = \mathscr{A} \sqcup \mathscr{W}$.

**4.1.4. Multiplicity systems.** Plumbing graphs with arrowhead vertices, in general, might carry an extra set of decorations: each arrowhead and non-arrowhead vertex has an additional *'multiplicity weight'*, denoted by $(m_v)$.

The typical example of a graph with arrowheads and multiplicity system is provided by an embedded resolution graph of an analytic function defined on a normal surface singularity, where the arrowheads correspond to the strict transforms of the zero set of the analytic function, the non-arrowheads to irreducible exceptional divisors, and the multiplicities are the vanishing orders of the pull-back of the function along the exceptional divisors and strict transforms; see 4.3.

More generally, the set of multiplicities represent a relative 2-cycle in the corresponding oriented plumbed 4-manifold, which in the homology group relative to the boundary represents zero. The *oriented plumbed 4-manifold* $P = P(\Gamma)$ is constructed in a similar way as the plumbed 3-manifold $M = M(\Gamma)$: one replaces the $S^1$-bundles with the corresponding disc-bundles $\mathscr{D}_v$ and one glues them by a similar procedure. Then $P$ is a 4-manifold with boundary such that $\partial P = M$. Each vertex $v$ determines a 2-cycle $C_v$ in $P$: If $v$ is an arrowhead then $C_v$ is an oriented generic disc-fiber of $\mathscr{D}_v$ – hence it is a relative cycle. If $v$ is a non-arrowhead vertex then $C_v$ is the oriented "core" (i.e. the zero section) of $\mathscr{D}_v$, – hence this is an absolute cycle. Their simultaneous orientations can be arranged compatibly with the graph.

In the present work we consider only those multiplicity systems $\{m_v\}_{v\in\mathscr{V}}$, which satisfy a set of compatibility relations. These relations are equivalent to the fact that the class of $C(\underline{m}) := \sum_{v\in\mathscr{V}} m_v C_v$ in $H_2(P, \partial P, \mathbb{Z})$ is zero. This can be rewritten as follows. Let $w$ be a fixed non-arrowhead vertex with Euler number $e_w$. Let $\mathscr{E}_w$ be the set of all adjacent edges, excluding loops supported by $w$. For each $e \in \mathscr{E}_w$, connecting $w$ to the vertex $v(e)$ (where $v(e)$ may be an arrowhead or not), let $\epsilon_e \in \{+,-\}$ be its sign-decoration. Then:

$$e_w m_w + \sum_{e\in\mathscr{E}_w} \epsilon_e m_{v(e)} = 0. \tag{4.1.5}$$

Indeed, this follows from the fact that $H_2(P, \mathbb{Z})$ is freely generated by the absolute classes of $C_w$ ($w \in \mathscr{W}$), hence the intersection numbers $(C(\underline{m}), C_w)$ vanish for all non-arrowhead vertices $w$ if and only if $[C(\underline{m})] = 0$ in $H_2(P, \partial P, \mathbb{Z})$.

**4.1.6. The intersection matrix and the multiplicity system.** The combinatorics and a part of the decorations of the graph $\Gamma$ can be codified into the *intersection matrix* of $\Gamma$. Furthermore, in the presence of arrowheads, the position of the arrowheads can be codified in an *incidence matrix*. Their definition is the following.

**Definition 4.1.7.** *The "intersection matrix" $A$ of $\Gamma$ is the symmetric matrix of size $|\mathscr{W}| \times |\mathscr{W}|$ whose entry $a_{wv}$ is the Euler number of $w$ if $w = v$, while for $w \neq v$ it is $\sum_e \epsilon_e$, where the sum is over all edges $e$ connecting $w$ and $v$, and $\epsilon_e \in \{+,-\}$ is the edge-decoration of $e$.*

*The graph $\Gamma$ is called negative definite if the intersection matrix $A$ is negative definite.*

*The "incidence matrix" $\mathfrak{I}$ of the arrows of $\Gamma$ is a matrix of size $|\mathcal{W}| \times |\mathcal{A}|$. For any $a \in \mathcal{A}$ let $w_a$ be that non-arrowhead vertex which supports the arrowhead $a$. Then, for each $a \in \mathcal{A}$, the entry $(a, w_a)$ is 1, all the other entries are zero.*

*The matrix $(A, \mathfrak{I})$ of size $|\mathcal{W}| \times (|\mathcal{W}| + |\mathcal{A}|)$ consists of the blocks $A$ and $\mathfrak{I}$.*

The matrix of the linear system of equations (4.1.5) in variables $\{m_w\}_{w \in \mathcal{V}}$ is the matrix $(A, \mathfrak{I})$. Hence, if $A$ is non-degenerate then from the position of the arrowheads, or from the incidence matrix, one recovers uniquely all the multiplicities.

Note also that if $P$ is the plumbed 4-manifold associated with the graph $\Gamma$, cf. 4.1.4, then $A$ can also be interpreted as the intersection form on $H_2(P, \mathbb{Z})$ associated with the basis $\{[C_w]\}_{w \in \mathcal{W}}$.

**4.1.8. The multiplicity system associated with an open book decomposition.** Consider a pair $(M, K)$ as in 4.1.3. $K$ is called a *fibered link* if it is the binding of an open book decomposition of $M$.

The case of a fibered link $K$ in a 3-manifold $M$ has a special interest in purely topological discussions too. Links provided by singularity theory are usually fibered. In such cases the pair $(M, K)$ has a plumbed representation provided by a plumbing graph (decorated with Euler numbers and genera) and arrows (representing $K$). Additionally, $p : M \setminus K \to S^1$ is a locally trivial fibration with a trivialization in a neighbourhood of $K$, cf. 2.1.7. In particular, $p$ sends any oriented meridian of any oriented component of $K$ to the positive generator of $H_1(S^1, \mathbb{Z})$.

In such a situation, we define a multiplicity system associated with the open book decomposition as follows.

First, we fix the link-components as distinguished fibers of the corresponding building blocks $B_w$. Then, for each non-arrowhead vertex $w$, let $\gamma_w$ be an oriented generic $S^1$-fiber of $B_w$, different from any fixed fiber corresponding to components of $K$. Here we use the same orientation of the $S^1$-bundle which was used in the plumbing construction. For $a \in \mathcal{A}$ we define $\gamma_a$ as the oriented meridian of the corresponding oriented component of $K$. For any loop $\gamma$ let $[\gamma]$ be its homology class.

**Definition 4.1.9.** *The multiplicity system associated with the fibration $p$ is the collection of integers $m_v := p_*([\gamma_w]) \in H_1(S^1, \mathbb{Z}) = \mathbb{Z}$, $v \in \mathcal{V}$. (Clearly, $m_a = 1$ for $a \in \mathcal{A}$.)*

The fact that this is indeed a multiplicity system can be seen as follows. Let $F$ be the oriented page of $p$ in $M$, with $\partial F = K$. By a homotopy one can push $F \setminus K$ in the interior of $P$ keeping $K = \partial F$ fixed. Then its relative homology class can be represented by the relative cycle $C(\underline{m})$. On the other hand, the corresponding relative homology class is zero, since $F$ sits in $\partial P$.

We wish to emphasize that if $M$ is a rational homology sphere, and $K$ is the binding of an open book decomposition of $M$, then the open book decomposition can be recovered from the pair $(M, K)$ by a theorem of Stallings. This means that there is a *unique* open book decomposition for any fixed binding whenever

$H_1(M, \mathbb{Q}) = 0$. The book of Eisenbud and Neumann in [33, page 34] also provides two different arguments for this fact, one of them based on [9], the other on [135].

On the other hand, in general, the information codified in the plumbing data of the pair $(M, K)$ together with the multiplicity system (that is the graph with arrows decorated with Euler numbers, genera and multiplicities), contains less information than the open book decomposition itself; see [86] for different examples, or 4.4 here.

**4.1.10. Arrowheads with multiplicity zero.** Assume that $K_1 \sqcup K_2$ is an oriented link in $M$, and the pair $(M, K_1 \sqcup K_2)$ has a plumbing representation as in 4.1.3. Here $K_1$ and $K_2$ consist of two disjoint sets of link components such that $(M, K_1)$ is a fibered link. In particular, its open book decomposition $p$ defines a multiplicity system on all the non-arrowheads and on all the arrows of $K_1$ as in 4.1.8. Then, we can put zero multiplicities on all the arrows of $K_2$. (In fact, $p_*([\gamma_a]) = 0$ for all meridians of $K_2$, hence this definition also works.) In this way, using zero-multiplicity arrowheads, we can identify link components of $M$ which are not components in the binding of the open book decomposition $(M, K_1)$.

**4.1.11. Manifolds with boundary.** Similarly, one can codify plumbed oriented 3-manifolds with boundary, where each boundary component is a torus. In general, one starts with an oriented closed 3-manifold $M$ with a link $K$ in it, cf. 4.1.3. Let $L$ be the collection of some of the components of $K$. Then after a small closed tubular neighbourhood $T(L)$ of $L$ is fixed, one deletes its interior $T^\circ(L)$ obtaining a manifold with boundary $M \setminus T^\circ(L)$. The other components of $K$, which are not in $L$, are kept as link components in $M \setminus T^\circ(L)$.

At the level of plumbing graphs, in the present article, this will be codified as follows. Assume that the arrowhead representing a connected component of $L$ is connected by an edge to the non-arrowhead vertex $v$. Then replace this supporting edge by a *dash-edge* (and delete its sign-decoration, or consider it irrelevant). An arrowhead that is supported by a dash-edge will be called *dash-arrow*. Therefore, the dash-arrows represent deleted solid tori containing as their core the components of the corresponding link.

Equivalently, if $r_w$ is the number of dash-arrows supported by the non-arrowhead vertex $w$, then one can also get the plumbed manifold with boundary using the plumbing construction by deleting $r_w$ solid tori, the inverse image of $r_w$ small open discs of the base space of $B_w$, from $B_w$. (This is codified in [94] by the decoration $[r_w]$ of $w$, instead of the $r_w$ dash-arrows of $w$ used here.)

Notice that in this way, (i.e. by replacing an arrow supported by a usual edge by an arrow supported by a dash-edge) one loses some information: for example, the Euler-number of the supporting non-arrowhead $w$ becomes irrelevant.

**4.1.12. Fibrations and multiplicities.** Additionally, if $M \setminus T^\circ(K)$ is a locally trivial fibration $p$ over $S^1$, one can define again a multiplicity system: $m_w := p_*([\gamma_w])$ for each non-arrowhead $w$, as above. Nevertheless, in this case, the arrowheads supported by dash-arrows will carry no multiplicity decorations (or, equivalently, they will be disregarded). This system satisfies the compatibility relations (4.1.5) in the following modified way: if $w$ is a non-arrowhead which supports no dash-arrow, then (4.1.5) is valid for that $w$. But, in general, no other relation holds.

(That is, (4.1.5) is valid for all non-arrowhead vertices $w$ with the convention that the multiplicity of the dash-arrows "can be anything".)

Notice that if $(M, K)$ has an open book decomposition, then the fibration of the complement of $K$ contains less information than the original open book decomposition: in general, a fibration $K \setminus T^{\circ}(K)$ cannot be extended canonically to an open book decomposition. Similarly, the multiplicity system associated with a fibration $p : M \setminus T^{\circ}(K) \to S^1$ contains less information than the multiplicity system associated with the original open book decomposition.

## 4.2 The Plumbing Calculus

**4.2.1.** The **plumbing calculus of oriented plumbed 3-manifolds and the corresponding plumbing graphs** targets the following classification result, cf. [94, (3.2)(i)]. According to this, there are 8 permitted operations of plumbing graphs. In Neumann's notation, seven of them are: R0(a), R1, R3, R5, R6, R7, R2/4. In Neumann's list an eighth operation appears as well, R0(b)'. Since it can be replaced by three consecutive applications of R0(a), we will omit it. (Note also that Neumann's list contains an additional operation R8; this one applies for graphs with dash-arrows, and it will appear below in 4.2.8.)

These operations satisfy the following two key properties:

1. (**Stability of the calculus**) Applying any of the above seven operations, or their inverses, to a plumbing graph $\Gamma$ does not change the oriented diffeomorphism type of $M(\Gamma)$.
2. (**Sufficiency of the calculus**) If $\Gamma_1$ and $\Gamma_2$ are two plumbing graphs and $M(\Gamma_1)$ and $M(\Gamma_2)$ are diffeomorphic by an orientation preserving diffeomorphism, then $\Gamma_1$ and $\Gamma_2$ are related by a sequence of the above operations or their inverses.

The "oriented calculus" is part of a larger set of operations, for which a similar statement is valid as above; it connects non-necessarily orientable plumbed 3-manifolds and their plumbed graphs. The larger class additionally contains those operations which reverse orientation, or which are valid for non-orientable manifolds too. For the complete list, from R0 to R8, see [94]. The oriented calculus selects exactly those operations which preserve the orientation of the orientable 3-manifold. As we are interested only in the oriented special class, we discuss only these ones.

For the completeness of the presentation we provide these operations, at least those which will be used in the present work. The operations below are applied for one of the connected components of the graph $\Gamma$.

**[R0](a)** Reverse the signs on all edges other than loops adjacent to any fixed vertex.

**[R1] (blowing down)** $\epsilon = \pm 1$ and the edge signs $\epsilon_0, \epsilon_1, \epsilon_2$ are related by $\epsilon_0 = -\epsilon\epsilon_1\epsilon_2$.

a.) 

b.) 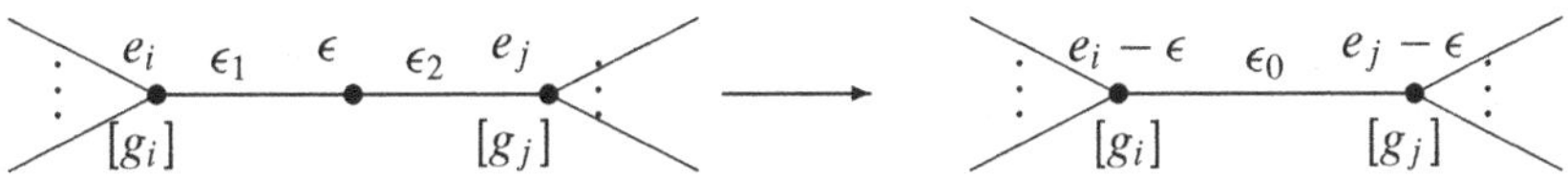


c.) 

**[R3] (0-chain absorption)** The edge signs $\epsilon_i'$ $(i = 1, ..., s)$ are related by $\epsilon_i' = -\epsilon\overline{\epsilon}\epsilon_i$ provided that the edge sign in question is not on a loop, and $\epsilon_i' = \epsilon_i$, if it is on a loop.

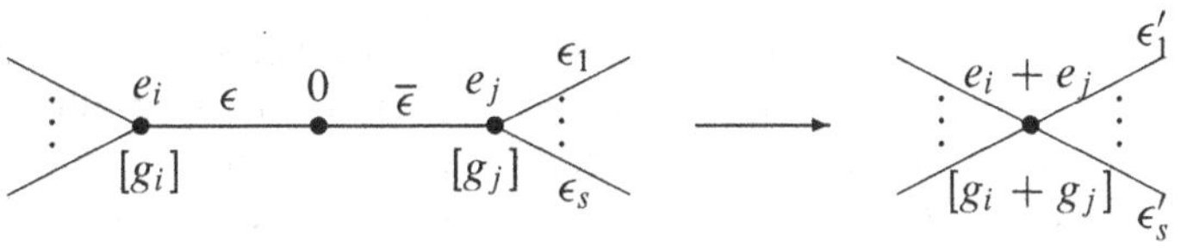


**[R5] (oriented handle absorption)**

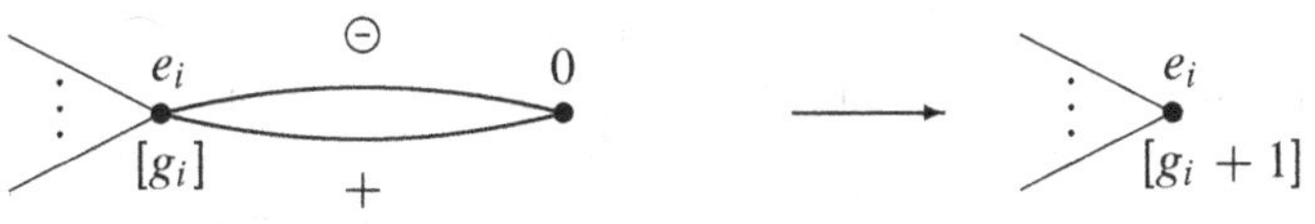


**[R6] (splitting)** If $\Gamma$ has the form

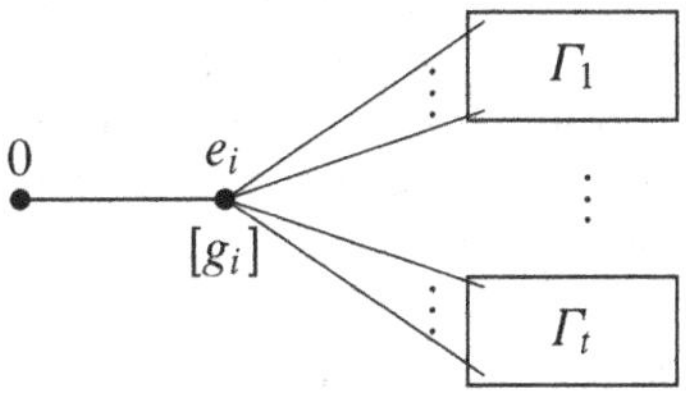

where each $\Gamma_j$ is connected and, for each $j \in \{1, \dots, t\}$, $\Gamma_j$ is connected to the vertex $i$ by $k_j$ edges, then replace $\Gamma$ by the disjoint union of $\Gamma_1, \dots, \Gamma_t$, and $2g_i + \sum_j (k_j - 1)$ copies of $\overset{0}{\bullet}$ .

**[R7] (Seifert graph exchange)** Replace

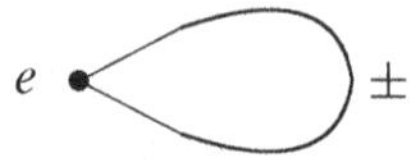


where $e \in \{-1, 0, +1\}$, by a star-shaped graph with all genera zero.

There are six cases, the pair $(e, \pm)$ can be $(-1, +)$, $(0, +)$, $(1, +)$, $(-1, \ominus)$, $(0, \ominus)$ and $(1, \ominus)$. The corresponding star-shaped graphs are (in Kodaira's notation, used in elliptic fibrations, or using the notations of extended $A$-$D$-$E$ graphs): $II$, $III(\tilde{A}_1)$, $IV(\tilde{A}_2)$, $II^*(\tilde{E}_6)$, $III^*(\tilde{E}_7)$, $IV^*(\tilde{E}_8)$.

Since they will not be used in the sequel, we omit the picture of the six graphs.

**[R2/4] (Unoriented handle absorption followed by two $\mathbb{R}P^2$-extrusions)** This operation will not be used in the sequel, hence again we will not give more details about it.

The interested reader can find details on both [R7] and [R2/4] in [94].

**4.2.2.** In the literature there are several special classes of graphs codifying special families of 3-manifolds, for which the graph calculus, that is the set of allowed operations, is more restrictive. Such special classes are for example, "spherical plumbing graphs", "orientable plumbing graphs with no cycles", or "star-shaped plumbing graphs". For more examples and for their calculus, see e.g. Theorem 3.2 in [94].

Besides the study of special families of 3-manifolds, there is another motivation to consider reduced sets of operations. If the class of plumbing graphs considered is the result of a special geometric construction, then they might carry some information in their shape or decorations which might be lost in the diffeomorphism type of $M(\Gamma)$. In such a situation, if we wish to preserve that extra information, then we must use only those operations which preserve it. Of course, in such a case, we cannot always expect the validity of the "sufficiency of the calculus", but we gain a stronger "stability".

For example, in the case of the plumbing calculus of *negative definite resolution graph* of a normal surface singularity, or the *dual graph* of any kind of complex curve configuration on a smooth complex surface, we prefer to use only $(-1)$-blow ups and its inverse instead of all possible operations of the smooth (oriented or non-oriented) calculus. In this way, we can make sure that the graph modified by the operation can again be realized in the corresponding complex analytical context. Moreover, the blow ups preserve the number of independent cycles and the genera of the graph (for their definition see below), which carry some analytic Hodge theoretical information, cf. 18.1.9.

**4.2.3. The "reduced plumbing calculus".** In the present book, guided by results of the present work, we also select a special set of operations from those used in the calculus of oriented plumbed 3-manifolds and their graphs. The collection of these operations is called *reduced set of operations*, and they generate the *reduced plumbing calculus*.

Our graphs are constructed from a singularity theoretical viewpoint. Using only the reduced set of operations allows for preserving features of these graphs inherited from their algebro-geometric/analytic construction, which might be lost if we run all the operations of the smooth calculus.

The principle by which we select the '*reduced set of operations*' is the following. For any decorated plumbing graph $\Gamma$ let $c(\Gamma)$ be the number of independent cycles in $\Gamma$ (i.e. the rank of $H_1(|\Gamma|)$, where $|\Gamma|$ is the topological realization of $\Gamma$). Furthermore, let $g(\Gamma)$ be the sum of the genus decorations of $\Gamma$, i.e $g(\Gamma) = \sum_{w\in\mathcal{W}(\Gamma)} g_w$. The point is that all the graphs provided by our main construction, associated with a fixed geometrical object (singularity) – but depending essentially on the choice of an embedded resolution –, share two properties: all of them are connected, and $c(\Gamma)+g(\Gamma)$ is the same for all of them (describing a geometric entity independent of the construction, cf. 13.6.10). Our *reduced set* contains exactly those operations of the oriented calculus which

$$\begin{array}{c}\textit{preserve connectedness and}\\ \textit{keep the integer } c(\Gamma)+g(\Gamma) \textit{ of the graphs fixed.}\end{array} \tag{4.2.4}$$

*In Neumann's notation this list is the following* : **R0(a)** (reverse the sign-decoration on all edges other than loops adjacent to a vertex), **R1** (blowing down $\pm 1$ vertices), **R3** (0-chain absorption), and **R5** (oriented handle absorption). The inverses of R1, R3 and R5 are called: blowing up, 0-chain extrusion, oriented handle extrusion.

**Remark 4.2.5.** There is one particular case of the splitting operation R6 which still satisfies the requirements (4.2.4). This operation has the following form, where the two left-edges might have any sign-decorations:

**[R6$^{naive}$] ("Naive" splitting)**

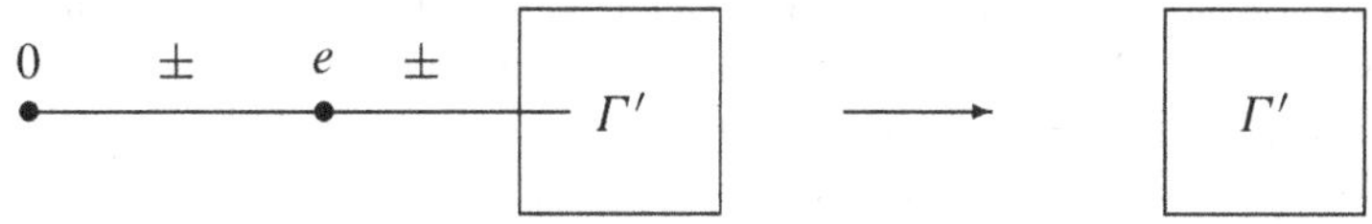


Hence, this operation can also be inserted in the list of reduced calculus; nevertheless, one can prove that it is a consequence of those already listed there. Indeed, by R0(a) we can assume that the signs of the left edges are $+$. By blowing up the left-edge, and blowing down the strict transform of the 0-vertex, we realize that $e$ can be replaced by $e \pm 1$. Hence by repeating this pair of operations, we can reduce $e$ to 0. Then a 0-chain absorption of this newly created 0-vertex finishes the argument.

In this book we will manipulate only the operations listed above. Nevertheless, if the reader wishes to use some other operations of the (oriented) plumbing calculus, this is perfectly fine if she or he wishes to focus only on $\partial F$. In fact, sometimes it is helpful to have in mind the "splitting operation" R6 too, since it helps to represent some of the manifolds as connected sums.

**Definition 4.2.6.** *We write $\Gamma_1 \sim \Gamma_2$ if $\Gamma_1$ can be obtained from $\Gamma_2$ by the above reduced plumbing calculus.*

**4.2.7. The strictly reduced calculus.** We can go further, and consider an even more restricted set of operations. It is based on the conjecture that under our construction of graphs, all the possible graphs associated with the same geometric object, a non-isolated singular germ $f$, share the same integers $c(\Gamma)$ and $g(\Gamma)$ (independently of the different choices in the construction). In fact, we conjecture that these integers are related to the weight filtration of the mixed Hodge structure on $H^1(\partial F, \mathbb{C})$, see Chap. 18.

Since these numbers are modified under the usual calculus, in fact even under the reduced calculus, we get that the weight filtration of the mixed Hodge structure is not a topological/smooth invariant of $\partial F$ (provided that the above mentioned conjecture is true). For concrete examples see 19.9.2 and 19.9.3.

This suggests, that if we would like to preserve this analytic information as well, we have to exclude the oriented handle absorption R5 from the list of operations of the "reduced calculus", and use an even more restrictive set, which is called *strictly reduced oriented calculus*. Hence, it only includes the operations R0(a), R1 and R3 and their inverses.

**4.2.8. Reduced oriented plumbing calculus of graphs with arrows.** If the graph has some arrows and/or dash-arrows, then all the above operations R0(a), R1, R3 and R5 of the reduced calculus have their natural analogs, complemented with some additional rules:

1. the vertex involved in R0(a), the ($\pm 1$)-vertex in R1, and the 0-vertex in R3 and R5 should be a non-arrowhead;
2. the ($\pm 1$)-vertex blown down in R1 can have at most two edges (including also those ones which support arrowheads); if the vertex has exactly one edge supporting an arrowhead, then we do not modify it by blow down;
3. the 0-vertex absorbed in R3 and R5, cannot support any kind of arrow;
4. by the operations, the arrows of the other vertices are naturally kept, and in the case of R3 the arrows supported by vertices $i$ and $j$ are summed;
5. if a vertex supports a dash-arrow then its Euler number is irrelevant.

Additionally, one has the following operation as well:

**[R8] (Annulus absorption for dash-arrows)**

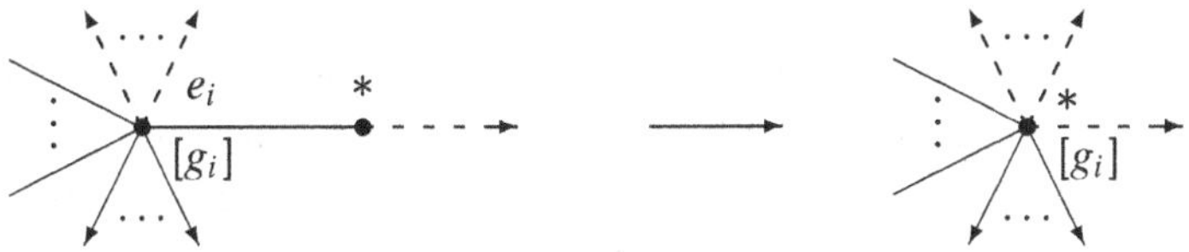


Here, if on the left hand side the vertex supports $s$ arrows and $t$ dash-arrows, then on the right hand side it supports $s$ arrows and $t + 1$ dash-arrows. The Euler number $*$ can be any integer.

A special, "degenerate" version of this is the operation

where both graphs represent a manifold with boundary obtained from $S^3$ by removing the tubular neighbourhoods of two Hopf link-components.

In the presence of a multiplicity system, all the above operations can be extended with taking the corresponding multiplicities into account in a natural and unique way such that the formulae (4.1.5) stay stable under the operations. We emphasize again, that the multiplicity of the dash-arrows is not well-defined, hence if a non-arrowhead vertex supports a dash-arrow, then its Euler number is not well-defined either.

**4.2.9. Changing the orientation.** If $\Gamma$ is an orientable plumbing graph, that is a graph with all $g_v \geq 0$, then let $-\Gamma$ be the same graph with the signs of all Euler and edge decorations reversed. Then $M(-\Gamma) = -M(\Gamma)$, that is, $-\Gamma$ provides the same manifold as $\Gamma$ but with opposite orientation.

## 4.3 Examples: Resolution Graphs of Surface Singularities

Let $(X, x)$ be a normal surface singularity and fix the germ $f : (X, x) \to (\mathbb{C}, 0)$ of an analytic function. In this section we review the definition of the embedded resolution graph $\Gamma(X, f)$ of $f$. More details can be found in the books of Laufer [57] and Eisenbud–Neumann [33], and also in the survey article of Lipman [65]. In Sect. 4.4 we also recall the basic topological properties of the link $K_X$ of $(X, x)$ and of the pair $(X, f^{-1}(0))$ including the representation $\arg_*(f)$ provided by the Milnor fibration associated with $f$.

We use the notation $(V_f, x) = (f^{-1}(0), x)$.

**4.3.1. The embedded resolution.** Let $(X, x)$ be a normal surface singularity and let $f : (X, x) \to (\mathbb{C}, 0)$ be the germ of an analytic function. An embedded resolution $\phi : (\mathscr{Y}, D) \to (U, V_f)$ of $(V_f, x) \subset (X, x)$ is characterized by the following properties. There is a sufficiently small neighborhood $U$ of $x$ in $X$, smooth analytic manifold $\mathcal{Y}$, and analytic proper map $\phi : \mathcal{Y} \to U$ such that:

1) if $E = \phi^{-1}(x)$, then the restriction $\phi|_{\mathcal{Y}\setminus E} : \mathcal{Y} \setminus E \to U \setminus \{x\}$ is biholomorphic, and $\mathcal{Y} \setminus E$ is dense in $\mathcal{Y}$;
2) $D = \phi^{-1}(V_f)$ is a divisor with only normal crossing singularities, i.e. at any point $P$ of $E$, there are local coordinates $(u, v)$ in a small neighbourhood of $P$, such that in these coordinates $f \circ \phi = u^a v^b$ for some non-negative integers $a$ and $b$.

If such an embedded resolution $\phi$ is fixed, then $E = \phi^{-1}(x)$ is called the *exceptional curve* associated with $\phi$. Let $E = \cup_{w\in\mathcal{W}} E_w$ be its decomposition in irreducible components. The closure $S$ of $\phi^{-1}(V_f \setminus \{0\})$ is called the strict transform of $V_f$. Let $\cup_{a\in\mathcal{A}} S_a$ be its decomposition into irreducible components. Obviously, $D = E \cup S$.

For simplicity, we will assume that $\mathcal{W} \neq \emptyset$, any two irreducible components of $E$ have at most one intersection point, and no irreducible exceptional component has a self-intersection. This can always be realized by some additional blow ups.

**4.3.2. The embedded resolution graph $\Gamma(X, f)$.** We construct the *dual embedded resolution graph* $\Gamma(X, f)$ of the pair $(X, f)$, associated with a fixed resolution $\phi$, as follows. Its vertices $\mathcal{V} = \mathcal{W} \sqcup \mathcal{A}$ consist of the nonarrowhead vertices $\mathcal{W}$ corresponding to the irreducible exceptional components, and arrowhead vertices $\mathcal{A}$ corresponding to the irreducible components of the strict transform $S$. If two irreducible divisors corresponding to $v_1, v_2 \in \mathcal{V}$ have an intersection point then we connect $v_1$ and $v_2$ by an edge in $\Gamma(X, f)$.

The graph $\Gamma(X, f)$ is decorated as follows. The edges are decorated by $+$. Any vertex $w \in \mathcal{W}$ is decorated with the self-intersection $e_w := E_w \cdot E_w$, which equals to the Euler number of the normal bundle of $E_w$ in $\mathcal{Y}$, and with the genus $g_w$ of $E_w$. Furthermore, the third decoration is the *multiplicity* (of $f$), defined for any $v \in \mathcal{V}$, which is the vanishing order of $f \circ \phi$ along the irreducible component corresponding to $v$. For example, if $f$ defines an isolated singularity, then for any $a \in \mathcal{A}$ one has $m_a = 1$.

**4.3.3. The resolution graph $\Gamma(X)$ of $(X, x)$.** We say that $\phi : \mathcal{Y} \to U$ is a resolution of $(X, x)$ if $\mathcal{Y}$ is a smooth analytic manifold, $U$ a sufficiently small neighbourhood of $x$ in $X$, $\phi$ is a proper analytic map, such that $\mathcal{Y} \setminus E$ (where $E = \phi^{-1}(x)$) is dense in $\mathcal{Y}$ and the restriction $\phi|_{\mathcal{Y}\setminus E} : \mathcal{Y} \setminus E \to U \setminus \{x\}$ is a biholomorphism.

If $E$ is a normal crossing curve, then the topology of the resolution and the combinatorics of the irreducible exceptional components $\cup_w E_w$ are codified in the *dual resolution graph* $\Gamma(X)$, associated with $\phi$. It is defined similarly as $\Gamma(X, f)$ in 4.3.2, but without arrowheads and multiplicities.

**4.3.4. Some properties of the graphs $\Gamma(X, f)$ and $\Gamma(X)$.**

(1) $\Gamma = \Gamma(X)$ can serve as a plumbing graph: the associated oriented plumbed 3-manifold $M(\Gamma)$ is diffeomorphic to the link $K_X$ of $X$, and the space $\mathcal{Y}$ of

the resolution can be identified with the plumbed 4-manifold $P(\Gamma)$ considered in 4.1.4.

Moreover, the multiplicity system of $\Gamma(X, f)$ satisfies the system of equations (4.1.5).

(2) The graphs $\Gamma(X, f)$ and $\Gamma(X)$ depend on the choice of $\phi$. Nevertheless, different dual graphs associated with different resolutions are connected by a sequence of blow ups and blow downs of $(-1)$-rational curves (operation R1 with $\epsilon = -1$).

By [94], from $K_X$ one can recover $\Gamma(X)$ up to this blow up ambiguity.

(3) The intersection matrix $A$ is negative definite; see [79], [57], or [40]. In particular, $A$ is non-degenerate, hence the multiplicities $\{m_w\}_{w\in\mathcal{W}}$ can be recovered from the Euler numbers and the multiplicities $\{m_a\}_{a\in\mathcal{A}}$, cf. 4.1.6.

(4) $m_v > 0$ for any $v \in \mathcal{V}$, hence the set of multiplicities determine the Euler numbers completely via (4.1.5). This "naive" property has a rather important technical advantage: a multiplicity can always be determined by a local computation, on the other hand the Euler number is a global characteristic class.

This principle will be used frequently in the present book.

(5) The graphs $\Gamma(X, f)$ and $\Gamma(X)$ are connected as follows from Zariski's Main Theorem, see [57] or [45].

**4.3.5. Examples.**

**Plane curve singularities.** If $(X, x)$ is smooth, then $(V_f, 0) \subset (X, x)$ can be resolved using only quadratic modifications. In this case, the graph $\Gamma(X, f)$ is a tree, and $g_w = 0$ for any $w \in \mathcal{W}$. See e.g. [16].

**Cyclic coverings.** Start with a normal surface singularity $(X, x)$ and a germ $f : (X, x) \to (\mathbb{C}, 0)$. Consider the covering $b : (\mathbb{C}, 0) \to (\mathbb{C}, 0)$ given by $z \mapsto z^N$, and construct the fiber product:

$$(X, x) \prod_{f,b} (\mathbb{C}, 0) = \{(x', z) \in (X \times \mathbb{C}, x \times 0) \ : \ f(x') = z^N\}.$$

By definition, $X_{f,N}$ is the normalization of $(X, x) \prod_{f,b} (\mathbb{C}, 0)$. The first projection induces a ramified covering $X_{f,N} \to X$ branched along $V_f$, with covering transformation group $\mathbb{Z}_N$. For more details see 5.3.

**Hirzebruch–Jung singularities [8, 47, 57, 109, 110].** For a normal surface singularity, the following conditions are equivalent. If $(X, x)$ satisfies either one of them, then it is called Hirzebruch–Jung singularity.

(a) The resolution graph $\Gamma(X)$ is a string, and $g_w = 0$ for any $w \in \mathcal{W}$. (In the terminology of low-dimensional topology, this is equivalent to the fact that the link $K_X$ is a *lens space*.)

(b) There is a finite proper map $\pi : (X, x) \to (\mathbb{C}^2, 0)$ such that the reduced discriminant locus of $\pi$, in some local coordinates $(u, v)$ of $(\mathbb{C}^2, 0)$, is $\{uv = 0\}$.

(c) $(X, x)$ is isomorphic with exactly one of the "model spaces" $\{A_{n,q}\}_{n,q}$, where $A_{n,q}$ is the normalization of $(\{xy^{n-q}+z^n = 0\}, 0)$, where $0 < q < n$, $(n, q) = 1$.

Usually, Hirzebruch–Jung singularities appear as in (b). If there is a map $\pi$ as in (b) with smooth reduced discriminant locus, then $(X, x)$ is automatically smooth. The following local situation is a prototype.

For any three *strictly positive* integers $a, b$ and $c$, with $\gcd(a, b, c) = 1$, let $(X, x)$ be the normalization of $(\{x^a y^b + z^c = 0\}, 0) \subset (\mathbb{C}^3, 0)$. Then the projection to the $(x, y)$-plane induces a map which satisfies (b). Let $z : (X, x) \to (\mathbb{C}, 0)$ be induced by $(x, y, z) \mapsto z$. Then the minimal embedded resolution graph of the pair $(X, z)$ is the following. (In the sequel sometimes we write $(a, c)$ for $\gcd(a, c)$.)

First, consider the unique $0 \leq \lambda < c/(a, c)$ and $m_1 \in \mathbb{N}$ with:

$$b + \lambda \cdot \frac{a}{(a,c)} = m_1 \cdot \frac{c}{(a,c)}. \tag{4.3.6}$$

If $\lambda \neq 0$, consider the continued fraction:

$$\frac{c/(a,c)}{\lambda} = k_1 - \cfrac{1}{k_2 - \cfrac{1}{\ddots - \cfrac{1}{k_s}}}, \quad k_1, \ldots, k_s \geq 2. \tag{4.3.7}$$

Then the next string is the embedded resolution graph of $z$:

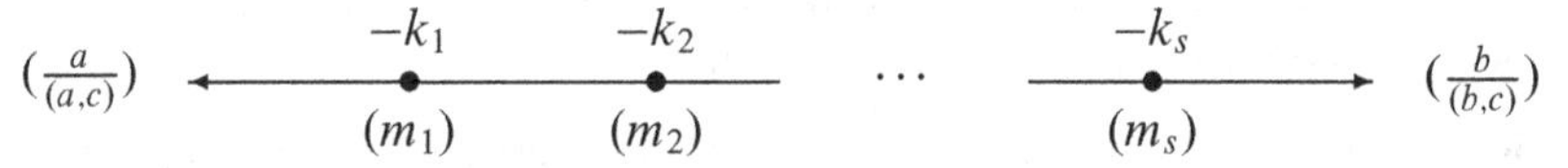


The arrow at the left (resp. right) hand side codifies the strict transform of $\{x = 0\}$ (resp. of $\{y = 0\}$). The first vertex has multiplicity $m_1$ given by (4.3.6); while $m_2, \ldots, m_s$ can be computed by (4.1.5) with all edge-signs $\epsilon = +$, namely:

$$-k_1 m_1 + \frac{a}{(a,c)} + m_2 = 0, \text{ and } -k_i m_i + m_{i-1} + m_{i+1} = 0 \text{ for } i \geq 2.$$

The same graph might also serve as the resolution graph of the germ $x$, induced by the projection $(x, y, z) \mapsto x$. The multiplicities of $x$ are given in the next graph, where the arrows codify *the same* strict transforms:

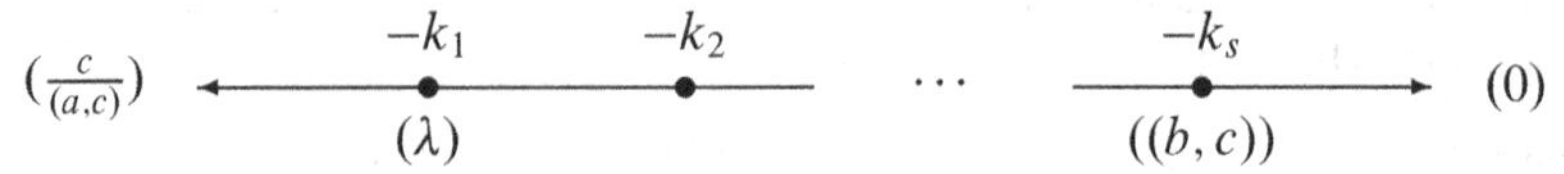


The other multiplicities can again be computed by (4.1.5), with all edge-signs +. Symmetrically, the graph of $y$ is:

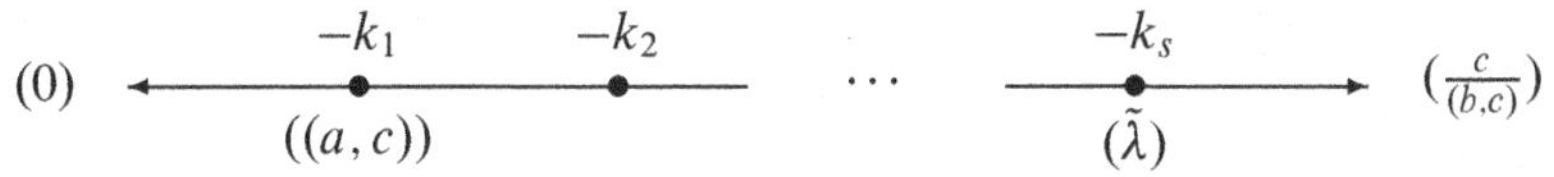


where $0 \leq \tilde{\lambda} < c/(c,b)$ and

$$a + \tilde{\lambda} \cdot \frac{b}{(b,c)} = m_s \cdot \frac{c}{(b,c)}.$$

Hence, the embedded resolution graph $\Gamma(X, x^i y^j z^k)$ of the germ $x^i y^j z^k$ defined on $X$ is the graph with the same shape, same Euler numbers and genera, and the multiplicity $m_v$ (for any vertex $v$) satisfying

$$m_v(x^i y^j z^k) = i \cdot m_v(x) + j \cdot m_v(y) + k \cdot m_v(z).$$

**Remark 4.3.8.** Note the negative sign in the (unusual) continued fraction expansion (4.3.7). Such an expression in the sequel will be called *Hirzebruch–Jung continued fraction*, and it will be denoted by $[k_1, k_2, \ldots, k_s]$. In the present work all continued fraction expansions are of this type.

**4.3.9. The strings** *Str***.** Next, we wish to define a string which will be used systematically in the main result. Strangely enough, its Euler numbers will be deleted. Before providing the precise definition, let us give a reason for it; see also 4.3.4(4).

As we already mentioned, once we know all the multiplicities and edge decorations, all the Euler numbers can be recovered by (4.1.5). In the main construction, we will glue together graphs whose multiplicity systems on common parts agree, but under the gluings the Euler numbers might change. In particular, as "elementary blocks" of the gluing construction we consider graphs with multiplicity decorations, but no Euler numbers: their "correct" Euler numbers will be determined last, after the gluing has been done and after deciding the edge-decorations.

**Definition 4.3.10.** *In the sequel, for positive integers $a$, $b$ and $c$,*

$$Str^{+}(a,b;c \,|\, i,j;k) \quad or \quad Str(a,b;c \,|\, i,j;k)$$

*denotes the string $\Gamma(X, x^i y^j z^k)$, together with its two arrowheads, all vertices (arrow-heads or not) weighted with multiplicities as above, and with all edge decorations $+$, but with all Euler numbers deleted. Moreover,*

$$Str^{\ominus}(a,b;c \,|\, i,j;k)$$

*denotes the same string but with all edges (connecting arrowheads and non-arrowheads) decorated with $\ominus$.*

*In particular, if $\lambda = 0$, then the corresponding string $Str^{\pm}(a,b;c\,|\,i,j;k)$ is a double arrow (decorated with $+$ or $\ominus$) having no non-arrowhead vertex.*

*The same $\pm$-double-arrow will be used for the string $Str^{\pm}(a,b;c\,|\,i,j;k)$ even if $a = 0$ or $b = 0$ (with the convention $a/(a,b) = 0$ whenever $a = 0$).*

## 4.4 Examples: Multiplicity Systems and Milnor Fibrations

**4.4.1. The homology of the link $K_X$ of $(X,x)$.** Let $(X,x)$ be a normal surface singularity, and let $K_X$ be its link, cf. 3.4. We fix a resolution $\mathscr{Y}$ with exceptional set $E = \{E_w\}_{w\in\mathscr{W}}$ and dual resolution graph $\Gamma = \Gamma(X)$. The intersection matrix $A$ can be identified with a $\mathbb{Z}$-linear map $A : \mathbb{Z}^{|\mathscr{W}|} \to (\mathbb{Z}^{|\mathscr{W}|})^*$, where for a $\mathbb{Z}$-module $M$, $M^*$ denotes its dual $Hom_{\mathbb{Z}}(M,\mathbb{Z})$. Since $A$ is non-degenerate, coker$(A)$ is a torsion group with $|\text{coker}(A)| = |\det(A)|$. Then from the homological long exact sequence of the pair $(\mathscr{Y}, K_X) = (P(\Gamma), M(\Gamma))$ one has

**Proposition 4.4.2.** *[48, 79, 112]*

$$H_1(K_X,\mathbb{Z}) = \text{coker}(A) \oplus H_1(E,\mathbb{Z}) = \text{coker}(A) \oplus \mathbb{Z}^{2g(\Gamma)+c(\Gamma)}.$$

**4.4.3. The topology of the pair $(K_X, V_f)$.** We consider an analytic germ $f : (X,x)\to(\mathbb{C},0)$; which is not necessarily an isolated singularity. Set $K_f := K_X \cap V_f$. In this section we wish to compare the multiplicity system of $\Gamma(X,f)$ and the generalized Milnor fibration $\arg := f/|f| : K_X \setminus K_f \to S^1$. If $f$ defines an isolated singularity, then this is an open book of $K_X$ with binding $K_f$.

The next Proposition is a general fact for compact 3-manifolds, see for example [33, page 34]:

**Proposition 4.4.4.** *The fibration* $\arg : K_X \setminus K_f \to S^1$ *is completely determined, up to an isotopy, by the induced representation* $\arg_* : H_1(K_X \setminus K_f,\mathbb{Z}) \to \mathbb{Z}$. *Moreover, if $\mathbb{Z}_n :=$ coker$(\arg_*)$, then the page of* arg *has $n$ connected components which are cyclically permuted by the monodromy.*

The map $\arg_*$ can be compared with the multiplicity system as follows.

Let $\mathbb{Z}^{\mathscr{V}}$ be the free abelian group generated by $\{\langle v\rangle\}_{v\in\mathscr{V}}$. Define the group $H_\Gamma$ as the quotient of $\mathbb{Z}^{\mathscr{V}}$ factorized by the subgroup generated by:

$$e_w\langle w\rangle + \sum_{v\in\mathscr{V}_w}\langle v\rangle \quad (\text{for all } w\in\mathscr{W}).$$

**Proposition 4.4.5.** *[48, 94, 112] One has the following exact sequences*

$$0 \to \mathbb{Z}^{\mathscr{A}} \xrightarrow{i} H_\Gamma \to \text{coker}(A) \to 0,$$

$$0 \to H_\Gamma \xrightarrow{j} H_1(K_X \setminus K_f,\mathbb{Z}) \xrightarrow{q} H_1(E,\mathbb{Z}) \to 0,$$

*where i is the composed map* $\mathbb{Z}^{\mathscr{A}} \hookrightarrow \mathbb{Z}^{\mathscr{V}} \to H_\Gamma$, *and* $j([\langle v\rangle]) = [\gamma_v]$. *(For the definition of* $\gamma_v$ *see 4.1.8.)*

Define $\mathbf{m} : H_\Gamma \to \mathbb{Z}$ by $\mathbf{m}([\langle v\rangle]) = m_v$. Then $\arg_* \circ j = \mathbf{m}$ inserts into the diagram:

$$\begin{array}{ccccccc} 0 \to H_\Gamma & \xrightarrow{j} & H_1(K_X \setminus K_f, \mathbb{Z}) & \xrightarrow{q} & H_1(E,\mathbb{Z}) \to 0. \\ & {\scriptstyle \mathbf{m}} \searrow & \downarrow {\scriptstyle \arg_*} & & \\ & & \mathbb{Z} & & \end{array}$$

If $K_X$ is a rational homology sphere, that is $H_1(E,\mathbb{Z}) = 0$, then $j$ is an isomorphism, and the set $\{m_a\}_{a\in\mathscr{A}}$ determines completely the Milnor fibration up to an isotopy.

In general one has the divisibilities, where $n$ is the order of coker($\arg_*$) as above,

$$n \,\big|\, \gcd\{m_v : v \in \mathscr{V}\} \,\big|\, \gcd\{m_a : a \in \mathscr{A}\}.$$

Nevertheless, it might happen that $n \neq \gcd\{m_v : v \in \mathscr{V}\}$, hence $n$ cannot be determined from $\mathbf{m}$. Or, even if $n = \gcd\{m_v : v \in \mathscr{V}\}$, in general, from the multiplicities one cannot recover $\arg_*$. We illustrate this by some examples borrowed from [86].

In these examples $(X,x)$ will be a Brieskorn singularity, or a cyclic covering. The reader might determine the corresponding graphs by his/her preferred method applicable for such cases; nevertheless, for such singularities, and for any of the coordinate functions, one can deduce the embedded resolution by using the algorithm of cyclic coverings presented in the Sect. 5.3.

**Example 4.4.6.** Set $(X,x) = (\{x^2 + y^7 - z^{14} = 0\}, 0) \subset (\mathbb{C}^3, 0)$ and take $f_1(x,y,z) = z^2$ and $f_2(x,y,z) = z^2 - y$. Then $\Gamma(X, f_1) = \Gamma(X, f_2)$ is the graph

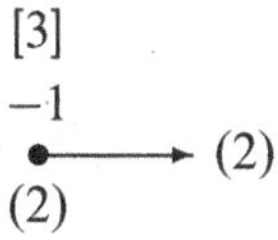


In both cases coker ($\arg_*$) is a factor of coker ($\mathbf{m}$) $= \mathbb{Z}_2$. In fact, coker ($\arg_*(f_1)$) $= \mathbb{Z}_2$ and $\arg_*(f_2)$ is onto. Indeed, the Milnor fibration of $z^2$ is the pullback by $z \mapsto z^2$ of the Milnor fibration of $z$, hence coker ($\arg_*$) $= \mathbb{Z}_2$. For the second statement it is enough to verify that the double covering $\{x^2 + y^7 - z^{14} = w^2 + y - z^2 = 0\} \subset \mathbb{C}^4$ is irreducible (notice that our equations are quasi-homogeneous, hence we can replace a small ball centered at the origin with the whole affine space). But this is true if its intersection with $y = 1$, that is $C := \{x^2 = z^{14} - 1; w^2 = z^2 - 1\} \subset \mathbb{C}^3$, is irreducible. The covering $C \to \mathbb{C}$, $(x,w,z) \mapsto z$, is a $\mathbb{Z}_2 \times \mathbb{Z}_2$ covering. The monodromy around $\pm 1$ is $(-1,-1)$, and around any $\alpha$ with $\alpha^{14} = 1$ and $\alpha^2 \neq 1$ is $(-1, +1)$. Hence the global monodromy group is the whole group $\mathbb{Z}_2 \times \mathbb{Z}_2$, in particular $C$ is irreducible.

Notice also that all the multiplicities of $f_2$ are even, nevertheless there is no germ $g : (X, x) \to (\mathbb{C}, 0)$ with $f_2 = g^2$.

**Example 4.4.7.** Set $(X, x) = (\{z^2 + (x^2 - y^3)(x^3 - y^2) = 0\}, 0)$ and $f_1 = x^2$ and $f_2 = x^2 - y^3$. Then $\Gamma(X, f_1) = \Gamma(X, f_2)$ is the following graph

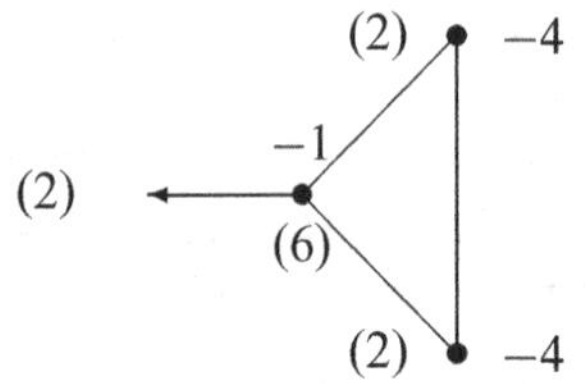


Then again, $\arg_*(f_1)$ has cokernel $\mathbb{Z}_2$, and $\arg_*(f_2)$ is onto.

Notice that in the above examples, $(X, x) = (\{z^2 + h(x, y) = 0\}, 0)$, $h$ is reduced, $f_2$ divides $h$ but it is not equal to $h$. For all such cases the monodromy argument given in 4.4.6 is valid. But all these examples define non-isolated singularities.

In order to construct examples of germs which define isolated singularities, we will use the well-known construction of series of singularities. Namely, assume that $f_1$ and $f_2$ have the same graph but have different representations $\arg_*$, and their zero sets have non-isolated singularities. Next, we find a germ $g$ such that the zero set of $f_i$ and $g$ have no common components ($i = 1, 2$). Then, for sufficiently large $k$, the germs $f_1 + g^k$ and $f_2 + g^k$, will define isolated singularities with the same embedded resolution graphs, but different representations $\arg_*$.

**Example 4.4.8.** Set $(X, x) = (\{x^2 + y^7 - z^{14} = 0\}, 0) \subset (\mathbb{C}^3, 0)$ and take $f_1(x, y, z) = z^2$ and $f_2(x, y, z) = z^2 - y$ as in 4.4.6. Let $P$ be the intersection point of the strict transform $S_a$ of $\{f_i = 0\}$ with the exceptional divisor $E$. Then, in some local coordinate system $(u, v)$ of $P$, $\{u = 0\}$ represents $E$, $\{v = 0\}$ represents $S_a$, and $f_i = u^2 v^2$. Consider $g = y$. Since $y$ in the neighborhood of $P$ can be represented as $y = u^2$ (modulo a local invertible germ), $f_i + g^k$ near $P$ has the form $u^2 v^2 + u^{2k}$. For example, if $k = 2$, then one needs one more blowing up in order to resolve $f_i + g^k$.

Therefore, $\Gamma(X, z^2 + y^k) = \Gamma(X, z^2 - y + y^k)$ for any $k \geq 2$; and for $k = 2$, the graphs have the following form:

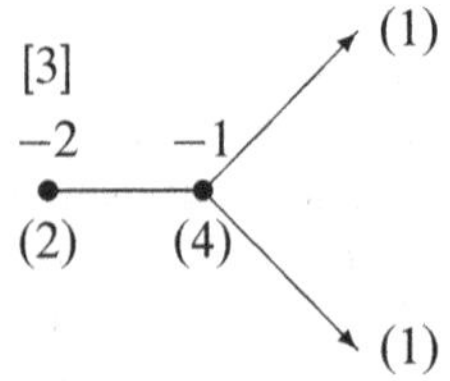

Notice that now **m** is onto, hence both $\arg_*(f_i)$ are onto. Nevertheless, $\arg_*(f_1) \neq arg_*(f_2)$ since their restrictions to a subgroup of $H_1(K_X \setminus K_f)$ (localized near the exceptional curve of genus 3) are different. This follows similarly as in 4.4.6.

**Example 4.4.9.** Set $(X, x) = (\{z^2 + (x^2 - y^3)(x^3 - y^2) = 0\}, 0)$ and $f_1 = x^2 + y^k$ and $f_2 = x^2 - y^3 + y^k$, where $k \geq 4$. Then by a similar argument as above, $\Gamma(X, f_1) = \Gamma(X, f_2)$, but $\arg_*(f_1) \neq \arg_*(f_2)$. The graph for $k = 4$ is

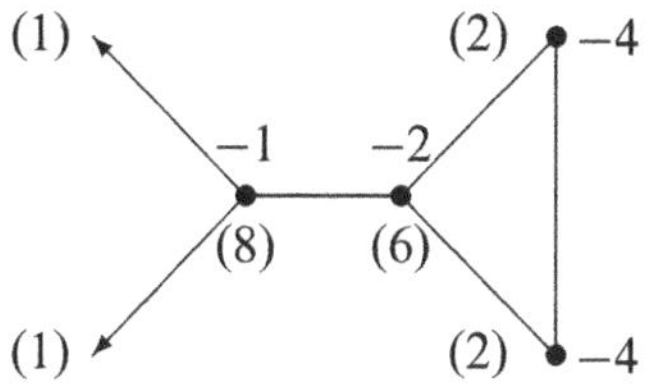

# Chapter 5
# Cyclic Coverings of Graphs

## 5.1 The General Theory of Cyclic Coverings

In this section we review a graph-theoretical construction from [86].

**Definition 5.1.1.** *A morphism of graphs $p : \Gamma_1 \to \Gamma_2$ consists of two maps $p_{\mathcal{V}} : \mathcal{V}(\Gamma_1) \to \mathcal{V}(\Gamma_2)$ and $p_{\mathcal{E}} : \mathcal{E}(\Gamma_1) \to \mathcal{E}(\Gamma_2)$, such that if $e \in \mathcal{E}(\Gamma_1)$ has end-vertices $v_1$ and $v_2$, then $p_{\mathcal{V}}(v_1)$ and $p_{\mathcal{V}}(v_2)$ are the end-vertices of $p_{\mathcal{E}}(e)$. If $p_{\mathcal{V}}$ and $p_{\mathcal{E}}$ are isomorphisms of sets, then we say that $p$ is an isomorphism of graphs.*

*If $\Gamma$ is a graph, we say that $\mathbb{Z}$ acts on $\Gamma$, if there are group-actions $a_{\mathcal{V}} : \mathbb{Z}\times\mathcal{V} \to \mathcal{V}$ and $a_{\mathcal{E}} : \mathbb{Z} \times \mathcal{E} \to \mathcal{E}$ of $\mathbb{Z}$ with the following compatibility property: if $e \in \mathcal{E}$ has end-vertices $v_1$ and $v_2$, then $a_{\mathcal{E}}(h, e)$ has end-vertices $a_{\mathcal{V}}(h, v_1)$ and $a_{\mathcal{V}}(h, v_2)$. The action is trivial if $a_{\mathcal{V}}$ and $a_{\mathcal{E}}$ are trivial actions.*

*If $\mathbb{Z}$ acts on both $\Gamma_1$ and $\Gamma_2$, then a morphism $p : \Gamma_1 \to \Gamma_2$ is equivariant if the maps $p_{\mathcal{V}}$ and $p_{\mathcal{E}}$ are equivariant with respect to the actions of $\mathbb{Z}$. If additionally $p$ is an isomorphism then it is called an equivariant isomorphism of graphs.*

Fix a finite graph $\Gamma$, and assume that $\mathbb{Z}$ acts on it in a trivial way.

**Definition 5.1.2.** *A $\mathbb{Z}$-covering, or cyclic covering of $\Gamma$ consists of a finite graph $G$, that carries a $\mathbb{Z}$-action, together with an equivariant morphism $p : G \to \Gamma$ such that the restriction of the $\mathbb{Z}$-action on any set of type $p^{-1}(v)$ ($v \in \mathcal{V}(\Gamma)$), respectively $p^{-1}(e)$ ($e \in \mathcal{E}(\Gamma)$), is transitive.*

Fix a cyclic covering $p : G \to \Gamma$. For any $v \in \mathcal{V}(\Gamma)$, let $\mathfrak{n}_v\mathbb{Z}$ be the maximal subgroup of $\mathbb{Z}$ which acts trivially on $p^{-1}(v)$. Similarly, for any $e \in \mathcal{E}(\Gamma)$ with end-vertices $\{v_1, v_2\}$, let $\mathfrak{d}_e \cdot \mathrm{lcm}(\mathfrak{n}_{v_1}, \mathfrak{n}_{v_2})\mathbb{Z}$ be the maximal subgroup of $\mathbb{Z}$ which acts trivially on $p^{-1}(e)$. These numbers define a system of strictly positive integers

$$(\mathbf{n}, \mathbf{d}) = \left\{\{\mathfrak{n}_v\}_{v\in\mathcal{V}(\Gamma)}; \{\mathfrak{d}_e\}_{e\in\mathcal{E}(\Gamma)}\right\}.$$

A. Némethi and Á. Szilárd, *Milnor Fiber Boundary of a Non-isolated Surface Singularity*, Lecture Notes in Mathematics 2037, DOI 10.1007/978-3-642-23647-1_5, 

**Definition 5.1.3.** $(\mathbf{n}, \mathbf{d})$ *is called the covering data of the covering* $p : G \to \Gamma$.

Sometimes we write

$$\mathfrak{n}_e := \mathfrak{d}_e \cdot \text{lcm}(\mathfrak{n}_{v_1}, \mathfrak{n}_{v_2}).$$

**Definition 5.1.4.** *Two cyclic coverings* $p_i : G_i \to \Gamma$ $(i = 1, 2)$ *are equivalent, denoted by* $G_1 \sim G_2$, *if there is an equivariant isomorphism* $q : G_1 \to G_2$ *such that* $p_2 \circ q = p_1$.

*The set of equivalence classes of cyclic coverings of* $\Gamma$, *all associated with a fixed covering data* $(\mathbf{n}, \mathbf{d})$, *is denoted by* $\mathscr{G}(\Gamma, (\mathbf{n}, \mathbf{d}))$.

**Theorem 5.1.5.** *[86]* $\mathscr{G}(\Gamma, (\mathbf{n}, \mathbf{d}))$ *has an abelian group structure and it is independent of* $\mathbf{d}$.

In general, $\mathscr{G}(\Gamma, (\mathbf{n}, \mathbf{d}))$ is non-trivial. Here are some examples.

**Example 5.1.6.** [86] For a covering data $(\mathbf{n}, \mathbf{d})$ one has:

1. Assume that $\Gamma$ is a tree. Then $\mathscr{G}(\Gamma, (\mathbf{n}, \mathbf{d})) = 0$ for any $(\mathbf{n}, \mathbf{d})$. Therefore, up to an isomorphism, there is only one covering $G$ of $\Gamma$. It has exactly $\gcd\{\mathfrak{n}_v\}_{v \in \mathscr{V}(\Gamma)}$ connected components.
2. Assume that $\Gamma$ is a cyclic graph, that is $\mathscr{V}(\Gamma) = \{v_1, v_2, \ldots, v_k\}$ and $\mathscr{E}(\Gamma) = \{(v_1, v_2), (v_2, v_3), \ldots, (v_k, v_1)\}$, where $k \geq 3$. Then $\mathscr{G}(\Gamma, (\mathbf{n}, \mathbf{d})) = \mathbb{Z}_n$, where $n = \gcd\{n_v : v \in \mathscr{V}(\Gamma)\}$.
3. For any subgraph $\Gamma' \subset \Gamma$ there is a natural surjection $pr : \mathscr{G}(\Gamma, (\mathbf{n}, \mathbf{d})) \to \mathscr{G}(\Gamma', (\mathbf{n}, \mathbf{d}))$.
4. Let $\Gamma$ be a graph with $c(\Gamma) = 1$, and let $\Gamma'$ be the unique minimal cyclic subgraph of $\Gamma$. If $n := \gcd\{n_v : v \in \mathscr{V}(\Gamma')\}$, then $\mathscr{G}(\Gamma, (\mathbf{n}, \mathbf{d})) = \mathbb{Z}_n$.

**Remark 5.1.7.** If $p : G \to \Gamma$ is a cyclic covering then $c(G) \geq c(\Gamma)$. Indeed, if $|G|$ and $|\Gamma|$ denote the topological realizations of the corresponding graphs, then the invariant subspace $H_1(|G|, \mathbb{Q})^{\mathbb{Z}}$ is isomorphic to $H_1(|\Gamma|, \mathbb{Q})$.

On the other hand, we have the following result which will be a key ingredient of the main construction in 10.2.8:

**Theorem 5.1.8.** *[86, (1.20)] Fix* $\Gamma$ *and* $(\mathbf{n}, \mathbf{d})$ *as above. Set* $\mathscr{V}^1 := \{v \in \mathscr{V}(\Gamma) : \mathfrak{n}_v = 1\}$. *Let* $\Gamma_{\neq 1}$ *be the subgraph of* $\Gamma$ *obtained from* $\Gamma$ *by deleting the vertices from* $\mathscr{V}^1$ *and their adjacent edges. If each connected component of* $\Gamma_{\neq 1}$ *is a tree, then* $\mathscr{G}(\Gamma, (\mathbf{n}, \mathbf{d})) = 0$.

**5.1.9. Variations.** One extends the set of coverings as follows (cf. [86, 1.25]).

1. Assume that we have two types of vertices: arrowheads $\mathscr{A}$ and non-arrowheads $\mathscr{W}$, i.e. $\mathscr{V} = \mathscr{A} \sqcup \mathscr{W}$. Then in the definition of a coverings $p : G \to \Gamma$ we add the following axiom: $\mathscr{A}(G) = p^{-1}(\mathscr{A}(\Gamma))$.
2. Assume that our graphs have some decorations. Then for a covering $p : G \to \Gamma$ we also require that the decorations of $G$ must be equivariant.
3. "Equivariant string insertion" means the following modification of $\Gamma$. One starts with the following data:

(a) a graph $\Gamma$ and a system $(\mathbf{n}, \mathbf{d})$ as above;
(b) a covering $p : G \to \Gamma$, as an element of $\mathscr{G}(\Gamma, (\mathbf{n}, \mathbf{d}))$;
(c) for any edge $e$ of $\Gamma$, we fix a string $Str^{\pm}(e)$ (with decorations):

$$Str^{\pm}(e) : \quad \overset{\pm}{\underset{(m_1)}{\bullet}} \quad \overset{\pm}{\underset{(m_2)}{\bullet}} \quad \overset{\pm}{\cdots} \quad \overset{\pm}{\underset{(m_s)}{\bullet}} \quad \overset{\pm}{\longrightarrow}$$

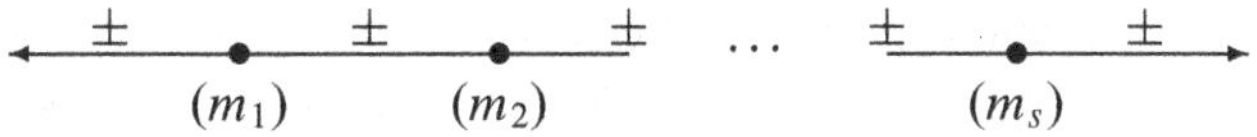

Then the new graph $G(\{Str^{\pm}(e)\}_e)$ is constructed as follows: we replace each edge $\tilde{e} \in p_{\mathscr{E}}^{-1}(e)$ (with end-vertices $\tilde{v}_1$ and $\tilde{v}_2$ and decoration $\pm$) of $G$ as shown below:

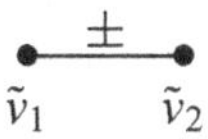

$$\underset{\tilde{v}_1}{\bullet} \overset{\pm}{\phantom{xx}} \underset{\tilde{v}_2}{\bullet}$$

is replaced by

$$\underset{\tilde{v}_1}{\bullet} \overset{\pm}{\phantom{x}} \underset{(m_1)}{\bullet} \overset{\pm}{\phantom{x}} \underset{(m_2)}{\bullet} \overset{\pm}{\cdots} \overset{\pm}{\phantom{x}} \underset{(m_s)}{\bullet} \overset{\pm}{\phantom{x}} \underset{\tilde{v}_2}{\bullet}$$

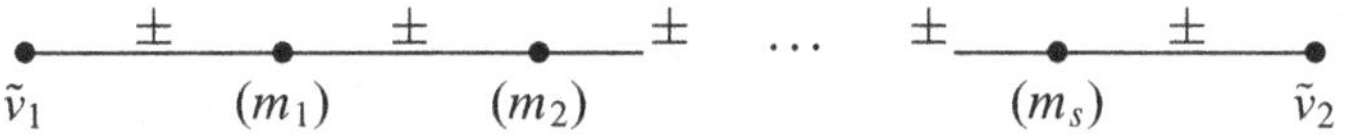

## 5.2   The Universal Cyclic Covering of $\Gamma(X, f)$

Fix a normal surface singularity $(X, x)$ and a germ $f : (X, x) \to (\mathbb{C}, 0)$ of an analytic function. Fix also a resolution $\phi : (\mathscr{Y}, D) \to (U, V_f)$ as in 4.3.1, and consider the associated embedded resolution graph $\Gamma(X, f)$. In this section we recall the construction of a canonical cyclic covering of this graph via the Milnor fibration of $f$. It is called the *universal cyclic covering of* $\Gamma(X, f)$. For more details see [28, 86].

There are several reasons why we include this construction in the present work:

- The universal cyclic covering is the prototype of all cyclic coverings provided by geometric constructions. For an analogous construction, which is an important ingredient in some proofs of the book, see 7.1.6.
- It shows, for a fixed graph $\Gamma(X, f)$, how one can codify graph-theoretically the possible differences of $\arg_*$. Thus, it is a necessary complement of Sect. 4.4.
- It guides all the geometry, in particular the resolution graphs, of cyclic coverings $X_{f,N}$.
- By examples which show that a graph can have several cyclic coverings, we emphasize the role and power of the key Theorem 5.1.8, which guarantees that the graph provided by the Main Algorithm in Chap. 10 is well-defined and unique.

**5.2.1. The construction of the covering** $p : G(X, f) \to \Gamma(X, f)$.
Let $T(E_w)$ ($w \in \mathscr{W}$) be a small tubular neighborhood of the irreducible curve $E_w$. By our assumption that any two irreducible components of the exceptional divisor have at most one intersection point (see 4.3.1), for any edge $e = (v, w)$ connecting two non-arrowheads, the intersection $T(E_w) \cap T(E_v)$ is a bidisc $T_e$. If $T(S_a)$, for $a \in \mathscr{A}$, is a small tubular neighborhood of the irreducible component $S_a$ of the strict transform $S$ (cf. 4.3.1), and $a$ is adjacent to $w_a \in \mathscr{W}$, then corresponding to the edge $e = (a, w_a)$ we introduce the bidisc $T_e = T(S_a) \cap T(E_{w_a})$. Set $T = (\cup_w T(E_w)) \cup (\cup_a T(S_a))$.

Next, we consider the smooth fiber $f^{-1}(\delta) \subset X$ lifted via $\phi$. For sufficiently small $\delta > 0$, the fiber $F := (f \circ \phi)^{-1}(\delta) \subset \mathscr{Y}$ is in $T$. Set $F_w = F \cap T(E_w)$ for any $w \in \mathscr{W}$, $F_a = F \cap T(S_a)$ for any $a \in \mathscr{A}$, and $F_e = F \cap T_e$ for any $e \in \mathscr{E}$.

It is possible to chose the geometric monodromy acting on $F$ in such a way that it preserves the subspaces $\{F_v\}_{v\in\mathscr{V}}$ and $\{F_e\}_{e\in\mathscr{E}}$. Then the connected components of $F_v$, respectively of $F_e$, are cyclically permuted by this action. Let $\mathfrak{n}_v$ and $\mathfrak{n}_e$ be the number of connected components of $F_v$ and $F_e$ respectively. Then, for any $e = (v_1, v_2)$, we have $\mathfrak{n}_e = \mathfrak{d}_e \cdot \operatorname{lcm}(\mathfrak{n}_{v_1}, \mathfrak{n}_{v_2})$ for some $\mathfrak{d}_e \geq 1$.

Now, we are able to construct the covering $p : G(X, f) \to \Gamma(X, f)$ associated with the resolution $\phi$. Above a vertex $v \in \mathscr{V}(\Gamma(X, f))$ there are exactly $\mathfrak{n}_v$ vertices of $G(X, f)$ which correspond to the connected components of $F_v$. The $\mathbb{Z}$-action is induced by the monodromy. If $v$ is an arrowhead in $\Gamma$ then by our agreement 5.1.9(1), all the vertices in $G$ above $v$ are arrowheads. Above an edge $e$ of $\Gamma$, there are $\mathfrak{n}_e$ edges of $G$. They correspond to the connected components of $F_e$. The $\mathbb{Z}$-action is again generated by the monodromy.

Fix an edge $\tilde{e}$ of $G$ (above the edge $e$ of $\Gamma$) which corresponds to the connected component $F_{\tilde{e}}$ of $F_e$. Similarly, take a vertex $\tilde{v}$ of $G$ (above the vertex $v$ of $\Gamma$) which corresponds to the connected component $F_{\tilde{v}}$ of $F_v$. Then $\tilde{e}$ has as an end the vertex $\tilde{v}$ if and only if $F_{\tilde{e}} \subset F_{\tilde{v}}$. In particular, $\tilde{v}_1$ and $\tilde{v}_2$ are connected in $G$ if and only if $F_{\tilde{v}_1} \cap F_{\tilde{v}_2} \neq \emptyset$: if $F_{\tilde{v}_1} \cap F_{\tilde{v}_2}$ has $\mathfrak{d}_e$ connected components, then $\tilde{v}_1$ and $\tilde{v}_2$ are connected exactly by $\mathfrak{d}_e$ edges.

Next, we list some basic properties of the universal covering graph complemented with several examples. For more details see [86].

**Example 5.2.2.** Assume that the link $K_X$ is a rational homology sphere. Then the covering data can be uniquely recovered from $\Gamma(X, f)$, hence, by 5.1.6(1), $G(X, f)$ itself is also determined. Indeed, for any $v \in \mathscr{V}(\Gamma)$ let $\mathscr{V}_v$ denote the set of vertices adjacent to $v$. Then $n_v = \gcd\{m_w : w \in \mathscr{V}_v \cup \{v\}\}$ for any $v \in \mathscr{V}(\Gamma)$, and $n_e := \gcd(m_{v_1}, m_{v_2})$ for any $e = (v_1, v_2) \in \mathscr{E}(\Gamma)$. Moreover, the number of connected components of $G(X, f)$ is exactly $\gcd\{m_v : v \in \mathscr{V}(\Gamma)\}$.

In general, one has the following connectedness result.

**Lemma 5.2.3.** *The number of connected components of the graph $G(X, f)$ is equal to the number of connected components of the Milnor fiber $F$ of the germ $f$. This number also agrees with* $|\operatorname{coker}(\arg_*(f))|$. *In particular, if $f$ defines an isolated singularity, then $G(X, f)$ is a connected graph.*

*Since* $|\mathrm{coker}\,(\arg_*(f))|$ *divides* $\gcd\{m_v : v \in \mathcal{V}\}$*, the graph* $G(X, f)$ *is connected whenever* $\gcd\{m_v : v \in \mathcal{V}\} = 1$*. (Nevertheless,* $G(X, f)$ *can be connected even if* $\gcd\{m_v : v \in \mathcal{V}\} \neq 1$*, see e.g. 5.2.4.)*

**Example 5.2.4.** Set $(X, x) = (\{x^2 + y^7 - z^{14} = 0\}, 0) \subset (\mathbb{C}^3, 0)$ and take $f_1(x, y, z) = z^2$ and $f_2(x, y, z) = z^2 - y$ as in 4.4.6. The next diagrams show the coverings $p : G(X, f_i) \to \Gamma(X, f_i)$ for $i = 1, 2$.

Note that the number of connected components of the graphs $G(X, f_i)$ is different.

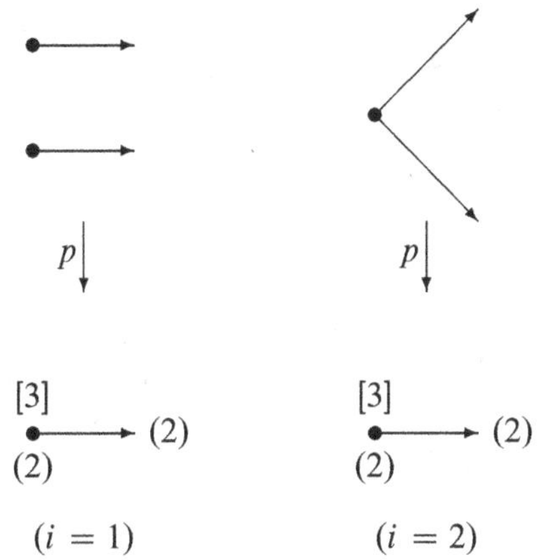


**Example 5.2.5.** Set $(X, x) = (\{x^2 + y^7 - z^{14} = 0\}, 0) \subset (\mathbb{C}^3, 0)$ and take $f_1(x, y, z) = z^2 + y^2$ and $f_2(x, y, z) = z^2 - y + y^2$ as in 4.4.8. The next diagrams show the coverings $p : G(X, f_i) \to \Gamma(X, f_i)$, where $i = 1, 2$.

In this case the number of independent cycles of the graphs $G(X, f_i)$ is different.

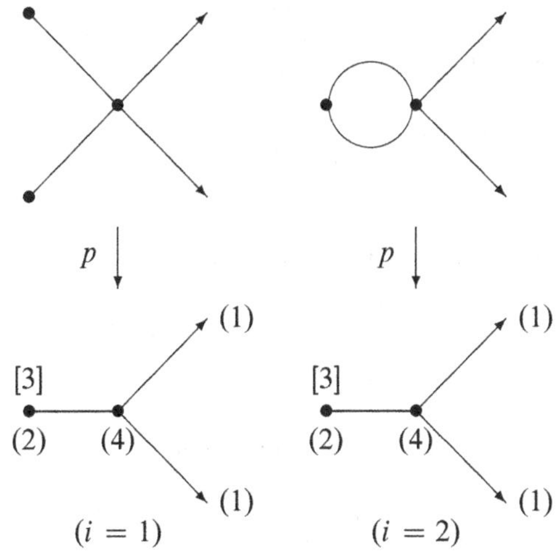

**Example 5.2.6.** Set $(X,x) = (\{z^2 + (x^2 - y^3)(x^3 - y^2) = 0\}, 0)$ and $f_1 = x^2 + y^4$ and $f_2 = x^2 - y^3 + y^4$ as in 4.4.9. Then the coverings $p : G(X, f_i) \to \Gamma(X, f_i)$ (for $i = 1, 2$) are:

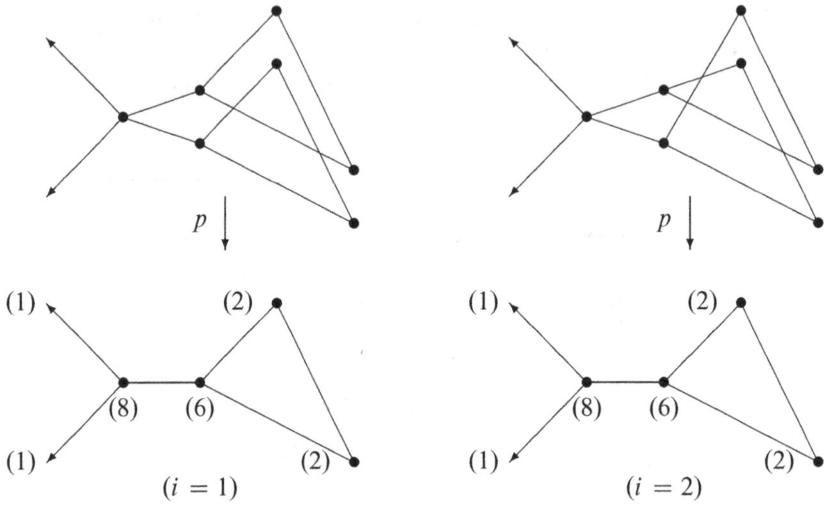


**Remark 5.2.7. Determining the complex monodromy.**
Consider a pair $(X, f)$ as above. Let $\arg : K_X \setminus K_f \to S_1$ be its generalized Milnor fibration as in 4.4.4. Assume that the Milnor fiber $F$ has $n$ connected components, and let $M : H_1(F, \mathbb{C}) \to H_1(F, \mathbb{C})$ be its algebraic monodromy acting on $H_1(F, \mathbb{C})$. Let $P(t)$ denote its characteristic polynomial $P(t) = \det(tI - M)$. Let $\{m_v\}_{v \in \mathscr{V}}$ be the multiplicities of the graph $\Gamma = \Gamma(X, f)$. In addition, for any non-arrowhead vertex $w \in \mathscr{W}$ of $\Gamma(X, f)$, denote by $\delta_w$ the number of adjacent vertices (with the notation of 5.2.2, $\delta_w = |\mathscr{V}_w|$). Then the following facts hold:

1. **A'Campo's formula** [2–4]

$$\frac{t^n - 1}{P(t)} = \prod_{w \in \mathscr{W}(\Gamma)} (t^{m_w} - 1)^{2 - 2g_w - \delta_w}. \tag{5.2.8}$$

Hence, for the Euler characteristic of the page $F$ one also has

$$\chi(F) = \sum_{w \in \mathscr{W}(\Gamma)} m_w \cdot (2 - 2g_w - \delta_w), \tag{5.2.9}$$

and the dimension of the generalized 1-eigenspace $(H_1(F, \mathbb{C}))_1$ is

$$(\dim H_1(F, \mathbb{C}))_1 = 2g(\Gamma) + 2c(\Gamma) + |\mathscr{A}(\Gamma)| - 1. \tag{5.2.10}$$

Note that if $f$ is an isolated singularity, then $\chi(F)$ and $P(t)$ can be determined from the pair $(K_X, K_f)$, that is, only from the binding of the open book decomposition, and the additional facts regarding $\arg_*(f)$ are not needed.

This is in contrast with the Jordan block structure of the operator $M$, see part 3 of this list.

2. By the Monodormy Theorem [2–4, 15, 22, 59, 61, 67], the size of a Jordan block of $M$ can at most be 2. Let $\#_\lambda^2$ be the number of Jordan blocks with eigenvalue $\lambda$ and size 2. Then from the Wang exact sequence applied to the Milnor fibration one gets (see e.g. [91]):

$$\dim\ker(M - I) = 2g(\Gamma) + c(\Gamma) + |\mathscr{A}(\Gamma)| - 1. \tag{5.2.11}$$

Therefore, using (5.2.10) too, $\#_1^2 = c(\Gamma(X))$, hence $\#_1^2$ depends only on $K_X$, and it is independent of the germ $f$.

3. [86, 3.25] Let $G = G(X, f)$ be the universal covering of $\Gamma(X, f)$, let $|G|$ be its topological realization. The cyclic action on $G$ induces an action on $|G|$. At homological level, this induces a finite morphism $M_{|G|}$ on $H^1(|G|, \mathbb{C})$. Then there is an isomorphism of pairs

$$(H^1(|G|, \mathbb{C}), M_{|G|}) = (\mathrm{im}(M^N - I), M), \tag{5.2.12}$$

where $N$ is an integer such that $\lambda^N = 1$ for any eigenvalue $\lambda$ of $M$. Hence, $\#_\lambda^2(f)$ equals the multiplicity of the root $\lambda$ in the characteristic polynomial of $M_{|G|}$.

For example, in the case of Example 5.2.5, the monodromy of $f_1$ is finite, while the monodromy of $f_2$ has a Jordan block of size 2, that is $\#_{-1}^2(f_2) = 1$.

4. Nevertheless, if $K_X$ is a rational homology sphere, each $\#_\lambda^k(f)$ can be determined from $\Gamma = \Gamma(X, f)$ [33, 86, 95]. Assume that $f$ defines an isolated singularity, and consider the integers $n_v$ ($v \in \mathscr{V}(\Gamma)$) and $n_e$ ($e \in \mathscr{E}(\Gamma)$) as in 5.2.2. For any fixed positive integer $N$ set $\mathfrak{n}_w := \gcd(n_w, N)$ and $\mathfrak{n}_e := \gcd(n_e, N)$. Then

$$\sum_{\lambda^N=1} \#_\lambda^2(f) = \sum_{e\in\mathscr{E}(\Gamma)} (\mathfrak{n}_e - 1) - \sum_{w\in\mathscr{W}(\Gamma)} (\mathfrak{n}_w - 1). \tag{5.2.13}$$

## 5.3 The Resolution Graph of $f(x, y) + z^N = 0$

**Remark 5.3.1.** Let $(X, x)$ be a normal surface singularity and $f$ the germ of an analytic function defined on $(X, x)$. Let us also fix a positive integer $N$.

The dual resolution graph of the cyclic covering $X_{f,N}$ depends essentially on the map $\arg_*(f)$. In Sect. 4.4 we emphasized that $\arg_*(f)$ cannot be determined from the dual embedded resolution graph $\Gamma(X, f)$. Hence, in general, the resolution graph of $X_{f,N}$ cannot be determined from $\Gamma(X, f)$ and $N$ either. Usually, from $\Gamma(X, f)$ and $N$ one can read the covering data of all the edges and all the vertices with $g_w = 0$, and obviously the graph $\Gamma = \Gamma(X, f)$ itself, but still missing are the covering data of vertices with $g_w > 0$ and the global "twisting data" codified in $\mathscr{G}(\Gamma, (\mathbf{n}, \mathbf{d}))$, whenever this covering group is non-trivial. For concrete examples, and a detailed discussion, see [86].

In order to determine the resolution graph of $X_{f,N}$, one needs to consider the *universal cyclic covering graph* of $\Gamma(X, f)$, which codifies the missing information, see [86]. On the other hand, if the link of $(X, x)$ is a rational homology sphere, then the covering data can be recovered from the multiplicity system of $\Gamma(X, f)$ and the integer $N$, furthermore, any cyclic covering of $\Gamma(X, f)$ can be determined from $\Gamma(X, f)$ and the covering data in a unique way, cf. 5.1. In particular, in such a case, $\Gamma(X_{f,N})$ can be recovered from $\Gamma(X, f)$ and $N$. This is definitely valid when $(X, x)$ is a smooth germ, that is when $X_{f,N} = \{f(x, y) + z^N = 0\}$ is the cyclic cover of a plane curve singularity.

Since this special case is what will be needed in the sequel, in this section it is all we recall. Different versions of this algorithm were used in several articles, see [6, 44, 57, 84–86, 97, 99, 103–105]. For its generalization to the Iomdin series, see [92]. It can also be viewed as the starting point of our Main Algorithm.

**5.3.2. The resolution graph of $X_{f,N}$ for a plane curve singularity $f$.** Assume that $f : (\mathbb{C}^2, 0) \to (\mathbb{C}, 0)$ is an isolated plane curve singularity, and let us fix a positive integer $N$. As above, $X_{f,N}$ denotes the germ of the isolated hypersurface singularity $\{f(x, y) + z^N = 0\}$. Germs of this type are also called *suspensions*. The projection $(x, y, z) \mapsto z$ induces a map $z : (X_{f,N}, 0) \to (\mathbb{C}, 0)$.

In fact, the algorithm provides a possible embedded resolution graph $\Gamma(X_{f,N}, z)$ from $\Gamma(\mathbb{C}^2, f)$ and $N$. By adding the germ $z$ we exploit fully the power of the multiplicity system and its "local nature", and we also exemplify the comment 4.3.4(4).

The graph $\Gamma(X_{f,N}, z)$ is a cyclic covering $p : \Gamma(X_{f,N}, z) \to \Gamma(\mathbb{C}^2, f)$ (with arrowheads and decorations, cf. 5.1.9) of $\Gamma(\mathbb{C}^2, f)$. The covering data is the following.

(a) For any $w \in \mathscr{W}(\Gamma(\mathbb{C}^2, f))$, let $\mathscr{V}_w$ be the set of all the vertices (arrowheads and non-arrowheads) adjacent to $w$. Furthermore, set

$$n_w := \gcd(m_v : v \in \mathscr{V}_w \cup \{w\}).$$

Above $w \in \mathscr{W}(\Gamma(\mathbb{C}^2, f))$ there are $\mathfrak{n}_w := \gcd(n_w, N)$ vertices of $\Gamma(X_{f,N}, z)$, each with multiplicity $m_w/\gcd(m_w, N)$ and genus $\tilde{g}$, where:

$$2 - 2\tilde{g} = \frac{(2 - |\mathscr{V}_w|) \cdot \gcd(m_w, N) + \sum_{v \in \mathscr{V}_w} \gcd(m_w, m_v, N)}{\mathfrak{n}_w}.$$

(b) Consider an edge $e = (w_1, w_2)$ of $\Gamma(\mathbb{C}^2, f)$

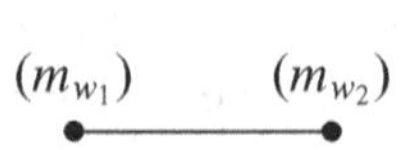

Set $n_e = \gcd(m_{w_1}, m_{w_2})$. Then $e$ is covered by $\mathfrak{n}_e := \gcd(n_e, N)$ copies of the following string (cf. 4.3.9)

$$Str(\frac{m_{w_1}}{\mathfrak{n}_e}, \frac{m_{w_2}}{\mathfrak{n}_e}; \frac{N}{\mathfrak{n}_e}|0, 0; 1).$$

(c) An arrowhead of $\Gamma(\mathbb{C}^2, f)$

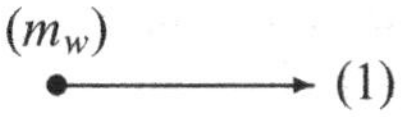


is covered by one string of type $Str(m_w, 1; N|0, 0; 1)$, whose right arrowhead will remain an arrowhead of $\Gamma(X_{f,N}, z)$ with multiplicity 1, and its left arrowhead is identified with the unique vertex above $w$.

(d) In this way, we obtain all the vertices, edges and arrowheads of $\Gamma(X_{f,N}, z)$, all the multiplicities of $z$, and some of the Euler numbers. The missing Euler numbers can be determined by (4.1.5). This ends the construction of $\Gamma(X_{f,N}, z)$.

If we drop the arrowheads and multiplicities of $\Gamma(X_{f,N}, z)$, we obtain $\Gamma(X_{f,N})$. The graphs $\Gamma(X_{f,N}, z)$ and $\Gamma(X_{f,N})$, in general, are not minimal. They can be simplified by blowing down operation.

**5.3.3. Brieskorn singularities.** Evidently, the above strategy can be applied for an arbitrary Brieskorn singularity $f(x, y, z) = x^{a_1} + y^{a_2} + z^{a_3}$ too. Nevertheless, in the next paragraph we present a much shorter procedure valid for this case, see [99].

For any cyclic permutation $(i, j, k)$ of $(1, 2, 3)$ take:

$$d_i := (a_i, [a_j, a_k]); \ \alpha_i := a_i/d_i; \ q_i := [a_j, a_k]/d_i;$$

and $0 \leq \omega_i < \alpha_i$ with $1 + \omega_i q_i \equiv 0 \pmod{\alpha_i}$. Here $[\cdot, \cdot] = \text{lcm}$ and $(\cdot, \cdot) = \gcd$.

For any $i \in \{1, 2, 3\}$, we construct a string $St_i$. If $\omega_i \neq 0$, take

$$\frac{\alpha_i}{\omega_i} = k_{i1} - \cfrac{1}{k_{i2} - \cfrac{1}{\ddots - \cfrac{1}{k_{is}}}}, \quad k_{i1}, \ldots, k_{is} \geq 2. \tag{5.3.4}$$

Then $St_i$ is the following graph (with a distinguished vertex **x**):

$St_i$ :

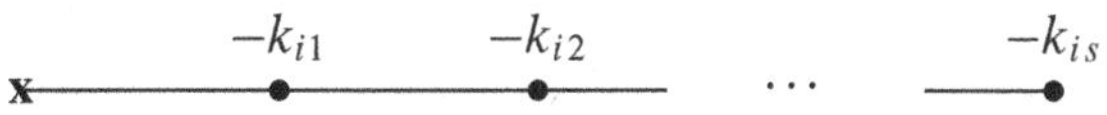


If $\omega_i = 0$, then $St_i$ contains only the distinguished vertex **x**, and it has no edges.

Then $\Gamma(V_f)$ is the star-shaped graph obtained using $(a_1, a_2)$ copies of $St_3$, $(a_2, a_3)$ copies of $St_1$, and $(a_3, a_1)$ copies of $St_2$, by identifying their distinguished vertices **x**. This vertex in $\Gamma(V_f)$ will have genus $g$ and Euler number $e$, where:

$$2-2g=\sum(a_i,a_j)-\frac{\prod(a_i,a_j)}{(a_1,a_2,a_3)}$$

and

$$-e=\frac{(a_1,a_2,a_3)}{\alpha_1\alpha_2\alpha_3}+\sum(a_i,a_j)\frac{\omega_k}{\alpha_k}.$$

Above $\sum$ and $\prod$ denotes the cyclic sum, respectively cyclic product.

**5.3.5. Seifert invariants. Orbifold Euler number.** [99] Those star shaped graphs whose Euler numbers on the legs are $\leq -2$ characterize the Seifert manifolds. Their topological invariants are codified as follows. Assume that $\{1,\dots,\nu\}$ is an index set for the legs. Each leg, via a continued fraction expansion as in (5.3.4), determines a pair $(\alpha_\ell,\omega_\ell)$, the corresponding "orbit invariant". Furthermore, the central vertex has a genus decoration $[g]$ and Euler number $e$. Then the collection $\{(\alpha_\ell,\omega_\ell),1\leq \ell\leq\nu;g,e\}$ is the *Seifert invariant* of the corresponding plumbed 3-manifold.

Usually, one also defines the *orbifold Euler number* by

$$e^{orb}:=e+\sum_\ell \omega_\ell/\alpha_\ell. \tag{5.3.6}$$

The intersection matrix (in the present normal form) is negative definite if and only if $e^{orb}<0$. The rational number $e^{orb}$ is also called "virtual degree", see [133].

One also has

$$-e^{orb}\cdot\prod_\ell \alpha_\ell=\det(-A)$$

and

$$e^{orb}(-M)=-e^{orb}(M). \tag{5.3.7}$$

# Chapter 6
# The Graph $\Gamma_{\mathscr{C}}$ of a Pair $(f, g)$: The Definition

## 6.1 The Construction of the Curve $\mathscr{C}$ and Its Dual Graph

**6.1.1. Introductory words.** The main tool of the present book is the weighted graph $\Gamma_{\mathscr{C}}$ introduced and studied in [92]. It has two types of vertices, non-arrowheads and arrowheads. The non-arrowhead vertices have two types of decorations: the first one is an ordered triple $(m; n, \nu)$ for some integers $m, \nu > 0$ and $n \geq 0$, while the second one is the "genus" decoration $[g]$, where $g$ is a non-negative integer. If $g = 0$ then we might omit this decoration. Any arrowhead has only one decoration, namely the ordered triple $(1; 0, 1)$. The edges are not directed and loops are accepted. Each edge has a decoration $\in \{1, 2\}$, which in some special situations can be omitted, since it can be recovered from the other decorations, cf. 6.2.4.

The graph $\Gamma_{\mathscr{C}}$ was introduced to study hypersurface singularities in three variables with 1-dimensional singular locus, and it was the main tool used getting resolution graphs of the members of the generalized Iomdin series.

More precisely, in that article we started with a hypersurface germ $f$ as above, and chose an additional germ $g : (\mathbb{C}^3, 0) \to (\mathbb{C}, 0)$ such that the pair $(f, g)$ formed an ICIS. The final output was the resolution graphs of the series of hypersurface singularities $f + g^k$, for $k$ large, determined in terms of $\Gamma_{\mathscr{C}}$ and $k$. Motivated by the fact that in addition $\Gamma_{\mathscr{C}}$ contains all the information needed to treat "almost all" the correction terms of the invariants of the series (see [92], or 1.1.3 and 7.1 here), we called $\Gamma_{\mathscr{C}}$ *"universal"*. Its power is reinforced by the present work as well.

Nevertheless, perhaps, the name *bi-colored relative graph associated with the pair* $(f, g)$ tells more about the geometry encoded in the graph. Here, the first attribute points out that the edges can be decorated by two "colors" (1 and 2), a key fact which has enormous geometrical effects and the source of pathological behaviors. By "relative" we wish to stress, that the graph codifies the $g$-polar geometry of $f$ concentrated near the singular locus of $V_f$. In particular, in $\Gamma_{\mathscr{C}}$ the functions $f$ and $g$ do not have a symmetric role. For more motivation and supporting intuitive arguments regarding $\Gamma_{\mathscr{C}}$, see Sect. 7.1 too.

A. Némethi and Á. Szilárd, *Milnor Fiber Boundary of a Non-isolated Surface Singularity*, Lecture Notes in Mathematics 2037, DOI 10.1007/978-3-642-23647-1_6, 

Geometrically, the graph $\Gamma_{\mathscr{C}}$ is the decorated dual graph of a special curve configuration $\mathscr{C}$ in the embedded resolution of the pair $(\mathbb{C}^3, V_f \cup V_g)$. Its presentation is the subject of the next paragraphs.

**6.1.2. The definition of $\mathscr{C}$ and $\Gamma_{\mathscr{C}}$ [92].** Consider an ICIS $(f, g) : (\mathbb{C}^3, 0) \to (\mathbb{C}^2, 0)$ and the local divisor $(D, 0) := (V_f \cup V_g, 0) \subset (\mathbb{C}^3, 0)$. Let

$$\Phi : \Phi^{-1}(D^2_\eta) \cap B_\epsilon \to D^2_\eta$$

denote a good representative of $(f, g)$ as in Sect. 3.1. Denote by $\Delta_\Phi \subset D^2_\eta$ its discriminant, as before.

Take an embedded resolution

$$r : V^{emb} \to U$$

of the pair $(D, 0) \subset (\mathbb{C}^3, 0)$. This means the following. The space $V^{emb}$ is smooth, $r$ is proper, $U$ is a small representative of $(\mathbb{C}^3, 0)$ of type $U = \Phi^{-1}(D^2_\eta) \cap B_\epsilon$, and the total transform $\mathbf{D} := r^{-1}(D)$ is a normal crossing divisor. Moreover, we assume that the restriction of $r$ on $V^{emb} \setminus r^{-1}(Sing(V_f) \cup Sing(V_g))$ is a biholomorphic isomorphism. (Note that $Sing(V_f) \cup Sing(V_g)$ is a smaller set than the "usual" singular locus $Sing(D)$ of $D$; nevertheless, since the intersection $V_f \cap V_g$ is already transversal off the origin, the above assumption can always be satisfied for a convenient resolution.)

In particular,

$$\widetilde{\Phi} = \Phi \circ r : \left(r^{-1}(\Phi^{-1}(D^2_\eta \setminus \Delta_\Phi) \cap B_\epsilon), r^{-1}(\Phi^{-1}(D^2_\eta \setminus \Delta_\Phi) \cap \partial B_\epsilon)\right) \to D^2_\eta \setminus \Delta_\Phi$$

is a smooth locally trivial fibration of a pair of spaces.

Note that the topology of $r$ is rather complicated, more complicated than the topology of a germ defined on a normal surface singularity. While in that case the exceptional locus is a curve, here the exceptional locus is a surface. The description and characterization of the embedding and intersection properties of the components of $\mathbf{D}$ can be a rather difficult task. The point is that in our next construction we will not need all these data, but only a special curve configuration. This curve is identified by the vanishing behaviour of the pullbacks of $f$ and $g$ on $\mathbf{D}$.

Denote by $\mathbf{D}_c$ those irreducible components of the total transform $\mathbf{D}$ along which *only* $f \circ r$ vanishes, that is

$$\mathbf{D}_c = \overline{r^{-1}(V_f \setminus V_g)}.$$

Here $\overline{\cdot}$ denotes the closure. Similarly, consider

$$\mathbf{D}_d = \overline{r^{-1}(V_g \setminus V_f)} \text{ and } \mathbf{D}_0 = r^{-1}(V_f \cap V_g).$$

**Definition 6.1.3.** *The curve configuration* $\mathscr{C}$ *is defined by*

$$\mathscr{C} = (\mathbf{D}_c \cap \mathbf{D}_0) \cup (\mathbf{D}_c \cap \mathbf{D}_d).$$

Thus, for each irreducible component $C$ of $\mathscr{C}$, there are exactly two irreducible components $B_1$ and $B_2$ of $\mathbf{D}$ for which $C$ is a component of $B_1 \cap B_2$. By the definition of $\mathscr{C}$, we can assume that $B_1$ is such that *only* $f \circ r$ vanishes on it, and $B_2$ is such that either only $g \circ r$ vanishes on it, or both $f \circ r$ and $g \circ r$.

Let $m_{f,B_i}$ (respectively $m_{g,B_i}$) be the vanishing order of $f \circ r$ (respectively of $g \circ r$) along $B_i$ ($i = 1, 2$). Then $m_{f,B_1} > 0$, $m_{g,B_1} = 0$, $m_{f,B_2} \geq 0$, and $m_{g,B_2} > 0$. To the component $C$ we assign the triple $(m_{f,B_1}; m_{f,B_2}, m_{g,B_2})$.

A component $C$ of $\mathscr{C}$ is either a compact (projective) curve or it is non-compact, isomorphic to a complex disc. The union of the non-compact components is the strict transform of $V_f \cap V_g$. Therefore, $(m_{f,B_1}; m_{f,B_2}, m_{g,B_2}) = (1; 0, 1)$ for them. The compact components are exactly those which are contained in $r^{-1}(0)$.

The graph $\Gamma_{\mathscr{C}}$ is the dual graph of the curve configuration $\mathscr{C}$.

The set of vertices $\mathscr{V}$ consists of non-arrowheads $\mathscr{W}$ and arrowheads $\mathscr{A}$. The non-arrowhead vertices correspond to the compact irreducible curves of $\mathscr{C}$ while the arrowhead vertices correspond to the non-compact ones.

In $\Gamma_{\mathscr{C}}$ one connects the vertices $v_i$ and $v_j$ by $\ell$ edges if the corresponding curves $C_i$ and $C_j \subset \mathscr{C}$ intersect in $\ell$ points. Moreover, if a compact component $C_i \subset \mathscr{C}$, corresponding to a vertex $v_i \in \mathscr{W}$, intersects itself, then each self-intersection point determines a loop supported by $v_i$ in the graph $\Gamma_{\mathscr{C}}$. The edges are not directed.

One decorates the graph $\Gamma_{\mathscr{C}}$ as follows:

1. Each non-arrowhead vertex $v \in \mathscr{W}$ has two weights: the ordered triple of integers $(m_{f,B_1}; m_{f,B_2}, m_{g,B_2})$ assigned to the irreducible component $C$ corresponding to $v$, and the genus $g$ of the normalization of $C$.
2. Each arrowhead vertex has a single weight: the ordered triple $(1; 0, 1)$.
3. Each edge has a weight $\in \{1, 2\}$ determined as follows. By construction, any edge corresponds to an intersection point of three local irreducible components of $\mathbf{D}$. Among them either one or two local components belong to $\mathbf{D}_c$. Correspondingly, in the first case let the weight of the edge be 1, while in the second case 2.

## 6.2 Summary of Notation for $\Gamma_{\mathscr{C}}$ and Local Equations

The next table and local coordinate realizations will be helpful in the further discussions and proofs.

**Vertices:**

The vertex $\overset{(m;n,\nu)}{\underset{[g]}{\bullet}}$ codifies a compact irreducible component $C$ of $\mathscr{C}$ of genus $g$.

There is a local neighbourhood $U_p$ of any generic point $p \in C$ with local coordinates $(u, v, w)$ such that $U_p \cap \mathbf{D} = B_1^l \cup B_2^l$, $B_1^l = \{u = 0\}$, $B_2^l = \{v = 0\}$, and

$$f \circ r|_{U_p} = u^m v^n, \text{ and } g \circ r|_{U_p} = v^\nu \text{ with } m, \nu > 0;\ n \geq 0.$$

Here $B_i^l$ are *local* components of $\mathbf{D}$ at $p$. A missing $[g]$ means $g = 0$.

The arrowhead vertex ➛ $(1; 0, 1)$ codifies a non-compact component. The local description goes by the same principle as above.

**Edges:** An edge corresponds either to an intersection point $p \in C_i \cap C_j$, or to a self-intersection of $C_i$ if $i = j$. There is a local neighbourhood $U_p$ of $p$ with local coordinates $(u, v, w)$ such that $U_p \cap \mathbf{D} = B_1^l \cup B_2^l \cup B_3^l$, and $B_1^l = \{u = 0\}$, $B_2^l = \{v = 0\}$, $B_3^l = \{w = 0\}$. Moreover, the local equations of $f$ and $g$ are as follows:

*An edge with decoration 1:*

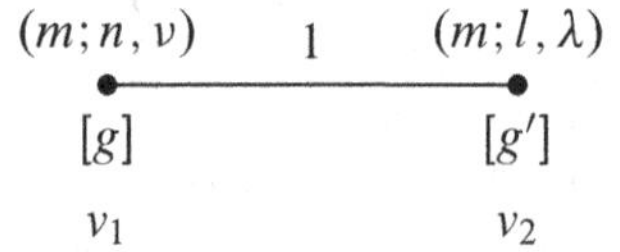


corresponds to local equations:

$$f \circ r|_{U_p} = u^m v^n w^l,\ g \circ r|_{U_p} = v^\nu w^\lambda,$$

where $m, \nu, \lambda > 0$ and $n, l \geq 0$.

One has similar equations with $m = \lambda = 1, l = 0, \nu > 0, n \geq 0$ if $v_2$ is an arrowhead:

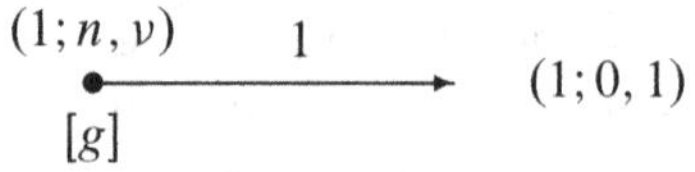


*An edge with decoration 2:*

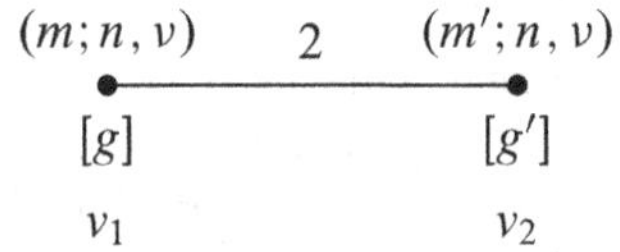


provides local equations:

$$f \circ r|_{U_p} = u^m v^{m'} w^n,\ \ g \circ r|_{U_p} = w^\nu$$

with $m, m', \nu > 0$ and $n \geq 0$.

One has similar equations with $m' = \nu = 1, n = 0$ and $m > 0$ if $v_2$ is an arrowhead:

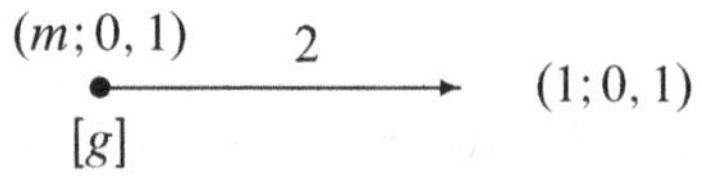


**Remark 6.2.4.** There is a compatibility between the weights that sometimes simplifies the decorations. Indeed, consider the following edge:

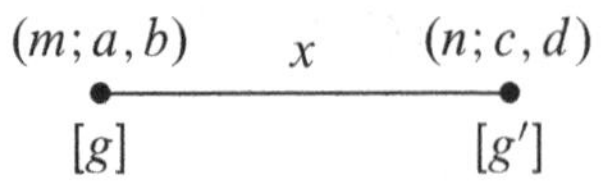


(a) if $m \neq n$, then $(a, b) = (c, d)$ and $x = 2$;
(b) if $(a, b) \neq (c, d)$, then $m = n$ and $x = 1$.

In particular, in the cases (a)–(b) above, the weight of the edge is determined by the weights of the vertices, hence, in these cases it can be omitted.

**Remark 6.2.5.** Clearly, different resolutions $r$ provide different curve configurations $\mathscr{C}$. In particular, the graph $\Gamma_{\mathscr{C}}$ is not unique. We believe that there is a "calculus" of such graphs connecting different graphs $\Gamma_{\mathscr{C}}$ coming from different embedded resolutions of $(D,0) \subset (\mathbb{C}^3,0)$, see the open problem 24.4.23 at the end of the book.

**Remark 6.2.6.** It is rather long and difficult to find a resolution $r$. In the literature there are very explicit resolution algorithms, but they are rather involved, and in general very slow, and usually with many irreducible exceptional components. Nevertheless, in the next Chapters we list many examples.

Finding the resolution of those examples, which do not have some specific form (which would help to find a canonical sequence of modifications or a direct resolution) we use a sequence of ad hoc blow ups following the naive principle: "blow up the worst singular locus", with the hope to obtain a more or less small configuration. Some of the computations are long, and are not given here. (Several of them were, in fact, done with the help of *Mathematica*.) Hence, we admit that for the reader the verification of some of the examples of $\Gamma_{\mathscr{C}}$ listed in the body of the book can be a really difficult job. Also, since the resolution procedure is not unique, an independent computation might lead to a different $\Gamma_{\mathscr{C}}$.

On the other hand, we will also list several families where we can find in a conceptual way resolutions which reflect the geometry and the structure of the singularities. The next chapters, starting from Chap. 8, contain an abundance of them.

Here we list some preliminary (specially chosen) examples in order to help the reader follow the first properties of $\Gamma_{\mathscr{C}}$ discussed in Chap. 7.

**Example 6.2.7.** Assume that $f = x^2y^2 + z^2(x+y)$ and take $g = x+y+z$. Then a possible $\Gamma_{\mathscr{C}}$ is:

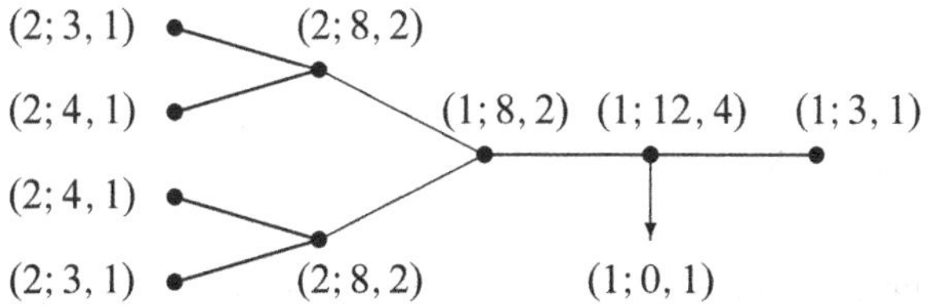


**Example 6.2.8.** Assume that $f = y^3 + (x^2 - z^4)^2$ and $g = z$. Then a possible $\Gamma_{\mathscr{C}}$ is:

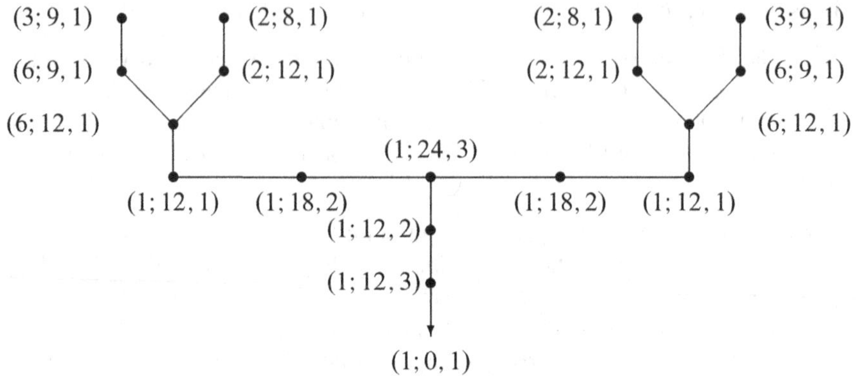


**Example 6.2.9.** Assume that $f = x^3y^7 - z^4$ and $g = x + y + z$. A possible graph $\Gamma_{\mathscr{C}}$ is:

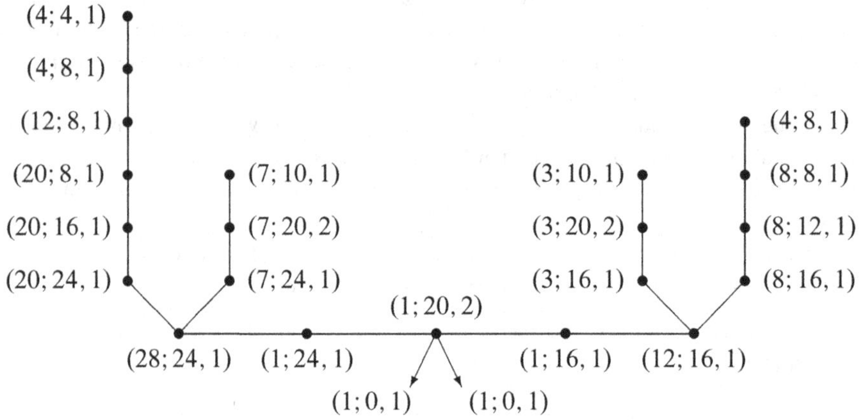


## 6.3 Assumption A

In the graph $\Gamma_{\mathscr{C}}$ a special attention is needed for edges of weight 2 with both end-vertices having first multiplicity $m = 1$ (including the case of loops too, when the two end-vertices coincide). Such an edge corresponds to a point $p$ which lies at the intersection of three locally irreducible components on exactly two of which only $f \circ r$ vanishes and that happens with multiplicity one. Thus the intersection of these two locally irreducible components is the strict transform of a component of $Sing(V_f)$ with transversal type $A_1$, which has *not* been blown up during the resolution procedure $r$.

Performing an additional blow up along this intersection, we obtain a new embedded resolution $r'$, whose special curve configuration $\mathscr{C}'$ will have an additional rational curve. The relevant edge of the dual graph $\Gamma_{\mathscr{C}}$ is changed to the dual graph $\Gamma_{\mathscr{C}'}$ with a new part via the following transformation:

$$
\begin{array}{ccc}
(1;n,v) \quad\;\; 2 \quad\;\; (1;n,v) & & (1;n,v) \quad 2 \quad (2;n,v) \quad 2 \quad (1;n,v) \\
\cdots\, \bullet \!\!-\!\!-\!\!-\!\!-\!\!-\!\!-\!\!-\!\!-\!\!-\!\! \bullet \cdots & \longrightarrow & \cdots\, \bullet \!\!-\!\!-\!\!-\!\! \bullet \!\!-\!\!-\!\!-\!\! \bullet \cdots \\
[g] \qquad\qquad\qquad [g'] & & [g] \qquad\qquad\qquad\qquad\qquad [g']
\end{array}
$$

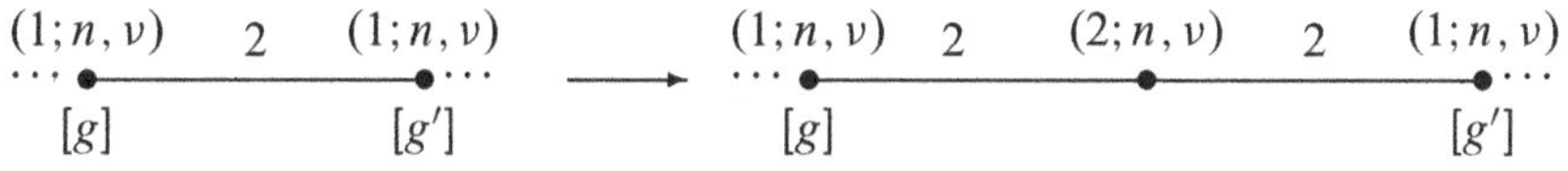

In order to avoid some pathological cases in the discussion, and to have a uniform treatment of the properties of $\Gamma_{\mathscr{C}}$ (e.g. of the "cutting edges" and subgraphs $\Gamma_{\mathscr{C}}^1$ and $\Gamma_{\mathscr{C}}^2$, see the next chapter), we will assume that $\Gamma_{\mathscr{C}}$ has no such edges, that is, we have already performed the extra blow-ups, if it was necessary. The same discussion/assumption is valid for loops and for the situation when one of the end-vertices above is replaced by an arrowhead.

This assumption is not crucial at all, the interested reader might eliminate it at the price of having to slightly modify the statements of the forthcoming Sects. 7.3 and 7.4. In fact, in the algorithm which provides $\partial F$, Assumption A is irrelevant.

# Chapter 7
# The Graph $\Gamma_{\mathscr{C}}$: Properties

## 7.1 Why One Should Work with $\mathscr{C}$?

From the definition 6.1.3 of the curve $\mathscr{C}$ it is not so transparent why exactly this configuration should play the crucial role in several results regarding non-isolated hypersurface singularities. In this section we wish to stress a universal property of $\mathscr{C}$, which motivates the definition, and will imply some immediate properties as well. We keep the notations of Chap. 6. In particular, $\Phi : \Phi^{-1}(D^2_\eta) \cap B_\epsilon \to D^2_\eta$ is a good representative, $(c, d)$ are the coordinates of the target, and $\Phi(\Sigma_f) = \Delta_1 = \{c = 0\}$.

We start with the definition of a special set "near" the discriminant component $\Delta_1$. For any integer $M > 0$, define the *wedge neighbourhood* of $\Delta_1$ by

$$\mathsf{W}_{\eta,M} = \{(c,d) \in D^2_\eta \mid 0 < |c| < |d|^M\}.$$

It is easy to verify that

$$\mathsf{W}_{\eta,M} \subset D^2_\eta \setminus \Delta_\Phi \ \text{ provided that } \ M \gg 0. \tag{7.1.1}$$

Hence, $\mathsf{W}_{\eta,M}$ is a small tubular neighbourhood of $\Delta_1 \setminus \{0\}$, not intersecting the other components of the discriminant. In particular, the restriction of $\Phi$ over $\mathsf{W}_{\eta,M}$ is a smooth locally trivial fiber bundle, equivalent with the restriction of the fibration over the torus $T_\delta$, containing all the information about the commuting monodromies $m_{\Phi,hor}$ and $m_{\Phi,ver}$ near $\Delta_1$, cf. Sect. 3.1.

The geometry of the fibration over $\mathsf{W}_{\eta,M}$, or of these two monodromies, were frequently used in the literature in connection with the "correction terms" of several singularity invariants. More precisely, if $i$ denotes a numerical invariant, then, usually $i(f+g^k)$, associated with the series $f+g^k$ approximating $f$ with $k$ large, are not easy to determine. Nevertheless, the correction term $i(f+g^k) - i(f)$, in many cases, depends only on the behaviour of $\Phi$ above $\mathsf{W}_{\eta,M}$. This is the source of several formulas: in the classical case of Iomdin $i$ is the Euler characteristic of

A. Némethi and Á. Szilárd, *Milnor Fiber Boundary of a Non-isolated Surface Singularity*, Lecture Notes in Mathematics 2037, DOI 10.1007/978-3-642-23647-1_7,

the Milnor fiber [49], in Siersma's article [117] $i$ is the zeta function, in M. Saito's paper [111] $i$ is the spectrum, while in [82,83] $i$ is the signature of the Milnor fiber.

Now, the restriction of $\Phi$ above $\mathsf{W}_{\eta,M}$ provides an alternative definition/characterization of the curve configuration $\mathscr{C}$, associated with the resolution $r$.

**Lemma 7.1.2. First characterization of** $\mathscr{C}$ *[92, (5.5)].*
*Fix a resolution $r$ and set $\widetilde{\Phi} = \Phi \circ r$. Then, for $M \gg 0$, one has*

$$\overline{\widetilde{\Phi}^{-1}(\mathsf{W}_{\eta,M})} \cap \widetilde{\Phi}^{-1}(0) = \mathscr{C}. \tag{7.1.3}$$

*Proof.* The proof follows by case-by-case local verification in the neighbourhood of different type of points of the resolution $r$, and from the properness of $\widetilde{\Phi}$.

For example, let us take a point $p \in \mathbf{D}_0 \setminus \mathscr{C}$. We have to show that $p \notin \overline{\widetilde{\Phi}^{-1}(\mathsf{W}_{\eta,M})}$. Assume the contrary, and take a local coordinate neighbourhood $U_p$ of $p$ with local coordinates $(u,v,w)$ such that $\widetilde{\Phi}|_{U_p} = (u^m, u^\mu)$ for some integers $\mu, m > 0$ (up to an invertible element). Then there exists a sequence $\{p_j = (u_j, v_j, w_j)\}_{j=1}^{\infty}$ in $\widetilde{\Phi}^{-1}(\mathsf{W}_{\eta,M}) \cap U_p$ such that $\lim_{j\to\infty} p_j = p$, hence $u_j \to 0$. Since $p_j \in \widetilde{\Phi}^{-1}(\mathsf{W}_{\eta,M})$ it follows that $|u_j|^m < |u_j|^{M\mu}$. This leads to a contradiction for $M$ sufficiently large.

The other local verifications are similar and are left to the reader. □

Therefore, one expects, that from the dual graph $\Gamma_{\mathscr{C}}$ of the special curve configuration $\mathscr{C}$, endowed with all the necessary multiplicity data (codifying the local behavior of $f$ and $g$ near $\mathscr{C}$), one is able to extract topological information about any set $\mathscr{S} \subset \Phi^{-1}(D_\eta^2) \cap B_\epsilon$ with $\Phi(\mathscr{S}) \subset \mathsf{W}_{\eta,M}$.

Note that for $\delta > 0$ sufficiently small, $\partial D_\delta \in \mathsf{W}_{\eta,M}$, hence $\Phi^{-1}(\partial D_\delta)$, the non-trivial part of the boundary of the Milnor fiber (by the discussion of Remark 3.1.11) is such a space. Hence, $\Phi$ over $\mathsf{W}_{\eta,M}$, or its lifting via $r$ codified in $\Gamma_{\mathscr{C}}$, should contain crucial information regarding $\partial F$ and its Milnor monodromy.

This fact is exploited in the present work.

Now we continue with some immediate consequences. From (7.1.3) one obtains the following.

**Corollary 7.1.4.** *a) For any open (tubular) neighbourhood $T(\mathscr{C}) \subset V^{emb}$, there exist a sufficiently large $M$ and a sufficiently small $\eta$ such that for any $(c,d) \in \mathsf{W}_{\eta,M}$ the "lifted Milnor fiber" $\widetilde{\Phi}^{-1}(c,d)$ is in $T(\mathscr{C})$.*

*b) For any $p \in \mathscr{C}$ and local neighbourhood $U_p$ of $p$, there exist a sufficiently large $M$ and a sufficiently small $\eta$ such that for any $(c,d) \in \mathsf{W}_{\eta,M}$ one has $U_p \cap \widetilde{\Phi}^{-1}(c,d) \neq \emptyset$.*

Therefore, the curve $\mathscr{C}$ can be regarded as the "limit" of the lifted Milnor fiber. In particular, from 7.1.4 and from the connectedness of the Milnor fiber, see 3.1.7, we obtain that

**Corollary 7.1.5.** *$\mathscr{C}$ is connected.*

**Remark 7.1.6.** We point out a natural decomposition of $\widetilde{F}_{\Phi}$, as an immediate consequence of the fact that the curve $\mathscr{C}$ is a "limit" of this fiber. Here $\widetilde{F}_{\Phi} = \widetilde{\Phi}^{-1}(c,d)$, with $(c,d) \in \mathsf{W}_{\eta,M}$ as in 7.1.4. The construction resonate with the construction of the universal cyclic covering graph presented in 5.2.1.

For each edge $e$ of $\Gamma_{\mathscr{C}}$, let $P_e$ be the corresponding intersection point of two components of $\mathscr{C}$. Let $U_{P_e}$ be a small neighbourhood of $P_e$. Then $\widetilde{F}_{\Phi}$ intersects $U_{P_e}$ in a tubular neighbourhood of some embedded circles of $\widetilde{F}_{\Phi}$. Consider the collection $B$ of all these circles. Then $\widetilde{F}_{\Phi}$ is a union

$$\bigcup_{v \in \mathscr{V}(\Gamma_{\mathscr{C}})} \widetilde{F}_v, \tag{7.1.7}$$

where each $\widetilde{F}_v$ is the closure of those components of $\widetilde{F}_{\Phi} \setminus B$ which are sitting in a tubular neighbourhood of the corresponding component $C_v$ of $\mathscr{C}$. Moreover, both horizontal and vertical monodromy actions over the torus $T_\delta$ (cf. Sect. 3.1) can be chosen in such a way that they preserve the cutting circles $B$ and each subset $\widetilde{F}_v$. Furthermore, their restrictions on each $\widetilde{F}_v$ are isotopic to a pair of commuting finite actions on $\widetilde{F}_v$. (This last fact follows from the fact that any $S^1$-bundle over a non-closed Riemann surface is trivial, and from the particular form of the local equations of $f$ and $g$ near $C_v$, cf. Sect. 6.2.)

The above limit procedure can be pushed further. Recall that $\Delta_1 = \{c = 0\} \cap D^2_\eta$, hence $\Delta_1$ is in the closure of $\mathsf{W}_{\eta,M}$. Hence, the limit procedure $(c,d) \mapsto (0,0)$ with $(c,d) \in \mathsf{W}_{\eta,M}$ can be done in two steps: first $(c,d)$ tends to some point of $\Delta_1 \setminus \{0\}$, then along this discriminant component we approach the origin. The analogues of 7.1.2 and 7.1.4 are the following:

**Lemma 7.1.8. Second characterization of $\mathscr{C}$** *[92, (5.8)].*
*With the same notations as in 7.1.2, one has*

$$\overline{\widetilde{\Phi}^{-1}(\Delta_1 \setminus \{0\})} \cap \widetilde{\Phi}^{-1}(0) = \mathscr{C}. \tag{7.1.9}$$

*In particular,*

a) *For any open tubular neighbourhood $T(\mathscr{C}) \subset V^{emb}$, if $|d|$ is sufficiently small then $\widetilde{\Phi}^{-1}((0,d)) \subset T(\mathscr{C})$.*
b) *For any $p \in \mathscr{C}$ and neighbourhood $U_p$ of $p$, if $|d|$ is sufficiently small then $U_p \cap \widetilde{\Phi}^{-1}((0,d)) \neq \emptyset$.*

This lemma will also have connectivity consequences, see 7.4.3.

## 7.2 A Partition of $\Gamma_{\mathscr{C}}$ and Cutting Edges

We start to review from [92] those properties of $\Gamma_{\mathscr{C}}$ that are needed in the sequel.

**Proposition 7.2.1.** *$\Gamma_{\mathscr{C}}$ is connected.*

*Proof.* Use 7.1.5 and the construction of $\Gamma_{\mathscr{C}}$. □

**Definition 7.2.2.** *The vertices of the graph $\Gamma_{\mathscr{C}}$ can be divided into two disjoint sets $\mathscr{V}(\Gamma_{\mathscr{C}}) = \mathscr{V}^1(\Gamma_{\mathscr{C}}) \cup \mathscr{V}^2(\Gamma_{\mathscr{C}})$, where $\mathscr{V}^1(\Gamma_{\mathscr{C}})$ (respectively $\mathscr{V}^2(\Gamma_{\mathscr{C}})$) consists of the vertices decorated by $(m; n, \nu)$ with $m = 1$ (respectively $m \geq 2$).*

*We will use similar notations for $\mathscr{W}(\Gamma_{\mathscr{C}})$ and $\mathscr{A}(\Gamma_{\mathscr{C}})$.*

*The set of edges $(v_1, v_2)$ with ends $v_1 \in \mathscr{V}^1(\Gamma_{\mathscr{C}})$ and $v_2 \in \mathscr{W}^2(\Gamma_{\mathscr{C}})$ will be called* cutting edges. *Their edge-decoration is always 2. We denote their index set by $\mathscr{E}_{cut}$.*

Note that $\mathscr{A}(\Gamma_{\mathscr{C}}) = \mathscr{A}^1(\Gamma_{\mathscr{C}})$, hence $\mathscr{V}^2(\Gamma_{\mathscr{C}}) = \mathscr{W}^2(\Gamma_{\mathscr{C}})$.

According to the decomposition $\mathscr{V} = \mathscr{V}^1 \cup \mathscr{V}^2$, we partition $\Gamma_{\mathscr{C}}$ into two graphs $\Gamma^1_{\mathscr{C}}$ and $\Gamma^2_{\mathscr{C}}$.

The description of the subgraphs $\Gamma^1_{\mathscr{C}}$ and $\Gamma^2_{\mathscr{C}}$ is the subject of the next sections.

## 7.3 The Graph $\Gamma^1_{\mathscr{C}}$

**7.3.1. The construction of $\Gamma^1_{\mathscr{C}}$.** The graph $\Gamma^1_{\mathscr{C}}$ is constructed in two steps.

First, consider the maximal subgraph of $\Gamma_{\mathscr{C}}$, which is spanned by the vertices $v \in \mathscr{V}^1(\Gamma_{\mathscr{C}})$ and has no edges of weight 2. Next, corresponding to each cutting edge – whose end-vertices $v_1$ and $v_2$ carry weights $(1; n, \nu)$ and $(m; n, \nu)$, $m > 1$ respectively – add an arrowhead decorated with the weight $(m; n, \nu)$ connected to $v_1$ by an edge. We will keep the decoration 2 of these "inherited cutting edges", although their type can be recognized by the principle 6.2.4.

In particular, $\Gamma^1_{\mathscr{C}}$ has two types of arrowheads: first, all the arrowheads of $\Gamma_{\mathscr{C}}$ remain arrowheads of $\Gamma^1_{\mathscr{C}}$, all of them with weight $(1; 0, 1)$; then, each cutting edge provides an arrowhead with weights of type $(m; n, \nu)$, with first entry $m > 1$.

We wish to provide more details regarding a special situation.

Assume that an edge of $\Gamma_{\mathscr{C}}$, decorated by 2, supports an arrowhead. Then, by Assumption A, cf. 6.3, the other vertex of the edge should automatically have weight $(m; 0, 1)$ with $m > 1$. In such a case, this edge becomes a double arrow of $\Gamma^1_{\mathscr{C}}$: an edge supporting two arrowheads, one with weight $(1; 0, 1)$, the other with $(m; 0, 1)$. This double arrow forms a connected component of $\Gamma^1_{\mathscr{C}}$.

**7.3.2. The simplified graph $G^1_{\mathscr{C}}$.** Deleting some of the information of $\Gamma^1_{\mathscr{C}}$, we obtain another graph $G^1_{\mathscr{C}}$, which looks like the weighted embedded resolution graph of a germ of an analytic function defined on a normal surface singularity; that is, $G^1_{\mathscr{C}}$ will be a plumbing graph as in 4.3.

The construction runs as follows.

First, keep the genus-decorations of all non-arrowheads. Next, for any non-arrowhead vertex, and for any arrowhead with decoration $(1; 0, 1)$, replace the weight $(1; n, \nu)$ by $(\nu)$. The weight $(m; n, \nu)$ of an arrowhead vertex with $m > 1$ is replaced by weight $(0)$. Furthermore, delete all old edge-decorations, and insert everywhere the new edge-decoration $+$ (hence, they can be even omitted by our convention from 4.3). Finally, we determine the Euler numbers via (4.1.5) using edge-decorations $\epsilon_e = +$.

**Proposition 7.3.3.** *[92, (5.27)]* $G^1_{\mathscr{C}}$ *is a possible embedded resolution graph of*

$$V_f^{norm} \xrightarrow{g\circ n} (\mathbb{C},0),$$

*where* $n : (V_f^{norm}, n^{-1}(0)) \to (V_f, 0)$ *is the normalization.*

*In particular, the number of connected components of* $\Gamma^1_{\mathscr{C}}$ *coincides with the number of irreducible components of the germ* $(V_f, 0)$. *A connected component of* $\Gamma^1_{\mathscr{C}}$ *which consists of a double arrow, corresponds to a smooth component of* $V_f^{norm}$ *on which* $g \circ n$ *is smooth, and which has not been modified during the resolution.*

*The arrowheads with multiplicity* (0) *represent the strict transforms of the singular locus* $\Sigma_f$.

If we do not wish to preserve the information about the position of the strict transforms of the singular locus $\Sigma_f$, we delete the arrowheads with weight (0) from $G^1_{\mathscr{C}}$. What remains is exactly the disjoint union of the embedded resolution graphs of the connected components of $(V_f^{norm}, g \circ n)$.

In a similar way, if we do not wish to keep any information about the germ $g$, we delete all multiplicity decorations and arrowheads with positive multiplicities. What remains is the collection of resolution graphs of the components of $V_f^{norm}$, where the (0)-arrowheads mark the strict transforms of $\Sigma_f$.

**Example 7.3.4.** Assume that $f = x^3y^7 - z^4$ and $g = x + y + z$ as in 6.2.9.

There are two cutting edges: the extreme edges of the horizontal string.

The graph $\Gamma^1_{\mathscr{C}}$ is the following:

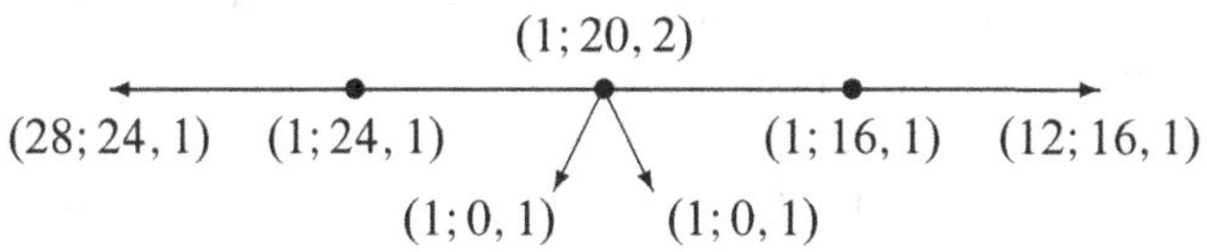


$G^1_{\mathscr{C}}$ is the graph:

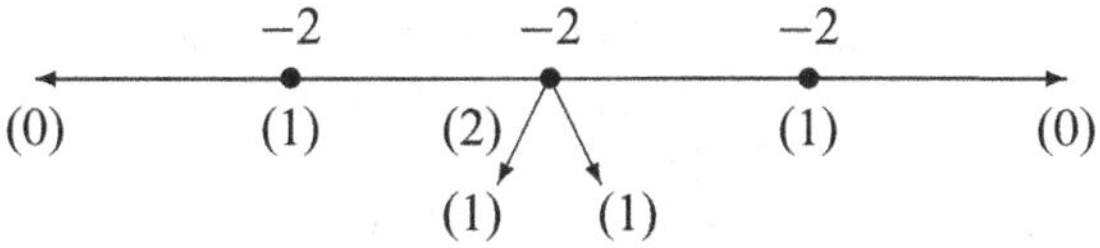


The two (0)-arrowheads correspond to the two components of the strict transform of the singular locus of $V_f$. The normalization of $V_f$ is an $A_3$ singularity.

## 7.4 The Graph $\Gamma^2_{\mathscr{C}}$

**7.4.1. The construction of $\Gamma^2_{\mathscr{C}}$.**
The "complementary" subgraph $\Gamma^2_{\mathscr{C}}$ is constructed in two steps as well.

First, consider the maximal subgraph of $\Gamma_{\mathscr{C}}$ spanned by the vertices $v \in \mathscr{V}^2(\Gamma_{\mathscr{C}})$. Then, corresponding to each cutting edge (cf. 7.2.2), whose end-vertices $v_1$ and $v_2$ carry weights $(1; n, \nu)$ and $(m; n, \nu)$ $(m > 1)$ respectively, regardless of $v_1$ being an arrowhead or not, glue an edge decorated by 2 to $v_2$, and make its other end-vertex an arrowhead weighted $(1; n, \nu)$ .

This ends the definition of $\Gamma^2_{\mathscr{C}}$. In order to understand its connected components, we need the following notation.

**Definition 7.4.2.** *For any $j \in \{1, \ldots, s\}$, we denote by $\mathbf{D}_{c,j}$ the collection of those components $B \subset \mathbf{D}_c$ which are projected via r* onto *$\Sigma_j$. Furthermore, define $\mathscr{C}_{\Sigma_j} \subset \mathscr{C}$ as the union of those irreducible components $C$ of $\mathscr{C}$ weighted $(m; n, \nu)$ with $m > 1$, for which $C \subset B$ for some component $B \subset \mathbf{D}_{c,j}$.*

Now we are ready to start the list of properties of $\Gamma^2_{\mathscr{C}}$.

**Proposition 7.4.3.** *[92, (5.32)] There is a one-to-one correspondence between the connected components of $\Gamma^2_{\mathscr{C}}$ and the irreducible components of $(Sing(V_f), 0) = (\Sigma, 0) = \cup_j \Sigma_j$.*

*More precisely, $\mathscr{C}_{\Sigma_j}$ is connected and its irreducible components correspond to the non-arrowhead vertices of one of the connected components $\Gamma^2_{\mathscr{C},j}$ of $\Gamma^2_{\mathscr{C}}$.*

*Proof.* Fix $j$. The connectedness of $\mathscr{C}_{\Sigma_j}$ follows from the next claim, similar to the limit properties 7.1.4 and 7.1.2.

a) For any open tubular neighbourhood $T(\mathscr{C}_{\Sigma_j}) \subset V^{emb}$ of $\mathscr{C}_{\Sigma_j}$, there exists a sufficiently small $\gamma > 0$ such that for any point $q \in \Sigma_j - \{0\} \subset \mathbb{C}^3$ with $|q| < \gamma$, $r^{-1}(q) \subset T(\mathscr{C}_{\Sigma_j})$.
b) For any $p \in \mathscr{C}_{\Sigma_j}$ and local neighbourhood $U_p$ of $p$, there exists a sufficiently small $\gamma > 0$ such that for any point $q \in \Sigma_j - \{0\}$ with $|q| < \gamma$, $U_p \cap r^{-1}(q) \neq \emptyset$.

These two statements can be verified by similar local computations as in 7.1.4. In fact, they can be deduced from 7.1.2 too.

This means that $\mathscr{C}_{\Sigma_j}$ is the "limit" of $r^{-1}(q) \cap \mathbf{D}_{c,j}$, where $q \in \Sigma_j \setminus \{0\}$ tends to 0. But the modification $r$, above any transversal slice $Sl_q$ at $q$ of $\Sigma_j$ (cf. Sect. 2.2), realizes an embedded resolution of $(Sl_q, Sl_q \cap V_f, q)$, hence $r^{-1}(q) \cap \mathbf{D}_{c,j}$ constitutes a collection of exceptional curves of an embedded resolution of this plane curve singularity. Hence, by Zariski's Main Theorem, cf. 4.3.4, it is connected. This implies that its limit $\mathscr{C}_{\Sigma_j}$ is connected as well. □

**Example 7.4.4.** Consider the example given in 6.2.9. In this case $\Gamma^2_{\mathscr{C}}$ has two components. One of them (situated at the left of the diagram) is $\Gamma^2_{\mathscr{C},1}$:

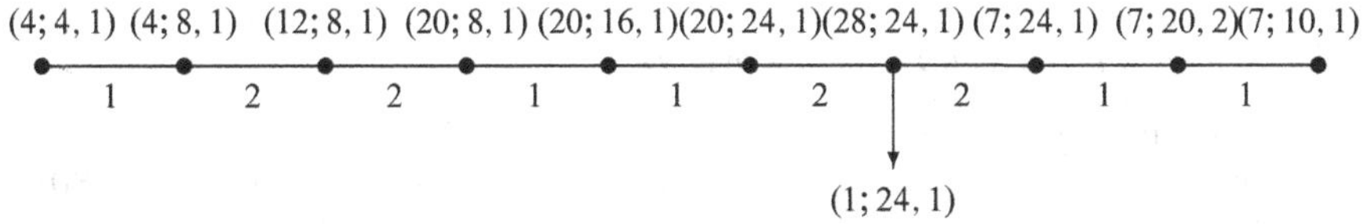


In fact, besides other data, $\Gamma^2_{\mathscr{C},j}$ contains all the information about the equisingularity type of the transversal singularity $T\Sigma_j$ of $\Sigma_j$. In order to make this statement more precise, we need some preparation.

**7.4.5. A partition of $\Gamma^2_{\mathscr{C},j}$.** We fix again the index $j$.

We introduce on $\mathscr{W}(\Gamma^2_{\mathscr{C},j})$ the following equivalence relation. First, we say that $w_1 \sim w_2$ if $w_1$ and $w_2$ are connected by an edge of weight 1, then we extend $\sim$ to an equivalence relation. If $K = \{w_{i_1}, \dots, w_{i_t}\}$ is an equivalence class, with decorations $(m; n_{i_l}, \nu_{i_l})$, then set $\nu(K) := \gcd(\nu_{i_1}, \dots, \nu_{i_t})$ and $m(K) := m$.

Each class $K$ defines a connected subgraph $G(K)$ of $\Gamma^2_{\mathscr{C},j}$ with vertices from $K$ and all the 1-edges connecting them. For the moment we keep all the decorations of the corresponding vertices of $G(K)$.

The equivalence classes $\{K_\ell\}_\ell$ determine a partition of $\mathscr{W}(\Gamma^2_{\mathscr{C},j})$. The subgraphs $G(K_\ell)$ are connected by 2-edges.

**7.4.6. Properties of the subgraphs $G(K)$.**

For a fixed equivalence class $K = \{w_{i_1}, w_{i_2}, \dots, w_{i_t}\}$, consider the corresponding irreducible curves $C_{i_1}, C_{i_2}, \dots, C_{i_t}$ of $\mathscr{C}$. By construction, there exists an irreducible component $B(K) \in \mathbf{D}_{c,j}$ which contains all of them. Moreover, the union $\mathscr{C}(K) := C_{i_1} \cup \dots \cup C_{i_t}$ is a connected curve. Let $T(K)$ denote a small tubular neighbourhood of $\mathscr{C}(K)$ in $B(K)$.

Note that the integer $m(K)$ is exactly the vanishing order of $f \circ r$ along $B(K)$. On the other hand, the local equations show that the restriction $g \circ r|_{T(K)} : T(K) \to \mathbb{C}$ provides the principal divisor $(g \circ r)^{-1}(0) = \sum_k \nu_{i_k} C_{i_k}$ in $T(K)$. Here, $T(K)$ can be changed to the inverse image of a small disc $D$ under $g \circ r|_{B(K)}$. Therefore, the divisor $\sum_k \nu_{i_k} C_{i_k}$ in $T(K)$ can be interpreted as a central fiber of the proper analytic map $g \circ r|_{T(K)} : T(K) \to D$.

**Lemma 7.4.7.** *The generic fiber of $(g \circ r)|_{T(K)}$ is a disjoint union of rational curves.*

*Proof.* Fix $d \neq 0$ with $|d|$ sufficiently small. Then $g^{-1}(d)$ intersects $\Sigma_j$ in $d_j$ points, say $\{q_i\}_i$, cf. 3.1.8. Then $((g \circ r)|T(K))^{-1}(d) = \cup_i (r^{-1}(q_i) \cap T(K))$. But $r^{-1}(q_i)$ is the singular fiber of an embedded resolution of the transversal singularity associated with $\Sigma_j$, and $r^{-1}(q_i) \cap T(K)$ is an irreducible curve of this exceptional locus, hence each $r^{-1}(q_i) \cap T(K)$ is a rational curve. □

Next, we recall a general property of a morphism whose generic fiber is rational, see [42], page 554:

**Fact.** *If $S$ is a minimal smooth surface, $D$ a disc in $\mathbb{C}$, and $\pi : S \to D$ any proper holomorphic map whose generic fiber $\pi^{-1}(d)$ $(0 \neq d \in D)$ is irreducible and rational, then $\pi$ is a trivial $\mathbf{P}^1$-bundle over $D$.*

If $S$ is not minimal, and the generic fiber of $\pi : S \to D$ is a disjoint union of (say, $N$) rational curves, then by the Stein Factorization theorem (see e.g. [45], page 280), and if necessary after shrinking $D$, there exists a map $b : D' \to D$ given by $z \mapsto z^N$, and $\pi' : S \to D'$ such that $\pi = b \circ \pi'$, and the generic fiber of $\pi'$ is irreducible and rational. Since the central fibers of $\pi$ and $\pi'$ are the same, it follows from the above fact that the central fiber of $\pi$ can be blown down successively until an irreducible rational curve is obtained. Being a principal divisor, its self-intersection is zero.

This discussion has the following consequences:

**Proposition 7.4.8. Properties of the graph $G(K)$.** *a) The graph $G(K)$ is a tree with all genus decorations $g_{w_k} = 0$. In particular, all the irreducible components of $\mathscr{C}_{\Sigma_j}$ are rational curves.*

*b) From the integers $\{\nu_{i_k}\}_{k=1}^t$ one can deduce the self-intersections $C_{i_k}^2$ of the curves $C_{i_k}$ in $B(K)$ as follows. First notice that the intersection matrix $(C_{i_k} \cdot C_{i_{k'}})_{k,k'}$, where the intersections are considered in $B(K)$, is a negative semi-definite matrix with rank $t-1$, and the central divisor $\sum_k \nu_{i_k} C_{i_k}$ is one element of its kernel (cf. [8], page 90). The intersections $C_{i_k} \cdot C_{i_{k'}}$ for $i_k \neq i_{k'}$ can be read from the graph $G(K)$ considered as a dual graph. Then the self-intersections can be determined from the relations $(\sum_k \nu_{i_k} C_{i_k}) \cdot C_{i_{k'}} = 0$. In particular, if $t = 1$, then $C_{i_1}^2 = 0$. If $t \geq 2$, then the graph is not minimal; if we blow down successively all the $(-1)$-curves we obtain a rational curve with self-intersection zero.*

*c) The number of irreducible (equivalently, connected) components of the generic fiber of $(g \circ r)|_{T(K)}$ is $\nu(K)$.*

*The fact that each irreducible component of the generic fiber is rational can be translated into the relation:*

$$2 \cdot \nu(K) = \sum_{k=1}^{t} \nu_{i_k} (2 - \delta_{w_{i_k}}^K),$$

*where $\delta_{w_{i_k}}^K$ is the number of vertices adjacent to $w_{i_k}$ in $G(K)$.*

**Example 7.4.9.** Consider the example 6.2.9 and its graph $\Gamma_{\mathscr{C},1}^2$ from 7.4.4. The graphs $G(K_l)$ are:

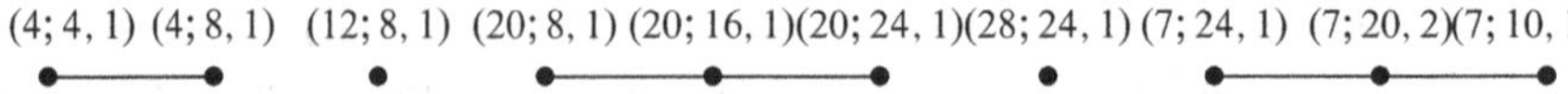

The corresponding central divisors and self-intersection numbers are the following:

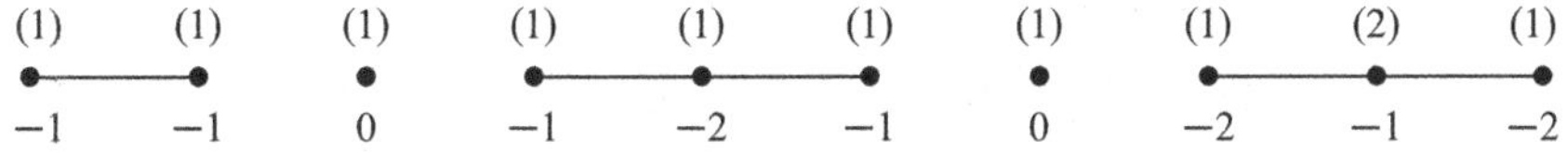


**7.4.10.** Now, we return to the connected component $\Gamma^2_{\mathscr{C},j}$ of $\Gamma^2_{\mathscr{C}}$ corresponding to $\Sigma_j$ $(1 \leq j \leq s)$. We consider the partition $\{G(K_\ell)\}_\ell$ of $\Gamma^2_{\mathscr{C},j}$; they are connected by 2-edges. The geometry behind the next discussion is the following.

Recall that $T\Sigma_j$ denotes the equisingular type of the transversal singularity associated with $\Sigma_j$, cf. Sect. 2.2, and $\deg(g|\Sigma_j) = d_j$, cf. 3.1.8. If $(Sl, q)$ is a transversal slice as in Sect. 2.2, then $r$ above $(Sl, q)$ determines a resolution of the transversal plane curve singularity $(Sl, Sl \cap V_f, q)$. We denote its weighted dual embedded resolution graph by $G(T\Sigma_j)$. Since in local coordinates it is easier to work with the pullback of $g$, it is convenient to replace the single point $q \in \Sigma_j - \{0\}$ by the collection of $d_j$ points $g^{-1}(d) \cap \Sigma_j$, where $|d|$ is small and non-zero. The dual weighted graph associated with the curves situated above these points consists of exactly $d_j$ identical copies of $G(T\Sigma_j)$, and it is denoted by $d_j \cdot G(T\Sigma_j)$.

Comparing the curves $r^{-1}(g^{-1}(d) \cap \Sigma_j)$ and $\mathscr{C}_{\Sigma_j}$ via the corresponding local equations, and using the results of Proposition 7.4.8, we obtain a cyclic covering of graphs

$$p : d_j \cdot G(T\Sigma_j) \to \{\text{a base graph}\}_j,$$

where the base graph and the covering data can be determined from $\Gamma^2_{\mathscr{C},j}$. This is given in the next paragraphs.

**The base graph** will be denoted by $\Gamma^2_{\mathscr{C},j}/\sim$. It is obtained from $\Gamma^2_{\mathscr{C},j}$ by collapsing it along edges of weight 1. More precisely, each subgraph $G(K_\ell)$ is replaced by a non-arrowhead vertex. If two subgraphs $G(K_\ell)$ and $G(K_{\ell'})$ are connected by $k$ 2-edges in $\Gamma^2_{\mathscr{C},j}$, then the corresponding vertices of $\Gamma^2_{\mathscr{C},j}/\sim$ are connected by $k$ edges. (In fact, in 7.4.12 we will see that each $k \leq 1$.) If the non-arrowhead vertices of $G(K_\ell)$ support $k$ arrowheads altogether, then on the corresponding non-arrowhead vertex of $\Gamma^2_{\mathscr{C},j}/\sim$ one has exactly $k$ arrowheads.

Since $\Gamma^2_{\mathscr{C},j}$ is connected, it is obvious that $\Gamma^2_{\mathscr{C},j}/\sim$ is connected as well.

**The covering data of** $p : d_j \cdot G(T\Sigma_j) \to \Gamma^2_{\mathscr{C},j}/\sim$.

Recall from 5.1 that the *covering data* of a projection $p : G \to \Gamma$ is a collection of positive integers $\{\mathfrak{n}_v\}_{v\in\mathscr{V}(\Gamma)}$ and $\{\mathfrak{n}_e\}_{e\in\mathscr{E}(\Gamma)}$, such that for each edge $e = (v_1, v_2) \in \mathscr{E}(\Gamma)$ one has $\mathfrak{n}_e = \mathfrak{d}_e \cdot \operatorname{lcm}(\mathfrak{n}_{v_1}, \mathfrak{n}_{v_2})$ for some integer $\mathfrak{d}_e$.

Now, we define a covering data for $\Gamma^2_{\mathscr{C},j}/\sim$. It is provided by the third entries $\nu$ of the weights $(m; n, \nu)$ of the vertices of $\Gamma^2_{\mathscr{C},j}$ and will be denoted by $\boldsymbol{\nu}$.

For any non-arrowhead vertex $w$ of $\Gamma^2_{\mathscr{C},j}/\sim$, which corresponds to $K_\ell$ in the above construction, set $\mathfrak{n}_w := \nu(K_\ell)$. For any arrowhead vertex $v$ of $\Gamma^2_{\mathscr{C},j}/\sim$, which corresponds to an arrowhead of $\Gamma^2_{\mathscr{C},j}$ with weight $(1; n, \nu)$, set $\mathfrak{n}_v := \nu$. For any edge of $\Gamma^2_{\mathscr{C},j}/\sim$, which comes from a 2-edge $e$ of $\Gamma^2_{\mathscr{C},j}$ with endpoints with weight $(*; n, \nu)$, set $\mathfrak{n}_e := \nu$.

The degeneration of $r^{-1}(g^{-1}(d) \cap \Sigma_j)$ into $\mathscr{C}_{\Sigma_j}$ provides the next result:

**Theorem 7.4.11. Characterization of the transversal singularities.**

*(a) For any $j$ there exists a cyclic covering of graphs*

$$p : d_j \cdot G(T\Sigma_j) \to \Gamma^2_{\mathscr{C},j}/\sim$$

*with covering data $\mathbf{v}$ and with the compatibility of the arrowheads: $\mathscr{A}(d_j \cdot G(T\Sigma_j)) = p^{-1}(\mathscr{A}(\Gamma^2_{\mathscr{C},j}/\sim))$, cf. 5.1.9.*

*(b) The decorations of $G(T\Sigma_j)$ can be recovered from the decorations of $\Gamma^2_{\mathscr{C},j}$ as follows: $m_w = m(K_\ell)$ for any $w \in \mathscr{W}(G(T\Sigma_j))$ sitting above a vertex corresponding to $K_\ell$; $m_v = 1$ for any arrowhead. The Euler numbers are determined via (4.1.5).*

*In particular, the weighted dual embedded resolution graph $G(T\Sigma_j)$ can be completely determined from the weighted graph $\Gamma^2_{\mathscr{C},j}$.*

**Corollary 7.4.12.** *1. $\Gamma^2_{\mathscr{C},j}$ is a connected tree.*

*2. With covering data $\mathbf{v}$, there is only one cyclic graph covering $p : G \to \Gamma^2_{\mathscr{C},j}/\sim$.*

*3. $d_j = \gcd\{\nu_w | w \in \mathscr{W}(\Gamma^2_{\mathscr{C},j})\}$, where $(m_w; n_w, \nu_w)$ is the weight of $C_w$*

*Proof.* The first part follows from the connectedness statement from 7.4.3, form the fact that $G(T\Sigma_j)$ is a tree, and from 5.1.7. The second and third parts follow from 5.1.6(1) and the connectedness of $G(T\Sigma_j)$. □

**Remark 7.4.13.** 1. For an example when the covering $p : d_j \cdot G(T\Sigma_j) \to \Gamma^2_{\mathscr{C},j}/\sim$ is not a bijection, see 7.5.5.

2. As we emphasized in 7.4.8(b–c), the collection of integers $\{\nu_w\}$ $(w \in \mathscr{W}(\Gamma^2_{\mathscr{C},j}))$ satisfies serious compatibility restrictions. Moreover, since in the cyclic covering $d_j \cdot G(T\Sigma_j) \to \Gamma^2_{\mathscr{C},j}/\sim$ the covering graph $d_j \cdot G(T\Sigma_j)$ has no cycles, this imposes some additional restrictions on the integers $\{\nu_w\}$ $(w \in \mathscr{W}(\Gamma^2_{\mathscr{C},j}))$.

3. Corollary 7.4.12(3) implies that, in fact, there is a graph covering of connected graphs

$$G(T\Sigma_j) \to \Gamma^2_{\mathscr{C},j}/\sim$$

whose covering data are those from $\nu$, where all integers are divided by $d_j$.

**Example 7.4.14.** Let us continue the example 7.4.9 (as the continuation of 6.2.9 and 7.4.4). In this case $d_1 = 1$ and the $G(T\Sigma_1) \to \Gamma^2_{\mathscr{C},1}/\sim$ is a bijection. Hence the algorithm gives for $G(T\Sigma_1)$ the embedded resolution graph

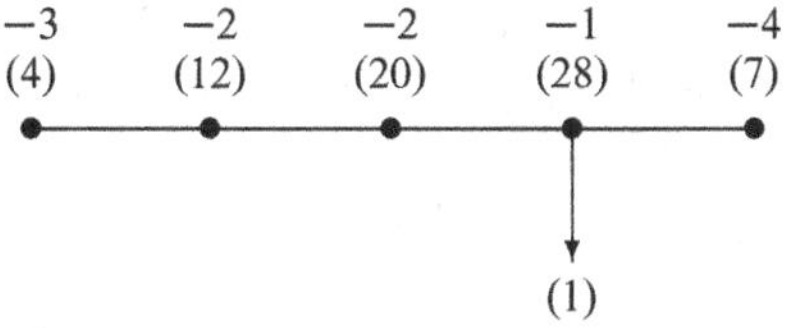


This is, of course, the minimal embedded resolution graph of $T\Sigma_1$ with local equation $u^7 - v^4 = 0$, as it is expected from the equations of $f = x^3y^7 - z^4$.

**Example 7.4.15.** Consider the graph in 6.2.8 for $f = y^3 + (x^2 - z^4)^2$ and $g = z$. $\Sigma = \{y = x^2 - z^4 = 0\}$ has two components, the transversal type of which are cusps of type $(2, 3)$. In the next diagram we put in dash-boxes the equivalence classes $K$ and the supports of the two components $\{\Gamma^2_{\mathscr{C},j}\}_{j=1,2}$:

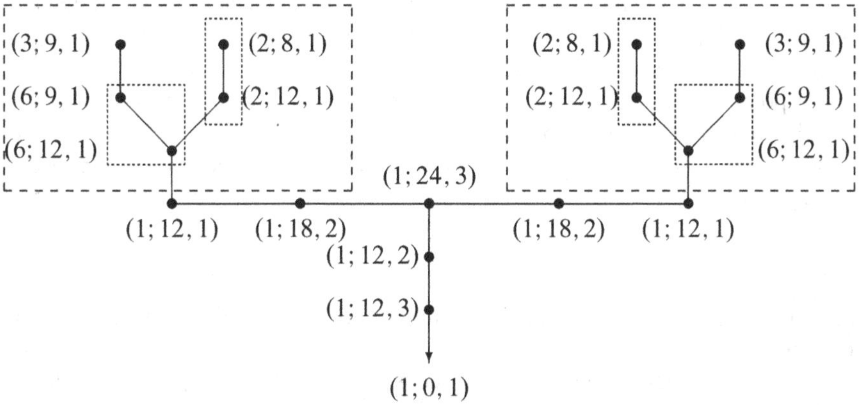


By the above algorithm, one can easily recover the transversal types, and the graph $E_6$ of the normalization of $V_f$.

The results of this chapter together with Theorem 5.1.8 culminate in the following corollary which is crucial for the algorithm presented in Chap. 10.

**Theorem 7.4.16.** *Up to isomorphism of cyclic coverings of graphs (with a fixed covering data), there is only one cyclic covering of the graph $\Gamma_{\mathscr{C}}$ provided that the covering data satisfies $\mathfrak{n}_v = 1$ for any $v \in \mathscr{V}^1(\Gamma_{\mathscr{C}})$.*

## 7.5 Cutting Edges Revisited

**7.5.1.** In this section we will analyze in detail the properties of cutting edges and we list some consequences.

Consider a cutting edge $e$ of $\Gamma_{\mathscr{C}}$, cf. 7.2.2. Recall that it always has edge decoration 2. Assume that the weights of the end-vertices have the form $(*\,; n, \nu)$. In order to indicate the dependence on $e$, we write $\nu = \nu(e)$.

Fix local coordinates as in Sect. 6.2, and set $C_e^* := B_1^l \cap B_2^l$. Here $B_1^l$ and $B_2^l$ are those two local components along which the restriction of the function $f \circ r$ vanishes, but $g \circ r$ does not vanish. Then $r$ projects $C_e^*$ onto a certain component $\Sigma_j$ of $\Sigma$. Moreover, $r|_{C_e^*} : C_e^* \to \Sigma_j$ is finite. Denote its degree by $d(e)$. Obviously

$$\deg(r|C_e^*) \cdot \deg(g|\Sigma_j) = \deg(g \circ r|C_e^*) = \nu(e),$$

hence

$$d(e) \cdot d_j = \nu(e). \tag{7.5.2}$$

Since $\nu(e)$ and $d_j$ can be obtained from $\Gamma^2_{\mathscr{C},j}$ (cf. 7.4.12), the degree $d(e)$ can also be recovered from $\Gamma^2_{\mathscr{C},j}$.

**7.5.3.** For every fixed $j \in \{1, \dots, s\}$, let $\mathscr{E}_{cut,j}$ be the set of cutting edges connecting $\Gamma^2_{\mathscr{C},j}$ with $\Gamma^1_{\mathscr{C}}$. Also, write $\#T\Sigma_j$ for the number of local irreducible components of the transversal singularity $T\Sigma_j$, which coincides with the number of connected components of $\partial F'_j$.

The point is that $\#(T\Sigma_j) \geq |\mathscr{E}_{cut,j}|$, and, in general, equality does not hold. Indeed, from the local equations and from the covering $r|_{C_e^*} : C_e^* \to \Sigma_j$ one deduces that each $e \in \mathscr{E}_{cut,j}$ is "responsible" for $d(e)$ local irreducible components of $T\Sigma_j$. In other words,

$$\#T\Sigma_j = \sum_{e \in \mathscr{E}_{cut,j}} d(e). \tag{7.5.4}$$

**Example 7.5.5.** Consider the example from 6.2.7. In this case $f = x^2y^2 + z^2(x+y)$, hence it has two singular components $\Sigma = \Sigma_1 \cup \Sigma_2 = \{xy = z = 0\}$, whose transversal type singularities are $A_1$, hence $\#T\Sigma_j = 2$ for $j \in \{1, 2\}$. The linear function $g$ induces on both $d_j = 1$. Furthermore, from the graph 6.2.7 we get that in both cases $\mathscr{E}_{cut,j} = 1$. This is compatible with the above discussion, since for the cutting edges $\nu(e) = 2$ in either case.

The graph $\Gamma^2_{\mathscr{C},j}$ has three vertices, all of them are in the same class, hence $\Gamma^2_{\mathscr{C},j}/\sim$ has only one vertex. The covering $G(T\Sigma_j) \to \Gamma^2_{\mathscr{C},j}/\sim$ is the following:

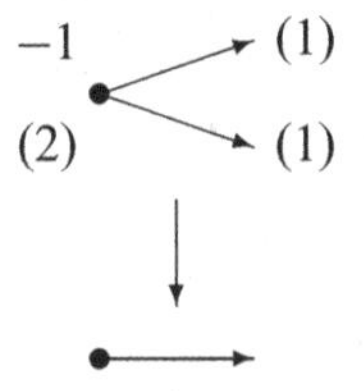


Hence, in general, the covering $G(T\Sigma_j) \to \Gamma^2_{\mathscr{C},j}/\sim$ is not a bijection.

**Example 7.5.6.** If $\#T\Sigma_j \neq 1$, then even if both graphs $\Gamma^1_{\mathscr{C}}$ and $\Gamma^2_{\mathscr{C}}$ are trees, it might happen that $\Gamma_{\mathscr{C}}$ has cycles. For example, in the case presented in 10.4.1,

$\Sigma = \Sigma_1$ is irreducible with $\#T\Sigma_j = 2$, both $\Gamma^1_{\mathscr{C}}$ and $\Gamma^2_{\mathscr{C}}$ are trees, and $\Gamma_{\mathscr{C}}$ has one cycle.

**7.5.7. Relationship with the resolution graph of the normalization $V_f^{norm}$.**
By Theorem 7.3.3 there is a one-to-one correspondence between the cutting edges and the irreducible components of the strict transform of $\Sigma_f$ in the normalization $V_f^{norm}$ of $V_f$. By the construction of the graph $\Gamma_{\mathscr{C}}$, and by the discussion 7.5.1, these are in bijection with components of type $C_e^*$. Moreover, via $r$ and Theorem 7.3.3, each $C_e^*$ can be identified with the corresponding strict transform component $St_e \subset V_f^{norm}$. In particular, the restriction of the normalization map $n$ satisfies:

$$\deg(n|_{St_e} : St_e \to \Sigma_j) = \deg(r|_{C_e^*} : C_e^* \to \Sigma_j) = d(e).$$

If $St(\Sigma_j)$ denotes the strict transform of $\Sigma_j$ in $V_f^{norm}$, then

$$St(\Sigma_j) = \bigcup_{e\in\mathscr{E}_{cut,j}} St_e,$$

and each $St_e$ contributes $d(e)$ components in $\#T\Sigma_j$, compatibly with (7.5.4). Looking at the local equations, specifically at the last multiplicities $\nu$, we obtain the next reinterpretation of the identity (7.5.4).

**Corollary 7.5.8.** *1. The vertical monodromy $m'_{j,ver}$ permutes the connected components of $\partial F'_j$; each orbit corresponds to a cutting edge $e \in \mathscr{E}_{cut,j}$, and the cardinality of the corresponding orbit is $d(e)$.*
*2. The vertical monodromy $m^{\Phi}_{j,ver}$ permutes the connected components of $d_j \cdot \partial F'_j$; each orbit corresponds to a cutting edge $e \in \mathscr{E}_{cut,j}$, and the cardinality of the corresponding orbit is $d_j d(e) = \nu(e)$.*

**7.5.9. The construction of the link $K$ of $f$ from $\Gamma_{\mathscr{C}}$.**
Consider the link $K^{norm} := K_{V_f^{norm}}$ of $V_f^{norm}$. It is the disjoint union of the (connected) links of the irreducible components of $V_f^{norm}$. In it consider the 1-dimensional sub-manifold

$$\bigcup_j (St_j \cap K^{norm}) = \bigcup_j \bigcup_{e\in\mathscr{E}_{cut,j}} (St_e \cap K^{norm}) \subset K^{norm}.$$

Assume that each component $St_e \cap K^{norm}$, denoted by $S_e^1$ (and which is diffeomorphic to $S^1$) is marked by two data, one of them is an element $j \in \{1,\dots,s\}$, the index $j = j(e)$ of $\Sigma_j$ onto which $St_e$ is mapped, the other is the degree $d(e)$ of $St_e \to \Sigma_j$.

We claim that from the data $(K^{norm}, \cup_e(S_e^1; j(e), d(e)))$ one can recover the link $K$ of $f$. Indeed, for each $j \in \{1,\dots,s\}$ fix a circle $S^1 = S_j^1$. Moreover, for each $e$ with $j(e) = j$ fix a cyclic covering $\phi_e : S_e^1 \to S_j^1$ of degree $d(e)$. Then $K$ is obtained from $K^{norm}$ by gluing its points via the maps $\phi_e$.

**Proposition 7.5.10.** *Introduce an equivalence relation on $K^{norm}$ as follows: $x \sim x'$ if and only if there exist $e$ and $e'$, with $x \in S^1_e$ and $x' \in S^1_{e'}$ (where $e = e'$ is allowed) such that $\phi_e(x) = \phi_{e'}(x')$ (and any other equivalence has the form $y \sim y$). Then*

$$K = K^{norm}/\sim .$$

Now, clearly, the above data $(K^{norm}, \cup_e(S^1_e; j(e), d(e)))$ can be deduced from $\Gamma_{\mathscr{C}}$. Recall that in Sect. 7.3 we provide the plumbing graph $G^1_{\mathscr{C}}$ for $K^{norm}$ from $\Gamma^1_{\mathscr{C}}$. One has only to modify this construction as follows. In the construction of $\Gamma^1_{\mathscr{C}}$ one has to decorate the 2-edges $e$ (or their arrowheads) by the extra decoration $(j(e), d(e))$, and keep this extra decoration for the (0)-arrowheads of $G^1_{\mathscr{C}}$ too. Then this enhanced $G^1_{\mathscr{C}}$ contains all the information needed to apply 7.5.10.

**7.5.11.** Here is a picture summarizing in a schematic form the essential features of the decomposition of the graph $\Gamma_{\mathscr{C}}$ into $\Gamma^1_{\mathscr{C}}$ and $\Gamma^2_{\mathscr{C}}$, the classes $K_\ell$, the types of the intersection points corresponding to 1- and 2- and cutting edges, and the degenerations (7.1.3), 7.1.4 and (7.1.9).

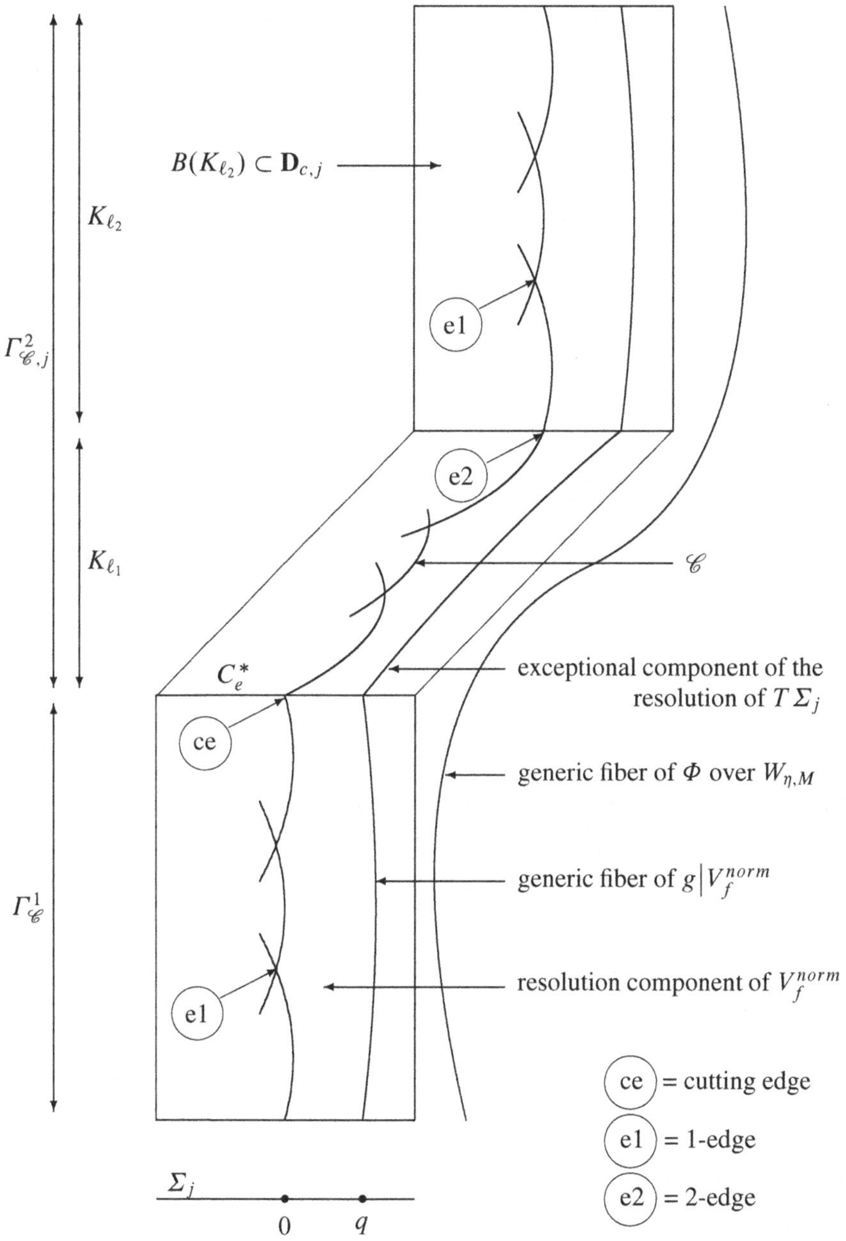

$B(K_{\ell_2}) \subset \mathbf{D}_{c,j}$
$K_{\ell_2}$
$K_{\ell_1}$
$\Gamma^2_{\mathscr{C},j}$
$\Gamma^1_{\mathscr{C}}$
e1
e2
ce
$\mathscr{C}$
$C_e^*$
exceptional component of the resolution of $T\Sigma_j$
generic fiber of $\Phi$ over $W_{\eta,M}$
generic fiber of $g|V_f^{norm}$
resolution component of $V_f^{norm}$
ce = cutting edge
e1 = 1-edge
e2 = 2-edge
$\Sigma_j$
0
$q$

# Chapter 8
# Examples: Homogeneous Singularities

## 8.1 The General Case

Assume that $f : (\mathbb{C}^3, 0) \to (\mathbb{C}, 0)$ is the germ of a homogeneous polynomial of degree $d$, and we choose $g$ to be a generic linear function with respect to $f$.

Let $C \subset \mathbb{CP}^2$ be the projective plane curve $\{f = 0\}$.

We show that a possible graph $\Gamma_{\mathscr{C}}$ can easily be determined from the combinatorics of the components and the topological types of the local singularities of $C$.

In this projective setting we use the following notations.

Let $C = \cup_{\lambda \in \Lambda} C_\lambda$ be the irreducible decomposition of $C$, and set $d_\lambda := \deg(C_\lambda)$. Hence $\sum_\lambda d_\lambda = d$. Furthermore, let $g_\lambda$ be the genus of the normalization of $C_\lambda$.

Let $\{p_j\}_{j \in \Pi}$ be the set of singular points of $C$. Assume that the local analytic irreducible components of $(C, p_j)$ are $(C_{j,i}, p_j)_{i \in I_j}$. Clearly, there is an "identification map" of global/local components $c : \cup_j I_j \to \Lambda$ which sends the index of a local component $C_{j,i}$ into the index $\lambda$ whenever $C_{j,i} \subset C_\lambda$.

Let $\Gamma_j$ be an embedded resolution graph of the local plane curve singularity $(C, p_j) \subset (\mathbb{C}^2, p_j)$. It has $|I_j|$ arrowheads, each with multiplicity (1).

These notations agree with some of the notations already considered for germs in three variables in the previous sections, for example, in Sect. 2.2. Indeed, the number of singular points of $C$ is the same as the number of irreducible components of $\Sigma_f$, hence $\Pi$ corresponds to $\{1, \dots, s\}$. Moreover, the local topological type of the plane curve singularity $(C, p_j) \subset (\mathbb{CP}^2, p_j)$ at a singular point $p_j$ of $C$ agrees with the corresponding $T\Sigma_j$, hence $|I_j| = \#T\Sigma_j$.

**Proposition 8.1.1.** *A possible $\Gamma_{\mathscr{C}}$ is constructed from the dual graphs $\{\Gamma_j\}_{j \in \Pi}$ and $|\Lambda|$ additional non-arrowhead vertices as follows:*

*First, for each $\lambda \in \Lambda$ put a non-arrowhead vertex $v_\lambda$ in $\Gamma_{\mathscr{C}}$ and decorate it with $(1; d, 1)$ and $[g_\lambda]$. Moreover, put $d_\lambda$ edges supported by $v_\lambda$, each of them decorated by 1 and supporting an arrowhead weighted by $(1; 0, 1)$.*

*Then, consider each graph $\Gamma_j$, keep its shape, but replace the decoration of each non-arrowhead with multiplicity $(m)$ by the new decoration $(m; d, 1)$, and*



A. Némethi and Á. Szilárd, *Milnor Fiber Boundary of a Non-isolated Surface Singularity*, Lecture Notes in Mathematics 2037, DOI 10.1007/978-3-642-23647-1_8,

*decorate all edges by 2. Furthermore, each arrowhead of $\Gamma_j$, corresponding to the local component $C_{j,i}$, is identified with $v_\lambda$, where $\lambda$ corresponds to $C_{j,i}$ via the local/global identification c.*

*Proof.* The following sequence of blow ups is performed. First the origin of $\mathbb{C}^3$ is blown up. This creates an exceptional divisor $E = \mathbb{CP}^2$ which intersects the strict transform $St(V_f)$ of $V_f$ along $C$. Moreover, the strict transform of $V_g$ (where $g$ is the chosen linear function) intersects the strict transform of each irreducible component $V(f)_\lambda$ in $d_\lambda$ discs.

The singular part of $St(V_f)$ consists of discs meeting $E$ in the singular points $p_j$ of $C$. The plane curve singularity $(C, p_j) \subset (\mathbb{CP}^2, p_j)$ can be resolved by a sequence of blow ups infinitely near points of $p$; this sequence is replaced in the present local product situation by blowing up infinitely near discs following the blowing up procedure of the corresponding plane curve singularity. Then the corresponding decorations follow easily. □

**Remark 8.1.2.** Notice that if a local graph $\Gamma_j$ is a double-arrow (representing a local singularity of type $A_1$ with local equation $xy = 0$) and both local irreducible components sit on the same global component $C_\lambda$, then by the above procedure the double arrow transforms into a loop supported on $v_\lambda$ decorated by 2. If the two local irreducible components sit on two different global components, then it becomes a 2-edge. In both cases, the corresponding edge will not satisfy Assumption A 6.3.

Nevertheless, if we consider embedded resolution graphs $\Gamma_j$ with at least one non-arrowhead vertex (e.g. the graphs of $A_1$ singularities will have one $(-1)$-vertex), then the graph $\Gamma_{\mathscr{C}}$ obtained in this way will satisfy Assumption A (and will be related with the previous graph by the moves of 6.3).

Usually, it is preferable to take for $\Gamma_j$ the minimal embedded resolution. Nevertheless, if we want to get $\Gamma_{\mathscr{C}}$ satisfying Assumption A, then we follow the convention that for an $A_1$ singularity $\Gamma_j$ contains one non-arrowhead.

In the next examples we ask the reader to determine for each case the graphs $\Gamma_j$. Several procedures are described in [16, 136]. We will provide only the output $\Gamma_{\mathscr{C}}$.

**Example 8.1.3.** If $f = z^d - xy^{d-1}$ with $d \geq 3$, then $\Gamma_{\mathscr{C}}$ is the following:

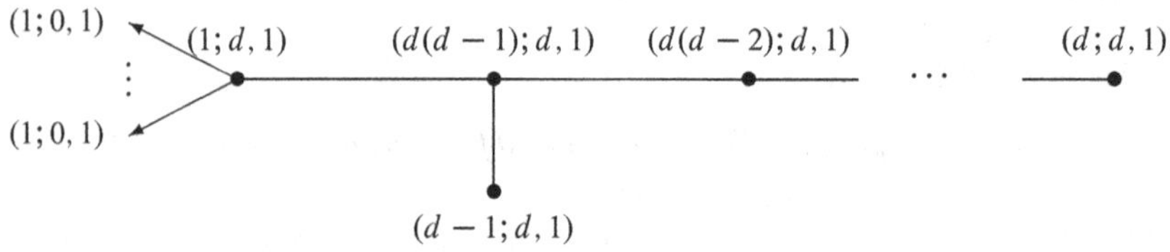


where there are $d$ arrowheads, and all the edges connecting non-arrowheads have decoration 2.

**Example 8.1.4.** Assume that $f = x^2y^2+y^2z^2+z^2x^2-2xyz(x+y+z)$. Then $C$ is an irreducible rational curve with three $A_2$ (ordinary cusp) singularities. Therefore, $\Gamma_{\mathscr{C}}$ is:

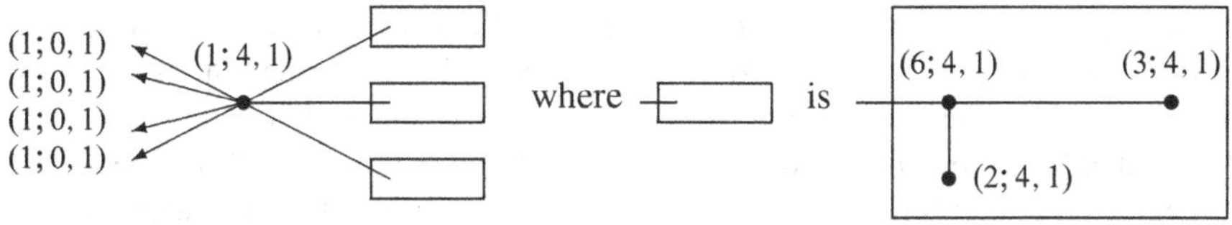


**Example 8.1.5.** Let $f = x^d + y^d + xyz^{d-2}$, where $d \geq 3$. Then a possible $\Gamma_{\mathscr{C}}$ which does not satisfy Assumption A is:

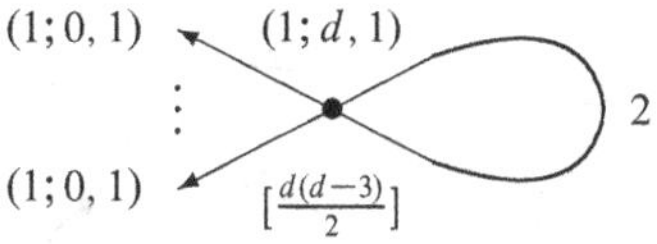


($d$ arrowheads)

Its modification as in 6.3, or as in (8.1.2), which satisfies Assumption A is:

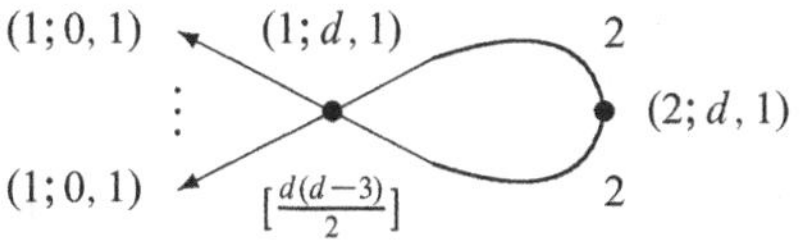


($d$ arrowheads)

**Remark 8.1.6.** Consider a **Zariski pair** $(C_1, C_2)$. This means that $C_1$ and $C_2$ are two irreducible projective curves that have the same degree and the topological type of their *local* singularities are the same, while their embeddings in the projective plane are topologically different. Then the two graphs $\Gamma_{\mathscr{C}}(C_1)$ and $\Gamma_{\mathscr{C}}(C_2)$ provided by the above algorithm will be the same. In particular, any invariant derived from $\Gamma_{\mathscr{C}}$ (e.g. $\partial F$) will not differentiate Zariski pairs.

**Remark 8.1.7.** Since $d_j = 1$ for any $j$, and $\nu(e) = d(e) = 1$ for any cutting edge $e$, one also has $\#T\Sigma_j = |\mathscr{E}_{cut,j}|$.

## 8.2 Line Arrangements

A special case of 8.1 is the case of line arrangements in $\mathbb{CP}^2$, that is, each connected component of $C$ is a line.

Having an arrangement, let $\{L_\lambda\}_{\lambda\in\Lambda}$ be the set of lines, and $\{p_j\}_{j\in\Pi}$ the set of intersection points. Write $|\Lambda| = d$, and for each $j$ set $m_j$ for the cardinality of $I_j := \{L_\lambda : L_\lambda \ni p_j\}$. Then $\Gamma_{\mathscr{C}}$ can be constructed as follows:

For each $\lambda \in \Lambda$ put a non-arrowhead vertex $v_\lambda$ with weight $(1; d, 1)$. For each $j \in \Pi$ put a non-arrowhead vertex $v_j$ with weight $(m_j; d, 1)$. Join the vertices $v_\lambda$ and $v_j$ with a 2-edge whenever $p_j \in L_\lambda$. Finally, put on each vertex $v_\lambda$ an edge with decoration 1, which supports an arrowhead with weight $(1; 0, 1)$.

Notice that $v_j$ is connected with $m_j$ vertices of type $v_\lambda$. Clearly, $v_j$ corresponds to the exceptional divisor obtained by blowing up an intersection point of $m_j$ lines. Notice that if in the special case of $m_j = 2$ – i.e. when $p_j$ sits only on $L_{\lambda_1}$ and $L_{\lambda_2}$ –, this blow up is imposed by Assumption A, cf. 6.3. Nevertheless, if we wish to neglect Assumption A, then this vertex $v_j$ can be deleted together with the two adjacent edges, and one can simply put a 2-edge connecting $v_{\lambda_1}$ with $v_{\lambda_2}$.

**Example 8.2.1.** In the case of the $A_3$ arrangement $f = xyz(x-y)(y-z)(z-x)$, the two graphs $\Gamma_{\mathscr{C}}$ (satisfying Assumption A or not) are:

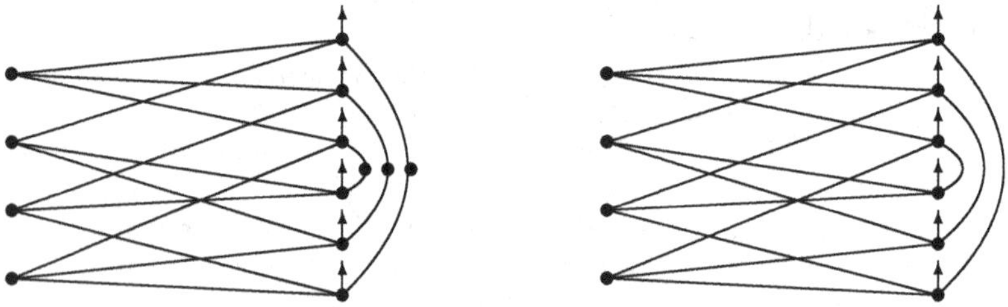

where on the left hand graph the four left-vertices are weighted by $(3; 6, 1)$, the next six by $(1; 6, 1)$, the remaining three by $(2; 6, 1)$, and the arrowheads by $(1; 0, 1)$. The edges supporting arrowheads are decorated by 1, the others by 2.

**Example 8.2.2.** The "simplified" graph $\Gamma_{\mathscr{C}}$, which does not satisfy Assumption A, for the generic arrangement with $d$ lines consists of $d$ vertices $v_\lambda$, each decorated with $(1; d, 0)$, each supporting an arrow $(1; 0, 1)$, and any pair of non-arrowheads is connected by a 2-edge.

# Chapter 9
# Examples: Families Associated with Plane Curve Singularities

## 9.1 Cylinders of Plane Curve Singularities

Consider $f(x, y, z) = f'(x, y)$ and $g(x, y, z) = z$, where $f' : (\mathbb{C}^2, 0) \to (\mathbb{C}, 0)$ is an isolated plane curve singularity. It is well-known (see e.g. [16, 45, 136]) that the embedded resolution of $(\mathbb{C}^2, V_{f'})$ can be obtained by a sequence of quadratic transformations. Replacing the quadratic transformations of the infinitely near points of $0 \in \mathbb{C}^2$ by blow ups along the infinitely near 1-dimensional axis of the $z$-axis, one obtains the following picture.

Let $\Gamma(\mathbb{C}^2, f')$ denote the minimal embedded resolution graph of the plane curve singularity $f' : (\mathbb{C}^2, 0) \to (\mathbb{C}, 0)$. Recall that, besides the Euler numbers and genera of the non-arrowheads, each vertex has a multiplicity decoration $(m)$, the vanishing order of the pull-back of $f'$ along the corresponding irreducible curve.

We say that $\{f = 0\}$ is the *cylinder* of the plane curve $\{f' = 0\}$.

In this situation, one can get a possible dual graph $\Gamma_{\mathscr{C}}$ from $\Gamma(\mathbb{C}^2, f')$ via the following conversion.

The shapes of the two graphs agree, only the decorations are modified: the Euler numbers are deleted, while for each vertex the multiplicity $(m)$ is replaced by $(m; 0, 1)$. The genus decorations in $\Gamma_{\mathscr{C}}$ – similarly as in $\Gamma(\mathbb{C}^2, f')$ – of all non-arrowheads are zero. Moreover, all edges in $\Gamma_{\mathscr{C}}$ have weight 2.

**Example 9.1.1.** Let $f(x, y, z) = f'(x, y) = (x^2 - y^3)(y^2 - x^3)$. Then $\Gamma(\mathbb{C}^2, f')$ is:

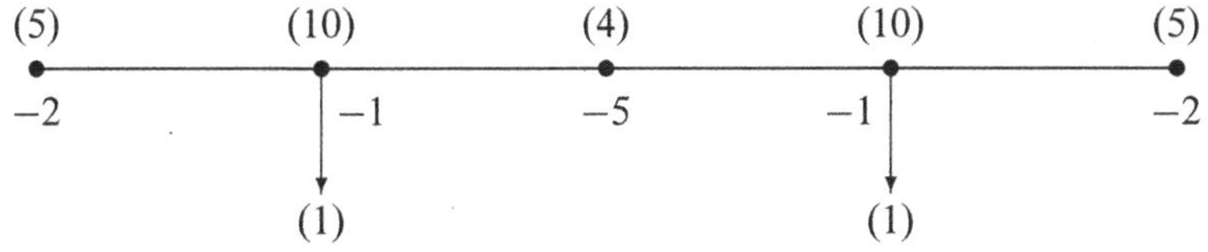


A. Némethi and Á. Szilárd, *Milnor Fiber Boundary of a Non-isolated Surface Singularity*, Lecture Notes in Mathematics 2037, DOI 10.1007/978-3-642-23647-1_9, 

which is transformed into $\Gamma_{\mathcal{C}}$ as:

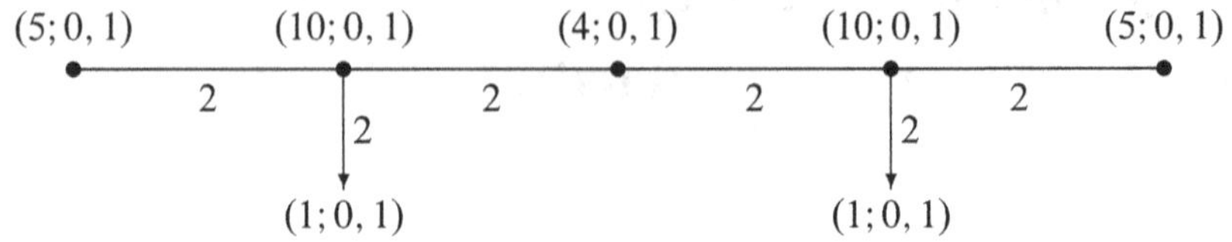


**9.1.2.** It is easy to verify that $\Gamma^2_{\mathcal{C}} = \Gamma_{\mathcal{C}}$, and $\Gamma^1_{\mathcal{C}}$ consists of $|\mathcal{A}|$ double arrows, where $|\mathcal{A}|$ is the number of irreducible components of $f'$. The statements of 7.3 and 7.4 can easily be verified.

## 9.2 Germs of Type $f = zf'(x, y)$

Here $f' : (\mathbb{C}^2, 0) \to (\mathbb{C}, 0)$ is an isolated plane curve singularity as above, $f(x, y, z) := zf'(x, y)$ and $g$ is a generic linear form in variables $(x, y, z)$.

For this family we found no nice uniform presentation of $\Gamma_{\mathcal{C}}$ with similar simplicity and conceptual conciseness as in Sect. 9.1, or in the homogeneous case. (We face the same obstruction as in the case of suspensions, explained in the second paragraph of 9.3.1). Since the 3-manifold $\partial F$ can be determined completely and rather easily for any $f = zf'(x, y)$ by another method, which will be presented in Chap. 21, we decided to omit general technical graph-presentations here. Nevertheless, particular testing examples can be determined without difficulty. For example, consider $f' = x^{d-1} + y^{d-1}$ when $f$ becomes homogeneous and $\Gamma_{\mathcal{C}}$ can be determined as in Chap. 8. Or, consider $f' = x^2 + y^3$, whose $\Gamma_{\mathcal{C}}$ is below. For more comments (and mysteries) regarding the possible graphs $\Gamma_{\mathcal{C}}$, see 21.1.8.

**Example 9.2.1.** Assume that $f = z(x^2 + y^3)$ and take $g$ to be a generic linear form. The "ad hoc blowing up procedure", using the naive principle to blow up the "worst singular locus", provides the following $\Gamma_{\mathcal{C}}$, where we only marked the 2-edges, and all unmarked edges are 1-edges:

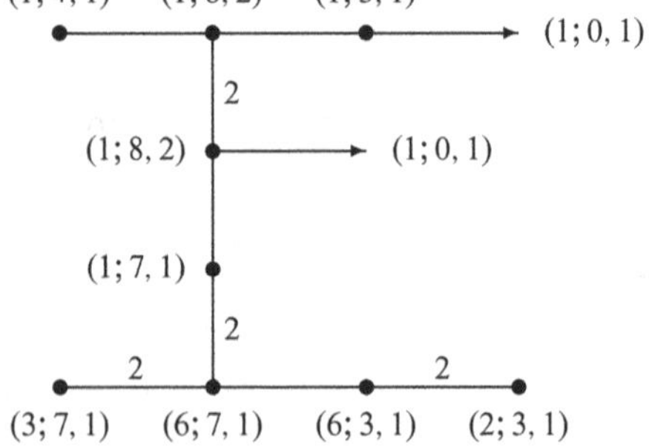

## 9.3 Double Suspensions

Suspension, or cyclic covering singularities are defined by functions of the form $f(x, y, z) = f'(x, y) + z^d$, where $f' : (\mathbb{C}^2, 0) \to (\mathbb{C}, 0)$ is plane curve singularity. If we wish to get $f$ non-isolated, we have to start with $f'$ non-isolated. When $d = 2$ the germ is called double suspension of $f'$. When $f'$ is not very complicated, one might find a convenient resolution by "ad hoc" blow ups, such as in the following case:

**Example 9.3.1.** Assume that $f = x^2y + z^2$ and $g = x + y$. Then a possible $\Gamma_{\mathscr{C}}$ is:

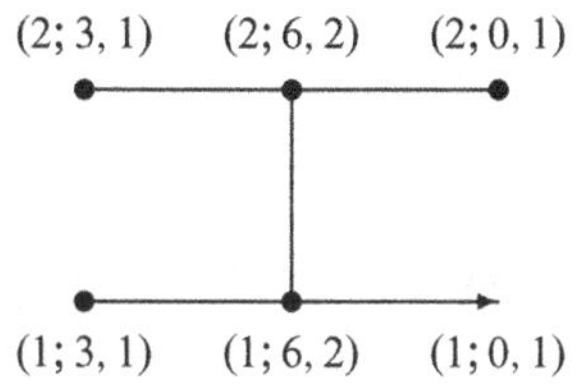


Of course, for the general family, we need a more conceptual and uniform procedure. In general, when determining $\Gamma_{\mathscr{C}}$, the construction of an embedded resolution $r$, as in 6.1, is not always simple, and it depends essentially on the choice of the germ $g$. Ideally, for any $f$, it would be nice to find a germ $g$ such that the pair $(f, g)$ would admit a resolution $r$ which reflects *only* the geometry of $f$, e.g. it is a "canonical", or "minimal" embedded resolution of $V_f$. For example, in the homogeneous case, resolving $f$ we automatically get a resolution which is good for the pair $(f, g)$ as well, provided that $g$ is a generic linear form. But, in general, "canonical" resolutions attached to $f$ by some geometric constructions used to resolve hypersurfaces do not have the extra property that they resolve a well-chosen $g$ as well (or, at least, the authors do not know such a general statement). Usually, the strict transform of $g$ may still have "bad contacts" with the created exceptional divisors even if we take for $g$ the generic linear form.

Nevertheless, for double suspensions $f = f' + z^2$, if one constructs a "canonical" resolution using the classical Jung construction fitting with the shape of $f$ (that is, based on the projection onto the $(x, y)$-plane, similarly as the methods described in 5.3), the obtained embedded resolution will be compatible with $g$ too, provided that we take for $g$ a generic linear form. We expect that a similar phenomenon is valid for arbitrary suspensions as well.

Since the embedded resolution of double suspensions is already present in the literature [7], this case can be exemplified without too much extra work. Nevertheless, the computations are not trivial, and their verification will require some effort from the reader, and familiarity with [7]. In the sequel we present the main steps needed to understand the procedure, we provide some examples, and we let the reader explore his/her favorite example.

We prefer to write $g$ as $g'(x, y) + z$, where $g'$ is a generic linear form (with respect to $f'$) in variables $(x, y)$.

The embedded resolution of $V_{fg} \subset (\mathbb{C}^3, 0)$ is constructed in several steps as in [7]. Although in that article $f'$ is isolated, the same procedure works in our case as well. We summarize the steps in the following diagram:

$$\begin{array}{ccccc} \tilde{X} & \xrightarrow{\psi} & X & \xrightarrow{\phi'} & U^3 \supset V_{fg} \\ & & {\scriptstyle p'}\downarrow & & \downarrow{\scriptstyle p} \\ & & Z & \xrightarrow{\phi} & U^2 \supset V_{f'g'} \end{array}$$

where

1. $U^3$ is a small representative of $(\mathbb{C}^3, 0)$ and $p : U^3 \to U^2$ is induced by the projection $(x, y, z) \to (x, y)$.
2. $\phi : Z \to U^2$ is an embedded resolution of $(V_{f'g'}, 0) \subset (\mathbb{C}^2, 0)$. We attach to each irreducible component $D$ of the exceptional divisor and to each strict transform component two nonnegative integers: the vanishing order $m(f')$, respectively $m(g')$ of $f'$, respectively $g'$, along that component.

   We take the minimal embedded resolution modified as in [7, (3.1)]: we assume that there are no pairs of irreducible components $D_{v_1}, D_{v_2}$ with $(D_{v_1}, D_{v_2}) \neq 0$ having both multiplicities $m_{v_1}(f'), m_{v_2}(f')$ odd. This can always be achieved from the minimal embedded resolution by an additional blow up at those intersection points where the condition is not satisfied.
3. $p' : X \to Z$ is the pull-back of $p : U^3 \to U^2$ via $\phi$, that is, $X$ is the product of $Z$ with the $z$-disc. By construction, in some local coordinates $(u, v, z)$ with $p'(u, v, z) = (u, v)$, any strict transform component of $V_f$ in $X$ has equation $u^{m_w(f')} + z^2$ above the generic point of an exceptional curve of $Z$, and $u^{m_w(f')} v^{m_v(f')} + z^2$ above an intersection point. The strict transform of $V_g$ is smooth; its local equations have similar form with the exponent of $z$ being one. Note that the contact of these two spaces along $z = 0$ is rather non-trivial.
4. $\psi$ is an embedded resolution of $(\phi')^{-1}(V_f) \subset X$, determined similarly as in [7, (3.4)]. This procedure constructs a "tower" of exceptional ruled surfaces over each exceptional divisor of $Z$. The algorithm of [7] constructs over each divisor of $Z$ a "minimal" tower, and the towers above divisors with even multiplicities are constructed first. Both these two conventions will be released now in order to get a resolution for the *pair* $V_f \cup V_g$.

The composed map $\phi' \circ \psi : \tilde{X} \to U^3$ serves for the modification $r$. Nevertheless, we wish to say here a word of warning. Usually we require that $r$ is an isomorphism above the complement of the singular locus of $V_f$. As it is explained in the Introduction of [7], or can be verified using the definitions, $\phi' \circ \psi$ fails to be an isomorphism above the union of $Sing(V_f)$ with the $z$-axis (because of the blow ups of the infinite near $z$-axis during the modification $\phi'$, as pull back of $\phi$). For example, the Milnor fiber of $f$ is not lifted diffeomorphically under this modification: it is blown up at its intersection points with the $z$-axis. Nevertheless, as the boundary $\partial F_{\epsilon,\delta}$ has no intersection points with the $z$-axis provided that $\delta \ll \epsilon$, this modification serves in this procedure as a genuine embedded resolution.

The above strategy leads to a combinatorial algorithm in two steps. In *Step 1* one determines the embedded resolution graph $\Gamma(f', g') := \Gamma(\mathbb{C}^2, f'g')$ (with the additional property mentioned in (2) above), but now weighted with both multiplicities $(m_v(f'), m_v(g'))$ of $f'$ and $g'$. In *Step 2* we determine the "towers" similarly as in [7], eventually constructed in a different order, or with extra blow ups. In the concrete examples below we will indicate the differences with [7]. Then, one reads from the "towers" the graph $\Gamma_{\mathscr{C}}$ of $(f, g)$. This appears as a "modified cover" of $\Gamma(f', g')$.

The following facts might be helpful in the construction of the above "modified cover" of graphs.

The non-arrowhead vertices of $\Gamma^1_{\mathscr{C}}$ cover the non-arrowhead vertices of $\Gamma(f', g')$ as follows. Fix $w \in \mathscr{W}(\Gamma(\mathbb{C}^2, f'g'))$. If $m_w(f')$ is even, *and* all the $f'$-multiplicities of the adjacent vertices are even, then $w$ is covered by 2 non-arrowhead vertices. In all other cases it is covered by only one vertex. This structure follows closely the structure and the position of the strict transform of $f$ in the resolution towers as it is described in Sect. 3.5 of [7].

The $(0, 1)$ arrowheads of $\Gamma(f', g')$ are covered by $(1; 0, 1)$-arrowheads of $\Gamma_{\mathscr{C}}$. If a non-arrowhead $w$ supports such an arrowhead in $\Gamma(f', g')$, and $m_w(f')$ is even, then it is covered by two arrowheads, otherwise only by one. Geometrically this is the only place where the strict transform of $g$ plays a role.

The two properties above describe behaviors common with $\mathbb{Z}_2$ graph coverings.

Next, above the arrowheads of $\Gamma(f', g')$ of type $(1, 0)$ we put nothing. The graph $\Gamma^2_{\mathscr{C}}$ appears above the arrowheads of $\Gamma(f', g')$ of type $(m, 0)$, $m > 1$. Fix such an arrowhead and the corresponding strict transform $St_a(f')$ of $f'$, which is supported by the exceptional component $E_w$. Then the entire tower above $St_a(f')$ is in $\mathbf{D}_c$ and all the intersection curves with the tower above $E_w$ enter in $\mathscr{C}$. Therefore, above the arrowheads of $\Gamma(f', g')$ of type $(m, 0)$ with $m > 1$ a lot of curves of $\mathscr{C}$ may appear, (and this part does not behave like a cyclic covering).

In certain cases, the genera of the projective irreducible components $C$ of the special curve configuration $\mathscr{C}$ may be difficult to determine from the *local* equations of $C$. However, if the link of the normalization of $V_f$ is a rational homology sphere, then we can be sure that in $\Gamma_{\mathscr{C}}$ all the genus decorations are zero, cf. 7.3.3 and 7.4.8. This is the case in all the examples worked out in this section.

Although in all the cases considered below the graph $\Gamma(f', g')$ is easy to determine, we provide them in order to emphasize the covering nature of the procedure.

**Example 9.3.2.** Assume that $f'(x, y) = x^a y^b$, where $a > 0$ and $b > 0$. The normalization is a Hirzebruch–Jung singularity, hence its link is a rational homology sphere.

We will distinguish three cases depending on the parity of $a$ and $b$.

**Case 1.** If both $a$ and $b$ are even, that is $f'(x, y) = x^{2n} y^{2m}$, and $g' = x + y$, then by *Step 1* the dual graph of the minimal good embedded resolution of $V_{f'g'}$, weighted by the vanishing orders of both $f'$ and $g'$, is

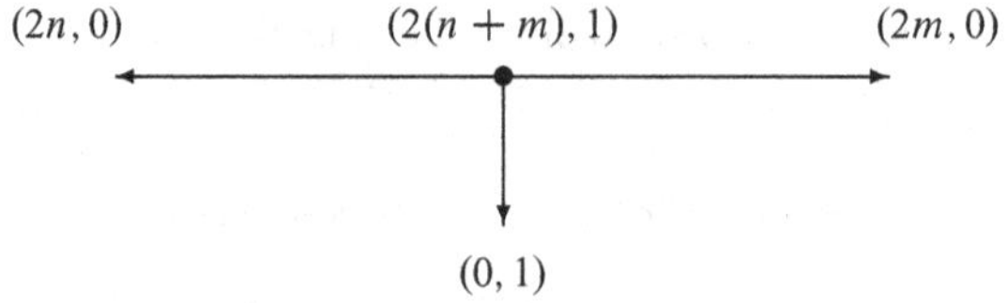


Here *Step 2* follows closely [7] with the following additional information: the tower above the exceptional divisor weighted $(2(n+m), 1)$ was constructed first, then the towers above the strict transforms of $f'$. A possible graph $\Gamma_{\mathscr{C}}$ of $(f, g)$ is:

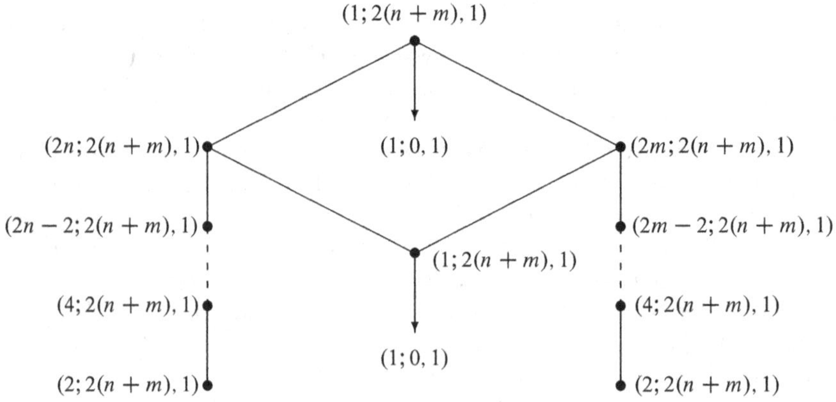


**Case 2.** If $a$ is even and $b$ is odd, that is $f'(x, y) = x^{2n} y^{2m+1}$, and $g' = x + y$, then by *Step 1* one gets the graph:

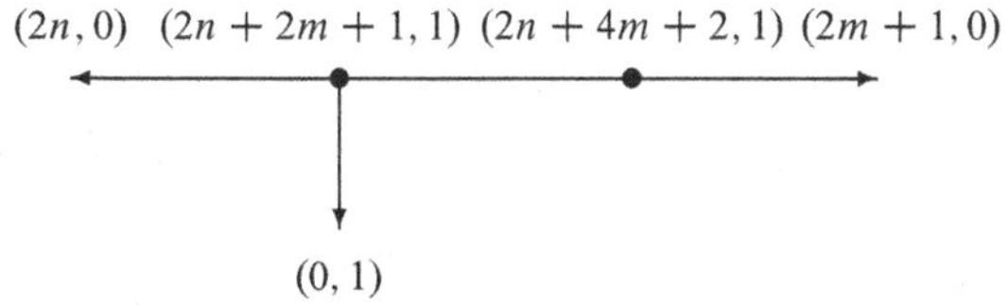


Here the assumption of not having adjacent irreducible components with both multiplicities $m_i(f), m_j(f)$ odd is taken into account.

By *Step 2*, a possible universal graph $\Gamma_{\mathscr{C}}$ of $(h, g)$ is:

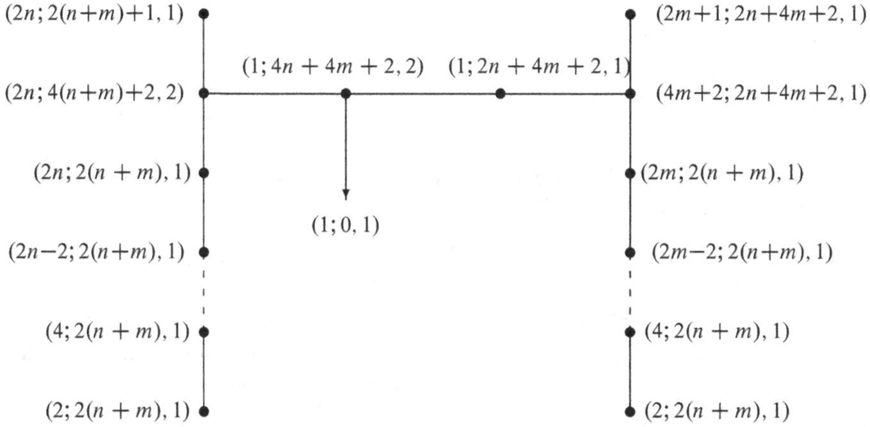


Here single blow-ups above the exceptional curves weighted $(2n + 4m + 2, 1)$ first, then $(2n + 2m + 1, 1)$ were used to ensure the strict transform of $g$ to be in transverse position with respect to the exceptional divisors appearing later. Then towers were constructed in the following order: first above the exceptional curve weighted $(2n+4m+2, 1)$ then $(2n+2m+1, 1)$, finally the strict transforms of $f'$.

**Case 3.** Finally, if both $a$ and $b$ are odd, that is $f'(x, y) = x^{2n+1}y^{2m+1}$, and $g' = x + y$, then *Step 1* produces the graph:

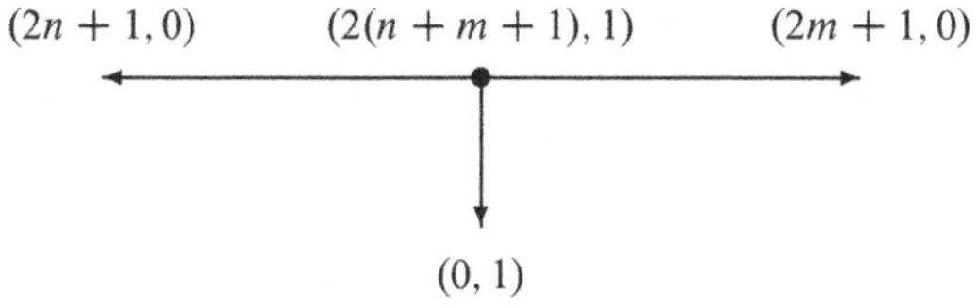

While a possible graph $\Gamma_{\mathscr{C}}$ is:

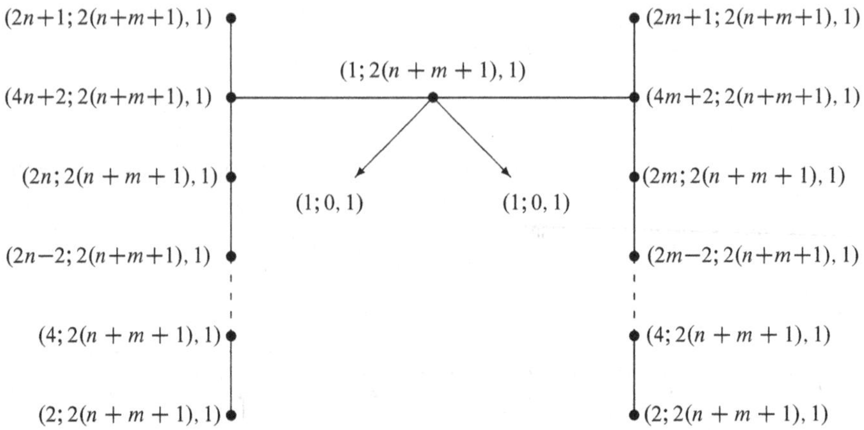


Here, a tower above the exceptional curve was constructed first, then towers above the strict transforms of $f'$.

**Example 9.3.3.** Consider the infinite family $T_{a,2,\infty}$ given by the local equation $f(x,y,z) = x^a + y^2 + xyz$.

If $a = 2$ then by a change of coordinates, $f$ can be rewritten as $f(x,y,z) = x^2 + y^2$, a case already treated in 9.1. Therefore, in this subsection we assume that $a \geq 3$. In this case, again, by completing the square and renaming variables, $f$ can be brought to the form $f(x,y,z) = x^2(x^{a-2} + y^2) + z^2$. In particular, the previous method can be used with $g(x,y,z) = x + y + z$.

The singular locus is $\Sigma_f = \{x = y = 0\}$ with transversal type $A_1$. Dividing the equation by $x^2$ and taking $t := y/x$, we get that $t^2 + zt + x^{a-2} = 0$, hence $t$ is in the normalization of the local ring, and the normalization is a hypersurface singularity of type $A_{a-3}$. In particular, the link of the normalization is a rational homology sphere.

First, we assume that $a$ is odd, that is $a = 2k + 3$.

*Step 1:* For $k \geq 1$ the graph $\Gamma(f', g')$ for $f' = x^2(x^{a-2} + y^2)$ and $g' = x + y$ is

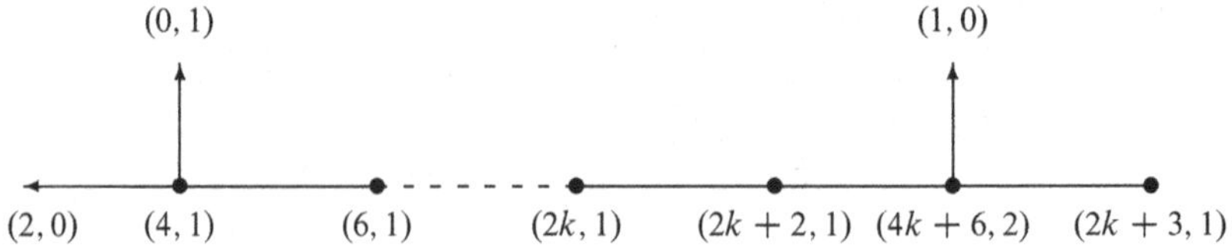

*Step 2:* A possible $\Gamma_{\mathscr{C}}$ for $(f, g)$ is:

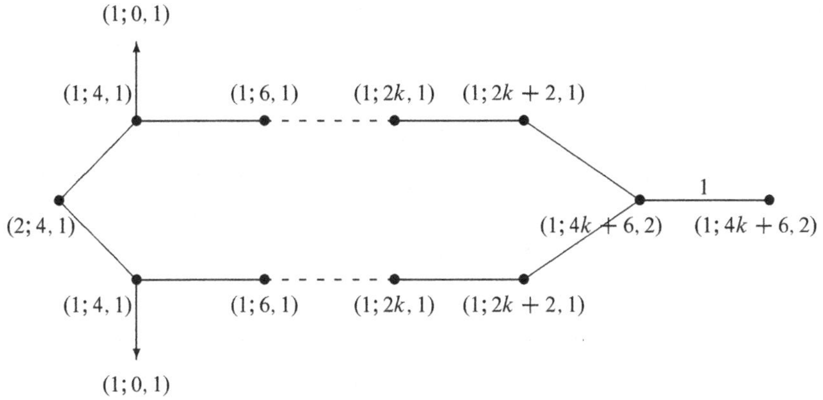


For $k = 0$ (i.e., for $a = 3$), the first graph is

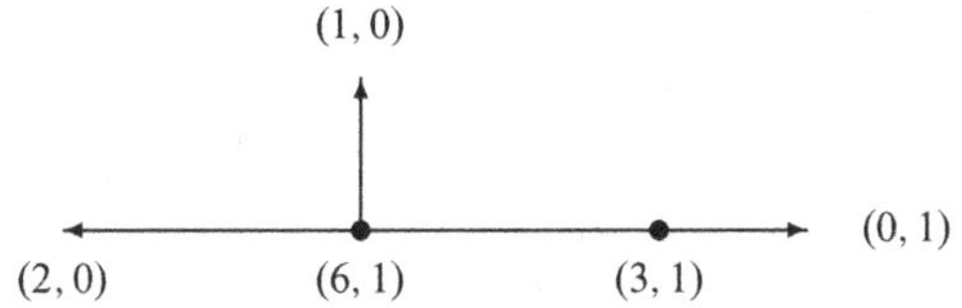


which is "covered" by the graph $\Gamma_{\mathscr{C}}$:

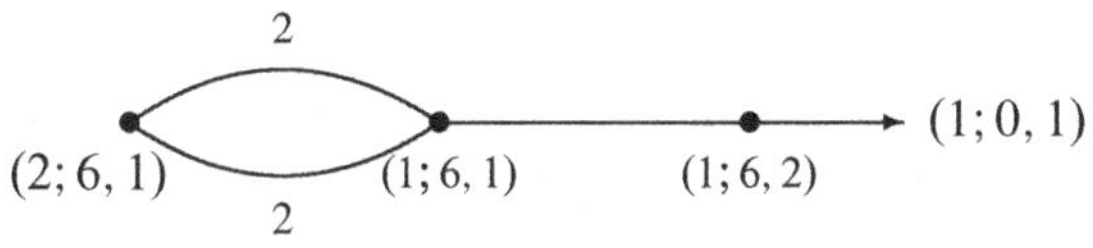


(For an alternative universal graph with different $g$, see 9.4.8.)

In case $a = 2k$, $k \geq 3$, *Step 1* provides

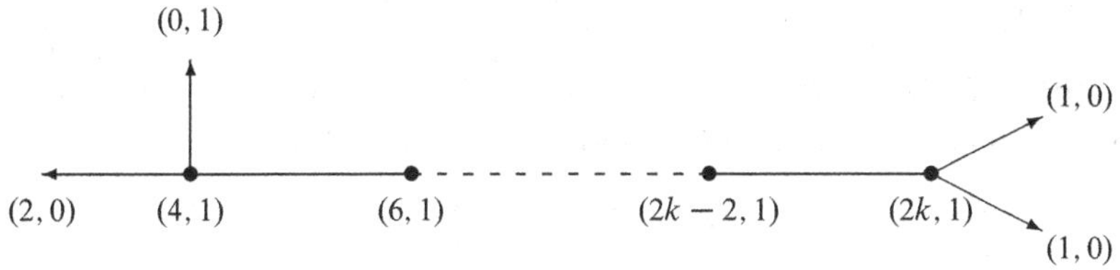

and $\Gamma_{\mathscr{C}}$ is

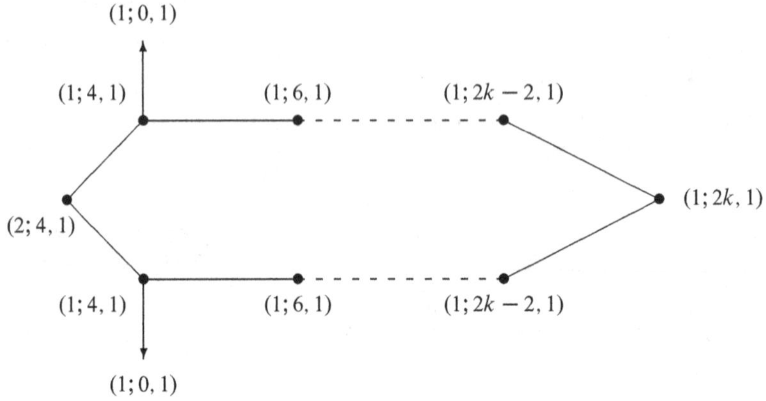


For $k = 2$ (that is $a = 4$) *Step 1* gives

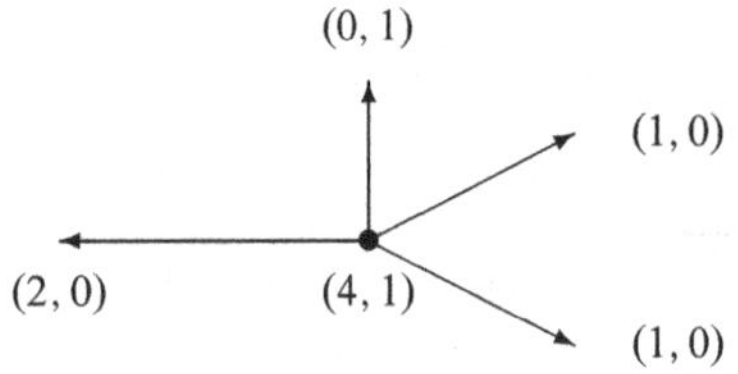


while *Step 2* provides

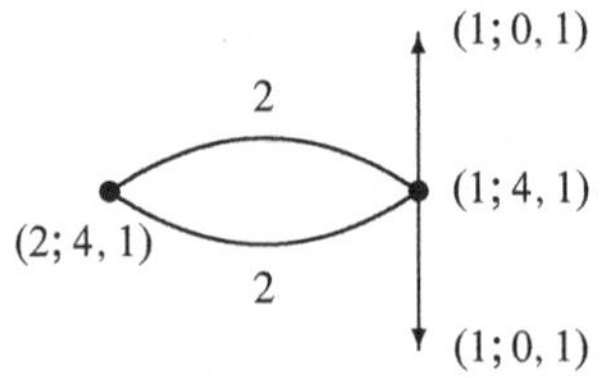


**Example 9.3.4.** Consider the family $f(x, y, z) = x^a y(x^2 + y^3) + z^2$ with $a \geq 2$ and $g = x + y + z$. Again, we need to consider two cases depending on parity of $a$. The equation of the normalization is $xy(x^2 + y^3) + z^2 = 0$ if $a$ is odd, and it is $y(x^2 + y^3) + z^2 = 0$ if $a$ is even. In both cases one can determine the plumbing graph of the link of the normalization using the algorithm 5.3. In particular, one gets that the link of the normalization is a rational homology sphere.

When $a$ is even, and taking into account, that no adjacent vertices can have odd $f'$-multiplicities, *Step 1* gives for $f'g'$:

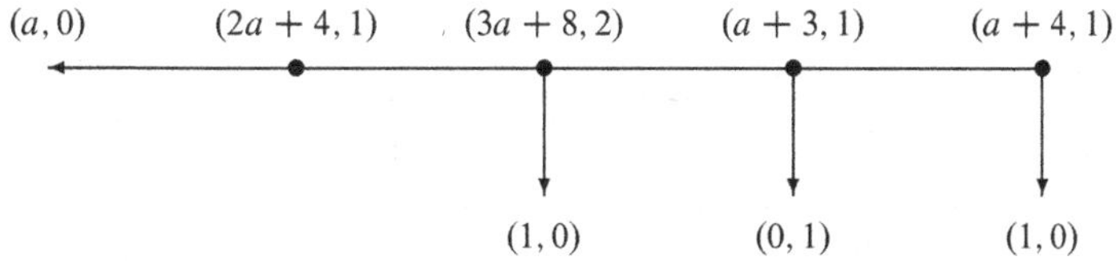


*Step 2* provides

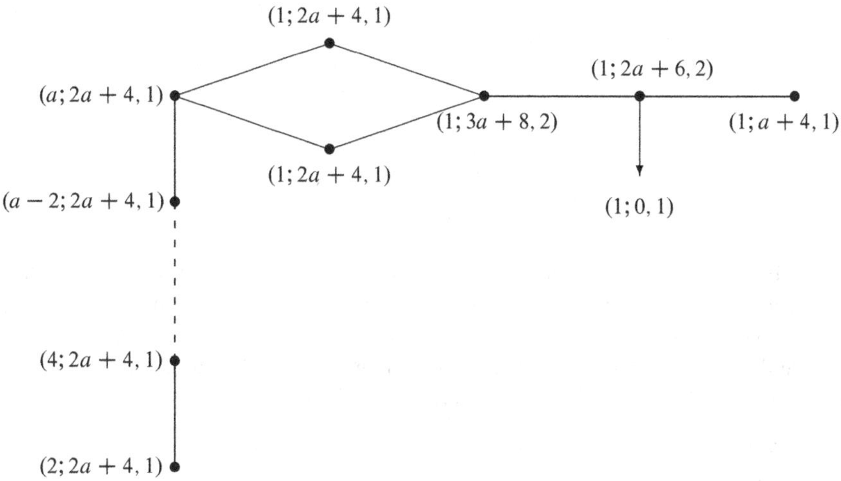


When $a$ is odd, the first graph is:

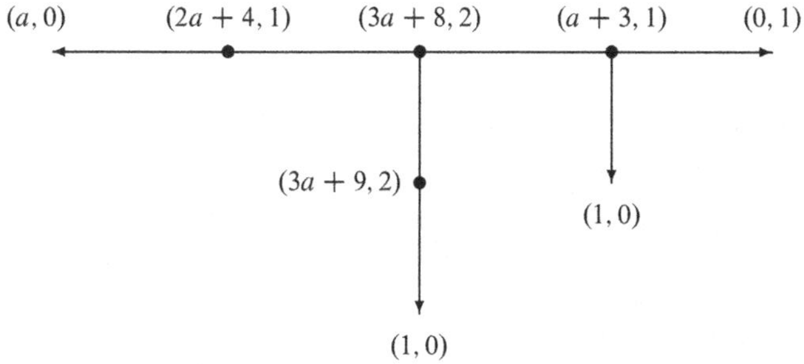

while *Step 2* provides

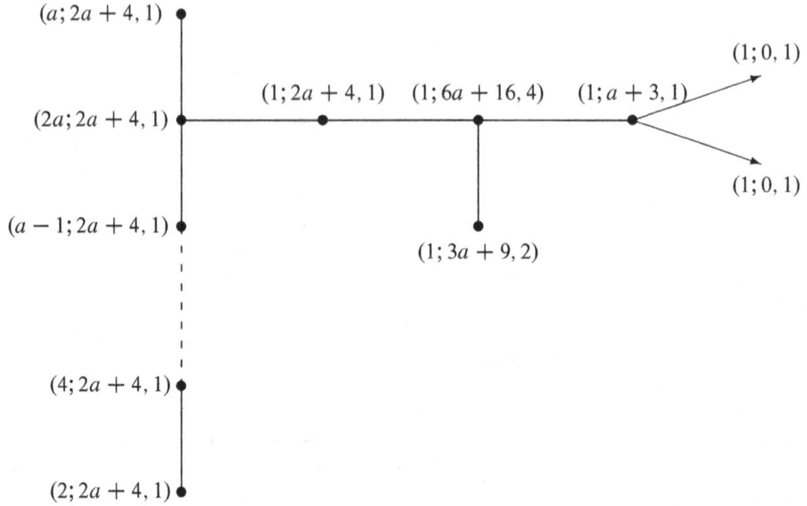


**Example 9.3.5.** The procedure 9.3.2 has a natural generalization for certain other suspensions too. For example, consider $f(x, y, z) = x^{md} y^{nd} + z^d$ with $\gcd(m, n) = 1$, and $g(x, y, z) = x + y + z$, hence $f'(x, y) = x^{dn} y^{dm}$, and $g' = x + y$.

Note that $f$ has $d$ local irreducible components, and each component is smooth. Hence, again we know that all the genus decorations of $\Gamma_{\mathscr{C}}$ are zero.

By *Step 1* the graph $\Gamma(f', g')$ is

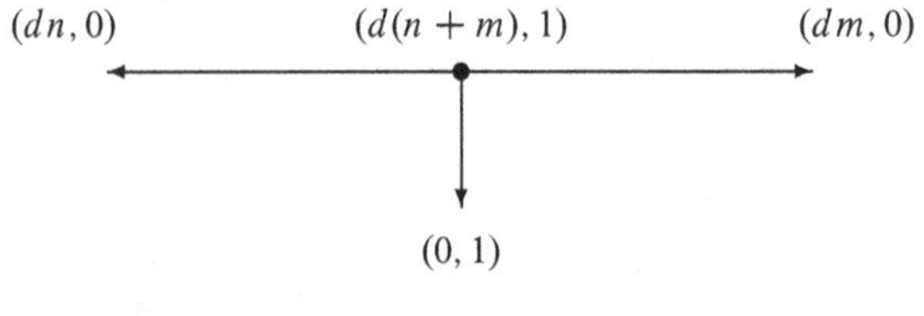


Finally, by *Step 2*, a possible graph $\Gamma_{\mathscr{C}}$ of $(f, g)$ is:

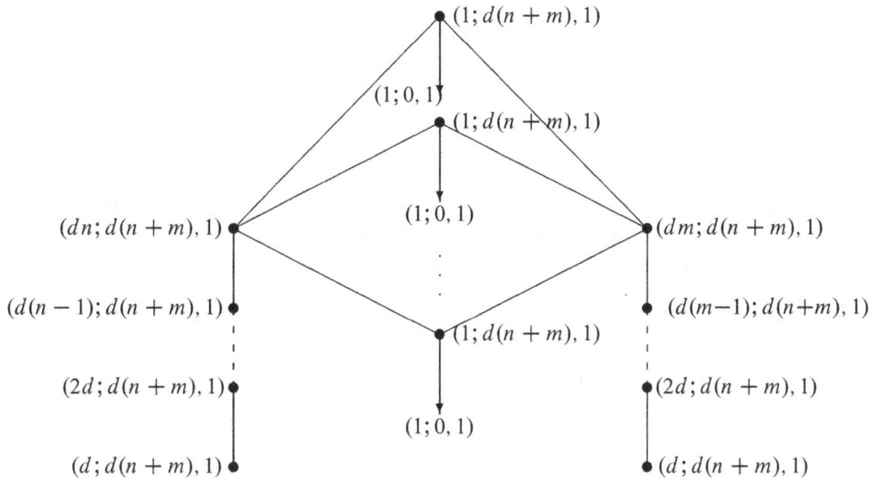


where in the middle column there are $d$ vertices.

For another suspension case, see 6.2.9.

## 9.4 The $T_{a,*,*}$-Family

**9.4.6. The $T_{a,\infty,\infty}$-family.** Let $f = x^a + xyz$ and set $g = x + y + z$.

If $a = 2k + 1$, $k > 1$, then a possible universal graph is:

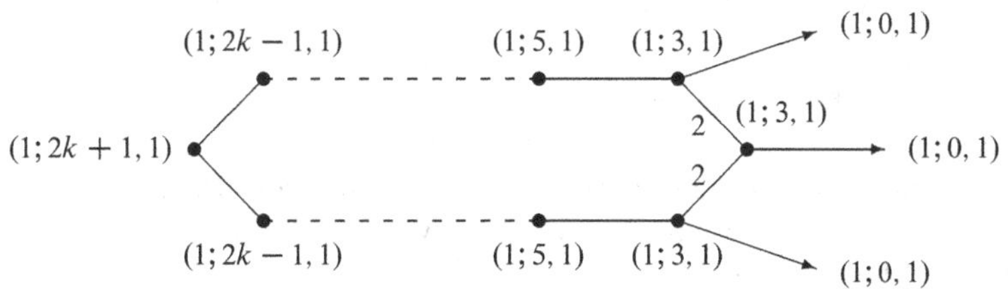


Above, all unmarked edges have weight 1.

The resolution process was started by a blow-up at the origin. Since $f = x^{2k+1} + xyz = x(x^{2k} + yz)$, there is only one singularity remaining, of the form $\{t^{2k-2} + sr = 0\}$. It is resolved by a series of blow-ups at infinitely near points, resulting in the above graph.

If $a = 3$ then $f$ is homogeneous. In particular, a single blow-up at the origin suffices and we get

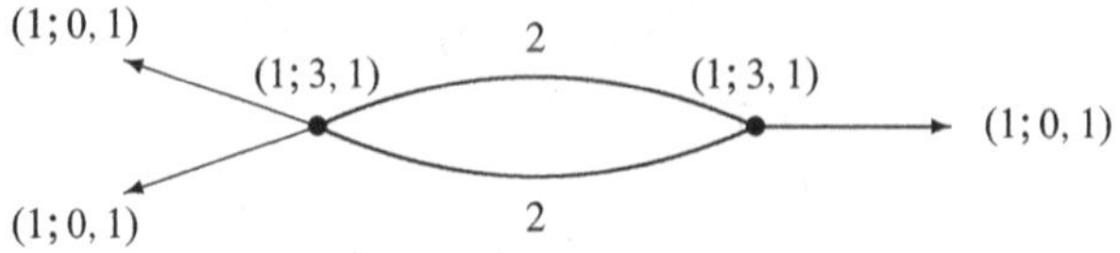


Set now $a = 2k$, $k > 2$. The same strategy for the resolution as above can be followed. However, when the strict transform of $f$ becomes smooth, it is not in normal crossing with the exceptional divisors. Two additional blow-ups along singular axes lead to the following universal graph:

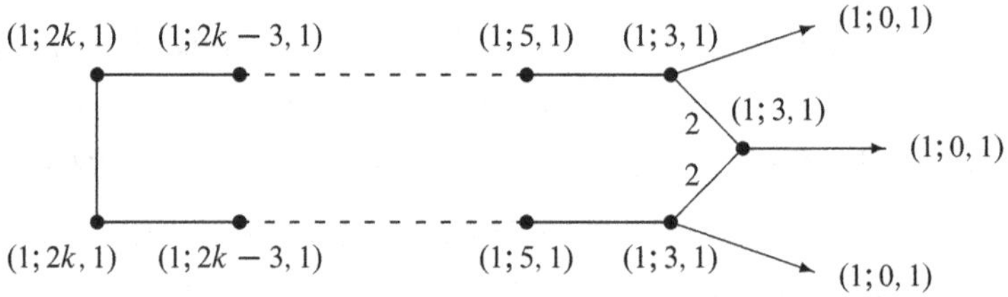


(All unmarked edges have weight 1, as before.)

Finally, in case $a = 4$ the previous resolution "strategy" leads to

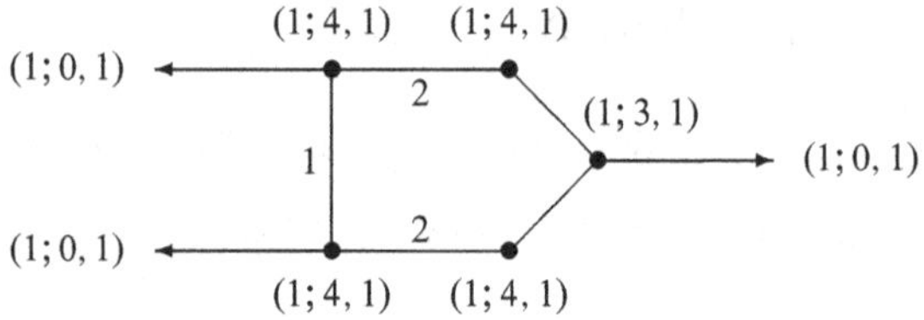


**9.4.7. The $T_{a,2,\infty}$-family (again).** Set $f(x, y, z) = x^a + y^2 + xyz$.

The cases $a = 3$ and $a = 5$ (with $g = z$) were already considered in [92], where $\Gamma_{\mathscr{C}}$ was obtained using an alternative/ad hoc resolution. The graphs $\Gamma_{\mathscr{C}}$ thus obtained will serve as clarifying examples for several geometric discussions in this work as well. The graphs are the following:

**Example 9.4.8.** If $f = x^3 + y^2 + xyz$ and $g = z$ then a possible $\Gamma_{\mathscr{C}}$ is:

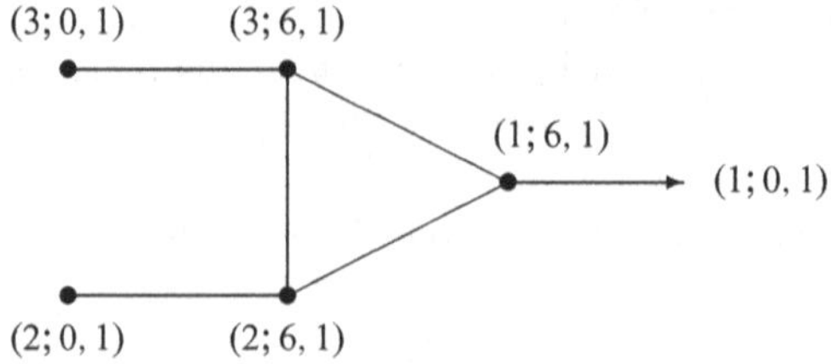

**Example 9.4.9.** If $f = x^5 + y^2 + xyz$ and $g = z$ then a possible $\Gamma_{\mathscr{C}}$ is:

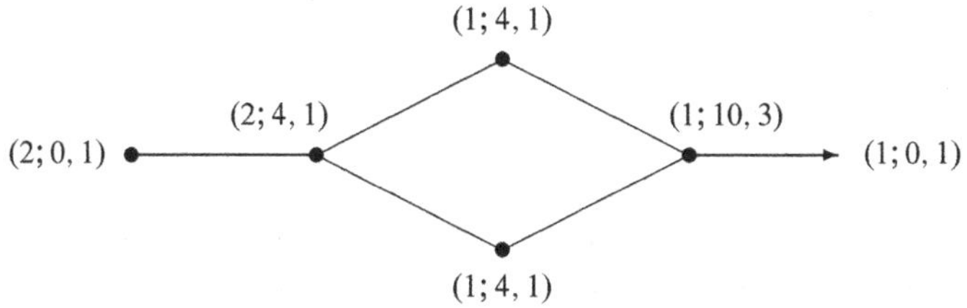


The general case (i.e. arbitrary $a \geq 3$), with $g = x + y + z$, is clarified in 9.3.3.

Notice that in the two examples above and in 9.3.3 we used different germs $g$ and different sequences of blow ups. Thus, the output graphs are also different.

# Part II
# Plumbing Graphs Derived from $\Gamma_{\mathscr{C}}$

# Chapter 10
# The Main Algorithm

## 10.1 Preparations for the Main Algorithm

**10.1.1. The goal of the chapter.** The algorithm presented in this chapter provides the plumbing representations of the 3-manifolds $\partial F$, $\partial_1 F$ and $\partial_2 F$, and the multiplicity systems of the open book decomposition of $(\partial F, V_g)$ as well as the generalized Milnor fibrations $\partial F \setminus V_g$, $\partial_1 F \setminus V_g$ and $\partial_2 F$ over $S^1$ induced by $\arg(g) = g/|g|$.

**10.1.2. Assumptions.** In 6.3 we imposed Assumption A on $\Gamma_{\mathscr{C}}$, which can always be realized by an additional blow up. Although in the Main Algorithm this restriction is irrelevant, in the geometric interpretations 7.3–7.4–7.5 it simplified and unified the presentation substantially.

In the next paragraph we introduce another restriction, Assumption B. In contrast with Assumption A, this new restriction plays a relevant role in the formulation of the algorithm and its proof. Nevertheless, in Chap. 12 we will formulate a new version of the algorithm from which Assumption B will be removed (but the proof of the new algorithm will still rely on the proof of the present original version).

**10.1.3. Assumption B.** In Chaps. 10 and 11 we will assume that $\Gamma_{\mathscr{C}}$ has no such edge decorated by 2 whose end-vertices would have the middle weights zero. This requirement is regardless of whether those end-vertices are arrowheads or not.

In the sequel, we call such an edge *vanishing 2-edge*.

Their absence can be assumed because of the following reason. Assume that $\Gamma_{\mathscr{C}}$ is associated with some resolution $r$ as in 6.1, and it has such a 2-edge $e$

$$\underset{v}{(m;0,\nu)} \overset{}{\underline{\qquad\quad 2 \qquad\quad}} \underset{v'}{(m';0,\nu)}$$

where the vertices $v$ and $v'$ correspond to the curves $C_v$ and $C_{v'}$ of $\mathscr{C}$. (If $v$ and $v'$ are non-arrowheads, they might have genus decorations as well.) Then the embedded

A. Némethi and Á. Szilárd, *Milnor Fiber Boundary of a Non-isolated Surface Singularity*, Lecture Notes in Mathematics 2037, DOI 10.1007/978-3-642-23647-1_10, 

resolution $r$ modified by an additional blow up with center $p \in C_v \cap C_{v'}$ provides a new graph $\Gamma'_{\mathscr{C}}$, where $e$ is replaced by

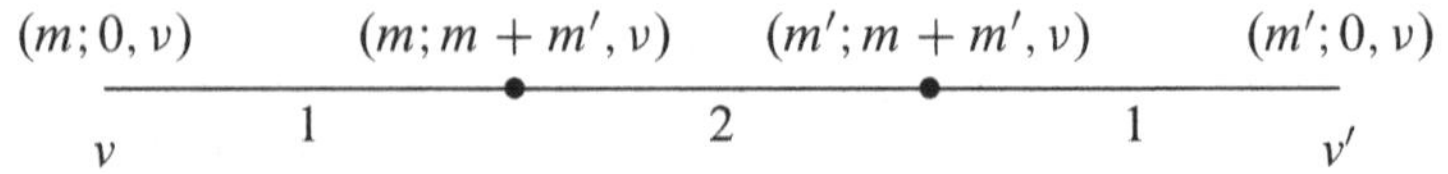


**10.1.4. Terminology –** ***legs*** **and** ***stars*****.** In the description of the algorithm we use the following expressions.

Fix a non-arrowhead vertex $v$ of $\Gamma_{\mathscr{C}}$ with weights $(m; n, \nu)$ and $[g]$. Then $v$ determines a *star* in $\Gamma_{\mathscr{C}}$, which keeps track of all the edges adjacent to $v$ along with their decorations and the weights $(k; l, \nu)$ of the vertices at the other end of the edges, but disregards the type of these vertices. The aim is to unify the different cases represented by loops and edges connecting non-arrowheads or arrowheads.

**Definition 10.1.5.** *A* **leg supported by the vertex** *$v$ has the form:*

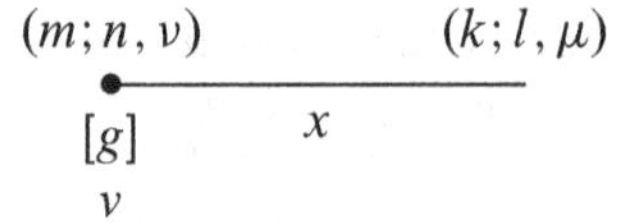


*where $x \in \{1, 2\}$ and the decorations satisfy the same compatibility conditions as the edges in 6.2.4. Then,* **a star***, by definition, consists of a vertex $v$ (together with its decorations) and a collection of legs supported by $v$.*

*Once $\Gamma_{\mathscr{C}}$ and $v$ are fixed,* the star of $v$ in $\Gamma_{\mathscr{C}}$ *is constructed as follows. Its "center" has the decorations $(m; n, \nu)$ and $[g]$ of the vertex $v$. Furthermore, any edge with decoration $x$, with end-vertices $v$ and $v'$ (where $v' \neq v$, and $v'$ is either an arrowhead or not) provides a leg with decorations $x$ and the weight (the ordered triple) of $v'$. In particular, if $v'$ is a non-arrowhead, and it is connected to $v$ by more than one edge, then each edge contributes a leg. Moreover, any loop supported by $v$ and weighted by $x$, provides* two *legs supported by $v$, both decorated by the same $x$ and their "free ends" by $(k; l, \mu) = (m; n, \nu)$.*

One has the following geometrical interpretation: regard $\Gamma_{\mathscr{C}}$ as the dual graph of the curve configuration $\mathscr{C}$, and let $v$ correspond to the component $C$. Then the legs of the star of $v$ correspond to the inverse images of the double points of $\mathscr{C}$ sitting on $C$ by the normalization map $C^{norm} \to C$.

## 10.2 The Main Algorithm: The Plumbing Graph of $\partial F$

Recall that in this section we assume that the graph $\Gamma_{\mathscr{C}}$ satisfies Assumption B.

First, we construct the plumbing graph of the open book decomposition of $\partial F$ with binding $V_g$ and fibration $\arg(g) : \partial F \setminus V_g \to S^1$, cf. 3.2.2 and 4.1.9. Hence, we have to determine the shape of the graph together with its arrows, and endow it with the genus and Euler number decorations and a multiplicity system. The graph will be determined as a covering graph $G$ of $\Gamma_{\mathscr{C}}$, modified with strings as in 5.1.9. In order to do this, we have to provide the *covering data of the graph-covering*, see 5.1.2.

**10.2.1. Step 1. – The covering data of the vertices $\{\mathfrak{n}_v\}_{v\in\mathscr{V}(\Gamma_{\mathscr{C}})}$.**

**Case 1.** Consider a non-arrowhead vertex $w$ of $\Gamma_{\mathscr{C}}$ decorated by $(m;n,\nu)$ and $[g]$. (In fact, by 7.4.8, $g=0$ whenever $m>1$.) Consider its star

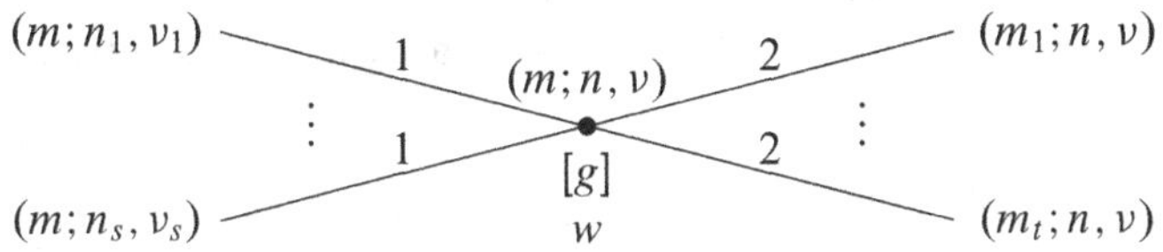


Let $s$ and $t$ be the number of legs weighted by $x=1$ and $x=2$ respectively. Then, in the covering procedure, above the vertex $w$ of $\Gamma_{\mathscr{C}}$ put $\mathfrak{n}_w$ non-arrowhead vertices, where

$$\mathfrak{n}_w=\gcd(m,n,n_1,\dots,n_s,m_1,\dots,m_t). \tag{10.2.2}$$

Furthermore, put on each of these non-arrowhead vertices the same multiplicity decoration $(\tilde{m})$, where

$$\tilde{m}=\frac{m\nu}{\gcd(m,n)}, \tag{10.2.3}$$

and the genus decoration $[\tilde{g}_w]$ determined by the formula:

$$\mathfrak{n}_w\cdot(2-2\tilde{g}_w)=(2-2g-s-t)\cdot\gcd(m,n) \tag{10.2.4}$$
$$+\sum_{i=1}^{s}\gcd(m,n,n_i)+\sum_{j=1}^{t}\gcd(m,n,m_j).$$

(In Step 3, the Euler number of each vertex will also be provided.)

**Case 2.** Consider an arrowhead vertex $v$ of $\Gamma_{\mathscr{C}}$, that is, $\to(1;0,1)$. Above the vertex $v$, in the covering graph $G$, put exactly one arrowhead vertex; i.e. set $\mathfrak{n}_v=1$. Let the multiplicity of this arrowhead be 1. In particular, all arrowheads of $G$ are: $\to(1)$.

**10.2.5. Step 2. – The covering data of edges $\{\mathfrak{n}_e\}_{e\in\mathscr{E}(\Gamma_{\mathscr{C}})}$ and the types of inserted strings.**

**Case 1.** Consider an edge $e$ in $\Gamma_{\mathscr{C}}$, with decoration 1:

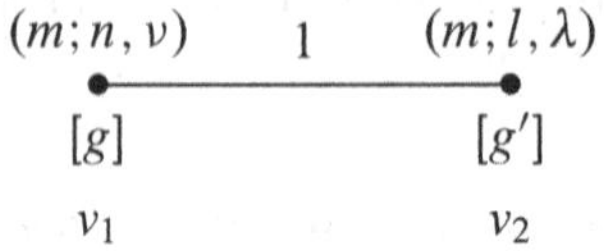


Define:

$$\mathfrak{n}_e = \gcd(m, n, l).$$

Notice that Step 1 guarantees that both $\mathfrak{n}_{v_1}$ and $\mathfrak{n}_{v_2}$ divide $\mathfrak{n}_e$.

Then, above the edge $e$ insert cyclically in $G$ exactly $\mathfrak{n}_e$ strings of type

$$Str\left(\frac{n}{\mathfrak{n}_e}, \frac{l}{\mathfrak{n}_e}; \frac{m}{\mathfrak{n}_e} \,\middle|\, \nu, \lambda; 0\right).$$

If the edge $e$ is a loop (that is, if $v_1 = v_2$), then the procedure is the same with the only modification that the end-vertices of the $\mathfrak{n}_e$ strings are identified cyclically with the $\mathfrak{n}_{v_1}$ vertices above $v_1$, hence they will form 1-cycles in the graph. In other words, on each vertex above $v_1$ one puts $\mathfrak{n}_e/\mathfrak{n}_{v_1}$ "closed" strings, that form loops.

If the right vertex $v_2$ is an arrowhead, that is, the edge $e$ is

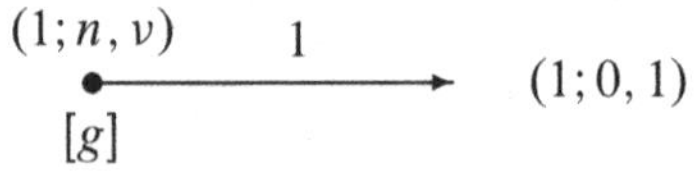


then complete the same procedure as above with $m = 1$ and $\mathfrak{n}_e = 1$: above such an edge $e$ put a single edge decorated by $+$, which supports that arrowhead of $G$ which covers the corresponding arrowhead of $\Gamma_{\mathscr{C}}$.

**Case 2.** Consider an edge $e$ in $\Gamma_{\mathscr{C}}$, with decoration 2:

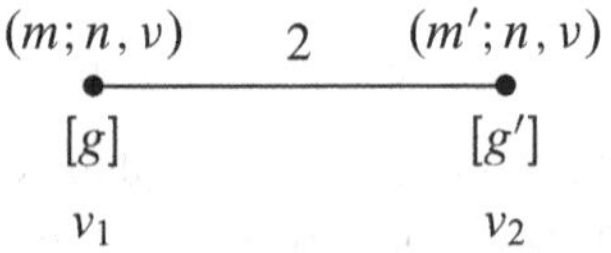


Notice that, by Assumption B (cf. 10.1.3), $n \neq 0$. For such an edge, define:

$$\mathfrak{n}_e = \gcd(m, m', n). \tag{10.2.6}$$

Notice again that both $\mathfrak{n}_{v_1}$ and $\mathfrak{n}_{v_2}$ divide $\mathfrak{n}_e$.

Then, above the edge $e$ insert cyclically in $G$ exactly $\mathfrak{n}_e$ strings of type

$$Str^{\ominus}\left(\frac{m}{\mathfrak{n}_e},\frac{m'}{\mathfrak{n}_e};\frac{n}{\mathfrak{n}_e}\,\Big|\,0,0;\nu\right).$$

If the edge is a loop, then we modify the procedure as in the case of 1-loops. Notice that by Assumption B, there are no 2-edges supporting arrowheads.

Note that above $\Gamma^1_{\mathscr{C}}$ and above the cutting edges the "covering degree" is always one.

**10.2.7. Step 3. – Determination of the missing Euler numbers.** The decorations provided by the first two steps are the following: the multiplicities of all the vertices, all the genera, *some* of the Euler numbers, and all the sign-decorations of the edges (those without $\ominus$ have decoration $+$). Then, finally, the missing Euler numbers are determined by formula (4.1.5).

**10.2.8. The output of the algorithm.** Notice that the set of integers $\{\mathfrak{n}_v\}_{v\in\mathscr{V}(\Gamma_{\mathscr{C}})}$ and $\{\mathfrak{n}_e\}_{e\in\mathscr{E}(\Gamma_{\mathscr{C}})}$ satisfy the axioms of a covering data. Furthermore, if $v\in\mathscr{V}^1(\Gamma_{\mathscr{C}})$ then $m=1$ hence $\mathfrak{n}_v=1$. Moreover, by Corollary 7.4.12, each $\Gamma^2_{\mathscr{C},j}$ is a tree. Therefore, by Proposition 5.1.8 and Theorem 7.4.16 we get that

*there is only one cyclic covering of $\Gamma_{\mathscr{C}}$ with this covering data (up to a graph-isomorphism).*

The graphs obtained by the above algorithm can, in general, be simplified by the operations of the oriented plumbing calculus (or their inverses), or by the reduced plumbing calculus. If we are interested only in the output oriented 3-manifold, we can apply this freely without any restriction. Nevertheless, if we wish to keep some information from the (analytic) construction which provides the graph (for example, if we wish to apply the results of Chaps. 13–18 regarding different horizontal and vertical monodromies), then it is better to apply only the *reduced plumbing calculus* of oriented 3-manifolds (with arrows), cf. 4.2. This is what we prefer to do in this book.

Moreover, even if we rely only on the reduced calculus, during the plumbing calculus, some invariants might still change. For example, the operation R5 modifies the sum $g(Gr)$ of the genus decorations and the number $c(Gr)$ of independent 1-cycles of a graph $Gr$. Since, in the sequel in some discussions these numbers will also be involved, by our graph notations we wish to emphasize that a certain graph is in the unmodified stage, or it was modified by the calculus. Hence, we will adopt the following notation:

**Definition 10.2.9.** *We write $G$, $G_1$ and $G_{2,j}$ for the graphs obtained by the original algorithm (associated with $\Gamma_{\mathscr{C}}$, $\Gamma^1_{\mathscr{C}}$ and $\Gamma^2_{\mathscr{C}}$, see below), while the general notation for the* modified graph *under the reduced plumbing calculus is $G^m$, $G^m_1$, $G^m_{2,j}$.*

Using these notations, one of the main results of the present work is the following.

**Theorem 10.2.10.** *The oriented 3-manifold $\partial F$ and the link $V_g \cap \partial F$ in it can be represented by an orientable plumbing graph (see 4.1 for the terminology).*

*More precisely, let $(f, g)$ be as in Sect. 3.1. Then, the decorated graph $G$ constructed above is a possible plumbing graph of the pair $(\partial F, \partial F \cap V_g)$, which carries the multiplicity system of the open book decomposition* $\arg(g)$ : $\partial F \setminus V_g \to S^1$. *If one deletes the arrowheads and the multiplicities, one obtains a possible plumbing graph of the boundary of the Milnor fiber $\partial F$ of $f$.*

The proof of Theorem 10.2.10 will be given in Chap. 11. Nevertheless, in the next paragraphs we wish to stress the main geometric idea of the proof.

If $f$ is an *isolated* hypersurface singularity, then the link $K = V_f \cap S^5_\epsilon$ is smooth, and there exists an orientation preserving diffeomorphism $\partial F \approx K$. Hence, $\partial F$ can be "localized", i.e. can be represented as a boundary of an arbitrary small representative of a (singular) germ. If that germ is resolved by a modification – whose existence is guaranteed by the existence of resolution of singularities –, then $K$ appears as the boundary of the exceptional locus, hence one automatically gets a plumbing representation for $K$. Its plumbing data can be read from the combinatorics of the exceptional set and the multiplicity system from the corresponding vanishing orders.

If $f$ is *not isolated* then $K$ is not smooth, and the above argument does not work. Even the fact that $\partial F$ has any kind of plumbing representation is not automatic at all. Nevertheless, the case of isolated singularities suggests that, if we were able to "localize" $\partial F$, as a link of a singular germ, then we would be able to extend the above procedure valid for isolated singularities to the non-isolated case as well. *This realization is the main point in the proof* of Theorem 10.2.10, but with the difference that the germ whose local link is $\partial F$ is *not holomorphic* (complex analytic), but it is *real analytic*. One has the following surprising result.

**Proposition 10.2.11.** *(See 11.3.3.) Let $f$ be a hypersurface singularity with a 1-dimensional singular locus. Take another germ $g$ such that $(f, g)$ forms an ICIS as in Sect. 3.1. For a sufficiently large even integer $k$ consider the real analytic germ*

$$\mathscr{S}_k := \{z \in (\mathbb{C}^3, 0) \ : \ f(z) = |g(z)|^k \}.$$

*Then $\mathscr{S}_k \setminus \{0\}$ is a smooth 4-manifold with a natural orientation whose link is independent of the choice of $g$ and $k$, and which, in fact, is orientation preserving diffeomorphic to $\partial F$.*

The proof of Theorem 10.2.10, in fact, describes the topology of a resolution of $\mathscr{S}_k$, and it shows that it is "guided" exactly by $\Gamma_{\mathscr{C}}$. Furthermore, the algorithm which provides $G$ from $\Gamma_{\mathscr{C}}$ extracts the combinatorics of the exceptional locus and its tubular neighbourhood from this resolution.

Notice that the above proposition is true for $k$ odd too, nevertheless, $\mathscr{S}_k$ for $k$ even has nicer analytic properties: for example, if $f$ and $g$ are polynomials, then $\mathscr{S}_k$ is a real algebraic variety.

**Remark 10.2.12.** (a) We would also like to stress that even though Proposition 10.2.11 is formulated and proved for germs in three variables, it is true for any germ $f : (\mathbb{C}^n, 0) \to (\mathbb{C}, 0)$ with 1-dimensional singular locus – the only modification in the statement and its proof is the replacement of $\mathbb{C}^3$ with $\mathbb{C}^n$.

(b) The power of Theorem 10.2.10 is not just the fact that it proves that $\partial F$ has a plumbing representation; for that already Proposition 10.2.11 is enough. Theorem 10.2.10 provides a very clear algorithm for the determination of the plumbing representation, which can be performed for any concrete example. Moreover, from the algorithm one can subtract essential theoretical information as well, as will be done in the next chapters.

(c) The investigation of the geometry of real analytic germs and their fibration properties is not new in the literature, see for example the results of A. Pichon and J. Seade regarding the germs of type $f\bar{g}$ [106, 107].

(d) Theorem 10.2.10 was obtained in 2004–2005; the Main Algorithm was presented at the Singularity Conference at Leuven, 2005. The material of the present book was posted on the Algebraic Geometry preprint server in 2009 [93].

The fact that the boundary of the Milnor fiber is plumbed was announced by F. Michel and A. Pichon in 2003 [73, 74]. Their proof appeared on the preprint server in 2010 [75].

The techniques prior to the present book were not sufficiently powerful to produce examples with cycles, and even to predict the necessity of edges with negative decorations. These are novelties of the present work.

## 10.3 Plumbing Graphs of $\partial_1 F$ and $\partial_2 F$

The above algorithm, which provides $\partial F$, is compatible with the decomposition of this space into its parts $\partial_1 F$ and $\partial_2 F$. In this section we make this statement precise.

**10.3.1. The graphs of $(V_f^{norm}, g\circ n)$ and $\partial_1 F$.** Consider the graph $\Gamma_{\mathscr{C}}^1$. Repeat Steps 1 and 2 from the Main Algorithm 10.2, but only for the vertices and edges contained in $\Gamma_{\mathscr{C}}^1$, *excluding* any edge inherited from a cutting edge. Replace any edge inherited from a cutting edge by an edge supporting an arrowhead with multiplicity (0). In this way we get a graph with all the multiplicities determined and with all edge-decorations $+$. Calculate the Euler numbers by (4.1.5). This graph will be denoted by $G_1$. (Note that it coincides with the graph $G_{\mathscr{C}}^1$ considered in 7.3.2 and Proposition 7.3.3.)

**Remark 10.3.2.** The vertices of $G_1$ can be identified with some of the vertices of $G$, hence if we disregard the decorations of the graphs, then $G_1$ is a subgraph of $G$. However, as decorated graph, $G_1$ is not a subgraph of $G$: that end-vertex of any cutting edge which is situated in $\Gamma_{\mathscr{C}}^1$ will have different Euler numbers in the two graphs $G$ and $G_1$. All the other Euler numbers evidently coincide.

The next theorem is essentially the same as Proposition 7.3.3; we consider it again to have a complete picture of the algorithm and its consequences.

**Theorem 10.3.3.** *$G_1$ is a possible embedded resolution graph of $(V_f^{norm}, g\circ n)$, where the arrows with multiplicity* (0) *represent the link-components determined by the strict transforms of* $Sing(V_f)$. *If we delete these 0-multiplicity arrows, we get a possible embedded resolution graph of $(V_f^{norm}, g\circ n)$. Furthermore, if we delete all the arrowheads and all the multiplicities, we get a possible resolution graph of the normalization of $V_f^{norm}$.*

*If in $G_1$ we replace the 0-multiplicity arrows by dash-arrows we get the plumbing representation of the pair $(\partial_1 F, V_g\cap\partial F)$, where the remaining arrows represent the link $V_g\cap\partial F$, and the multiplicities are the multiplicities of the local trivial fibration* $\arg(g): \partial_1 F\setminus V_g\to S^1$. *In particular, if all remaining non-dash-arrows and all multiplicities are deleted as well, we get the plumbing graph of the 3-manifold with boundary $\partial_1 F$.*

*If in $G_1$ all arrows are replaced by dash-arrows and the multiplicities are deleted, we get the plumbing graph of the manifold with boundary $\partial_1 F\setminus T^\circ(V_g)$.*

**10.3.4. The graph of $\partial_2 F$.** Let $G_2$ be the graph obtained from $G$ as follows. Delete all vertices and edges of $G$ that are above the vertices and edges of $\Gamma^1_{\mathscr{C}}$, and replace the unique string above any cutting edge by a dash-arrow (putting no multiplicity decoration on it). Obviously, $G_2$ has $s$ connected components $\{G_{2,j}\}_{1\le j\le s}$; where $G_{2,j}$ is related with $\Sigma_j$ as in 7.4. Clearly, $G_{2,j}$ can be determined from $\Gamma^2_{\mathscr{C},j}$ by a similar procedure as $G$ is obtained from $\Gamma_{\mathscr{C}}$, and by adding dash-arrows above the arrowheads.

**Theorem 10.3.5.** *For each $j = 1,\dots,s$, $G_{2,j}$ is a possible plumbing graph for the 3-manifold with boundary $\partial_{2,j}F$, where the set of multiplicities consists of the multiplicity system associated with the fibration* $\arg(g): \partial_{2,j}F\to S^1$. *If all multiplicities are deleted then obviously we get a plumbing graph of the 3-manifold with boundary $\partial_{2,j}F$.*

**10.3.6. The gluing tori.** Each connected component $\partial_{2,j}F$ ($j=1,\dots,s$) is glued to $\partial_1 F$ along $\partial\partial_{2,j}F$, which is a union of tori. Since for each cutting edge $e$ one has $\mathfrak{n}_e = 1$ (i.e. in the Main Algorithm exactly one string is inserted above $e$), the number of these tori is exactly the cardinality of $\mathscr{E}_{cut,j}$, the number of cutting edges adjacent to $\Gamma^2_{\mathscr{C},j}$.

Let $e$ be such a cutting edge, and use the notations of 7.5 regarding this edge. Let $T_e$ be the torus component of $\partial(\partial_{2,j}F)$ corresponding to $e$. Furthermore, consider the fibration $T_e\to L_j$ from 2.3.1. Then the fiber of this projection consists of $d(e)$ circles, the corresponding orbit of the (permutation) action of ${m'}_{j,ver}$ on $\partial F'_j$. In particular, the number of connected components of $\partial F'_j$ is $\sum_{e\in\mathscr{E}_{cut,j}} d(e)$, as it was already noticed in (7.5.4). Recall also that ${m'}_{j,hor}$ acts on $T_e$ trivially, cf. 2.3.1(2). For more details see Sect. 7.5.

These gluing tori appear in the link $K$ of $V_f$ as well. Indeed, consider $L_j = K\cap\Sigma_j$ as in Sect. 2.1. Let $T(L_j)$ be a tubular neighbourhood of $L_j$ in $S^5_\epsilon$ as in 2.3. Then $\partial T(L_j)$ intersects $K$ in $|\mathscr{E}_{cut,j}|$ tori, which can be identified with the gluing tori of $\partial F$, see also Remark 2.3.3.

**Remark 10.3.7.** From the plumbing graphs of $\partial_1 F$ and $\partial_2 F$ it is impossible to recover the graph of $\partial F$, since the *gluing* information (an automorphism of the gluing tori) cannot be read from the partial information contained in the graphs of $\partial_1 F$ and $\partial_2 F$. (See e.g. examples 10.4.1 and 10.4.2.) This gluing information is exactly one of the main advantages of the graph $\Gamma_{\mathscr{C}}$ and of the Main Algorithm, which provides the full $\partial F$.

**Remark 10.3.8.** In fact, on $G_{2,j}$ one can put even more information/decoration inherited from $G$. If one introduces a "canonical" framing (closed simple curve) of the boundary components of $\partial_{2,j} F$, then one can define a well-defined multiplicity of the dash-arrows as well (inherited from $G$). Usually, such a framing is needed when one wishes to "close" with solid tori a 3-manifold that has toric boundary components.

Here we will make this completion via the following construction. Consider the graph $\overline{G_{2,j}}$ obtained as follows.

From the graph $G$ delete all those vertices which are vertices of $G_1$. All the remaining vertices are non-arrowheads; keep their genus, Euler number and multiplicity decorations. Keep all the edges which connect these vertices, and keep their decorations as well. Finally, keep any edge which connects a vertex $v$ in $G_1$ with another vertex $w$ not in $G_1$, keep its decoration $\ominus$, and replace $v$ with an arrowhead having the same multiplicity as $v$ has in $G_1$. This graph is denoted by $\overline{G_2}$. Its connected components are indexed by $\{1, \dots, s\}$ and there is a natural bijection (induced by inclusion) with the graphs $G_{2,j}$. The connected component $\overline{G_{2,j}}$ of $\overline{G_2}$ which contains the vertices of $G_{2,j}$ is called the *canonical closure* of $G_{2,j}$. $\overline{G_{2,j}}$ contains $|\mathscr{E}_{cut,j}|$ arrowheads.

It is clear that $\overline{G_{2,j}}$ can be obtained from $\Gamma^2_{\mathscr{C},j}$ as well.

If we delete all the multiplicities of $\overline{G_{2,j}}$, but we keep the arrowheads, we get a plumbing graph of a closed 3-manifold (without boundary) and a link in it. This 3-manifold, denoted by $\overline{\partial_{2,j} F}$, will be called the *canonical closure* of $\partial_{2,j} F$, since it can be obtained from $\partial_{2,j} F$ by gluing some solid tori to its boundary components in a canonical way dictated by the above construction. The corresponding link in it is denoted by $L_{cut,j}$. The manifold $\partial_{2,j} F$ is obtained from $\overline{\partial_{2,j} F}$ by deleting the interior of a tubular neighbourhood $T_j$ of $L_{cut,j}$.

For each component of $L_{cut,j}$ consider the oriented meridian $\gamma_e$ in $T_j$. The collection $\{\gamma_e\}_{e \in \mathscr{E}_{cut,j}}$ serves as a framing in $\partial(\partial_{2,j} F)$. Using this framing $\partial_{2,j} F$ can be closed in a canonical way to get $\overline{\partial_{2,j} F}$.

**Remark 10.3.9.** Using the graph $G$ one can decorate both the graphs $G_1$ and $G_2$ even more so that all boundary components of $\partial_1 F$ and $\partial_2 F$ will be canonically identified with $S^1 \times S^1$ in such a way that gluing them provides $\partial F$.

Since a complete description of $\partial F$ is already provided by the Main Algorithm, we omit the description of these decorations. But, definitely, the interested reader might consider and add this data to the picture as well.

## 10.4 First Examples of Graphs of $\partial F$, $\partial_1 F$ and $\partial_2 F$

**Example 10.4.1.** Assume that $f = x^3 + y^2 + xyz$ and $g = z$ as in 9.4.8. Then the output of the Main Algorithm is the following graph $G$:

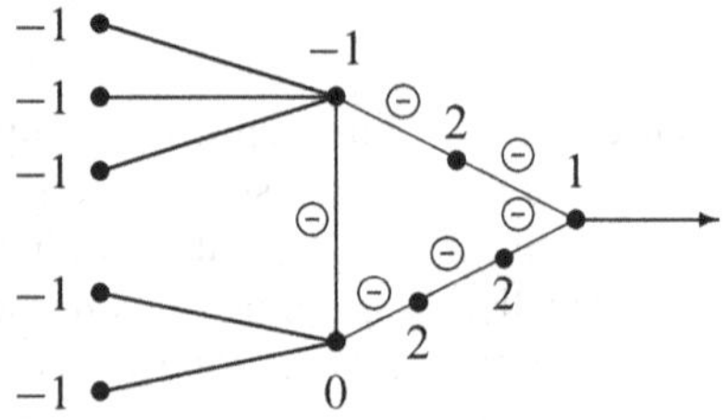

and all the multiplicities are (1).

There is only one non-arrowhead vertex in $G$ with multiplicity 1. Therefore, one has the following graphs for $(V_f^{norm}, g \circ n)$, $\partial_1 F \setminus T(V_g)$, and $\partial_1 F$:

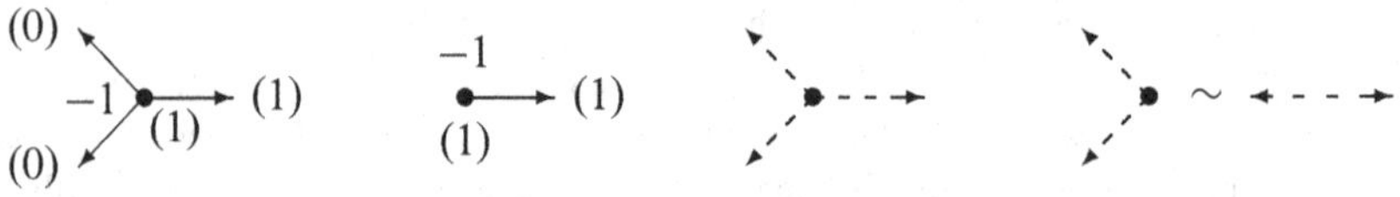

$G_1$ $\quad (V_f^{norm}, g \circ n) \quad \partial_1 F \setminus T(V_g) \quad \partial_1 F$

Moreover, by plumbing calculus, the graph of $\partial_2 F$ is also the double dash-arrow ◂- - -▸. Notice that both parts $\partial_1 F$ and $\partial_2 F$ are extremely simple 3-manifolds with boundary, namely, both are isomorphic to $S^1 \times S^1 \times [0, 1]$. The main information in $\partial F$ is exactly how these parts are glued.

By calculus starting from $G$, the boundary $\partial F$ is represented by the graph:

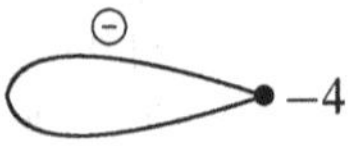

The multiplicity system of the open book decomposition $(\partial F, V_g)$ is given by:

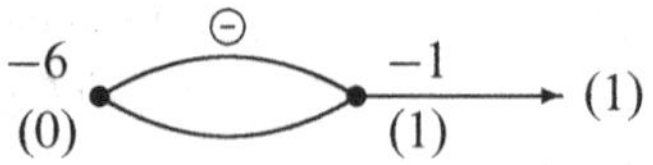

**Example 10.4.2.** Assume that $f = x^2y + z^2$ and $g = x + y$, cf. 9.3.1. Then, by the Main Algorithm and plumbing calculus we get that the (minimal) plumbing graph of $\partial F$ consists of a unique vertex with genus zero and Euler number $-4$, i.e. $\partial F$ is the lens space $L(4, 1)$. Moreover,

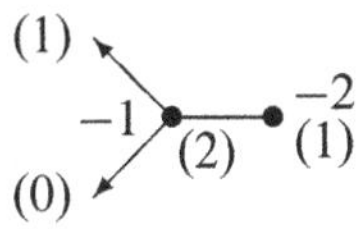


$G_1$

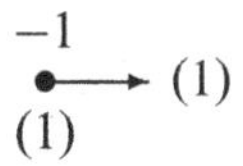


$(V_f^{norm}, g \circ n)$

$\partial_1 F$

and

$\partial_2 F$ :

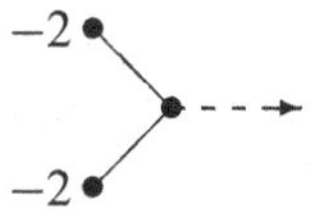


Notice that for the unique cutting 2-edge $e$ in 9.3.1 one has $d_1 = 1$, hence $\nu = d(e) = 2$. Therefore, although the transversal singularity has two local irreducible components, $\partial_1 F$ and $\partial_2 F$ are glued by only one torus. The point is that the transversal type is $A_1$, and the two local irreducible components of the transversal singularity are permuted by the vertical monodromy, see 7.5 and 10.3.6 (and compare also with the next example and Example (3.1) of [119]).

The open book decomposition of $(\partial F, V_g)$ is given by:

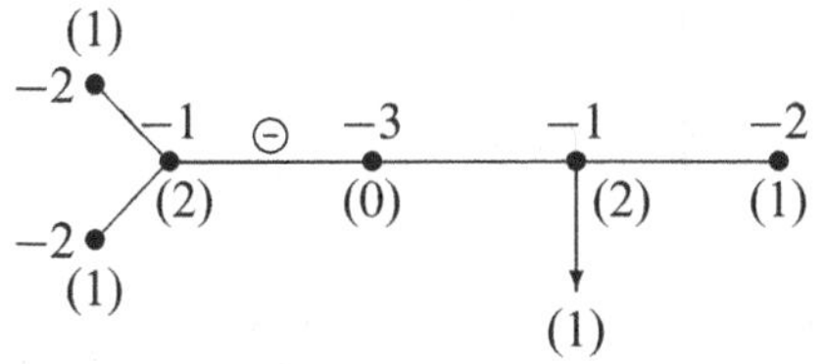


**Example 10.4.3.** In both cases of 9.4.8 and 9.4.9, the singular locus $\Sigma$ of $V_f$ is irreducible and consists of the $z$-coordinate axis. The transversal type is an $A_1$ singularity, hence $\#T(\Sigma) = 2$. Furthermore, $|\mathscr{E}_{cut,1}| = 2$ and for both cutting edges $d(e) = 1$. Therefore (see 7.5 and 10.3.6), the action of the vertical monodromy does not permute the two local components, and in both cases 9.4.8 and 9.4.9, there are two gluing tori.

On the other hand, it might happen that the two local components of a transversal $A_1$ singularity are permuted by the vertical monodromy, see e.g. 10.4.2.

Similarly, in the case of 6.2.7 (compare also with 7.5.5), $\Sigma = \Sigma_1 \cup \Sigma_2$, and for both $\Sigma_j$ the transversal type is $A_1$, hence $\#T(\Sigma_j) = 2$ $(j = 1, 2)$. Moreover, for both $j$, $|\mathscr{E}_{cut,j}| = 1$, $d_j = 1$, and $d(e) = 2$; hence $\partial_{2,j} F$ is glued to $\partial_1 F$ by exactly one torus.

**Example 10.4.4.** Assume that $f = x^d + y^d + xyz^{d-2}$, $d \geq 3$. For $\Gamma_{\mathscr{C}}$ see the second graph of 8.1.5 (which satisfies Assumption A). In this case $\Sigma$ is irreducible with transversal type $A_1$. The gluing data are $|\mathscr{E}_{cut,1}| = 2$ and $d_1 = 1$. For both cutting edges $d(e) = 1$, hence one has two gluing tori. The graph of $\partial F$ is:

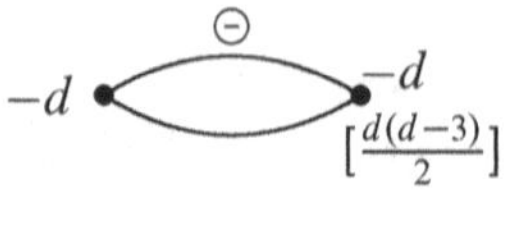


Recall that if a normal surface singularity is weighted homogeneous, then its link is a Seifert 3-manifold, and it can be represented by a star-shaped plumbing graph (or, in the degenerate case, by a string). Note that this is not true in the present situation: the above equation is homogeneous, nevertheless, the graph has a cycle. The same remark is valid for the weighted homogeneous equation from 10.4.1.

Note also that the above graph has a *negative definite intersection matrix* $A$, nevertheless there is no sequence of modifications by plumbing calculus which would eliminate the negative edge-decoration. Hence, $\partial F$ cannot be the link of a normal surface singularity.

**Example 10.4.5.** For $f = x^3y^7 - z^4$ the graph $\Gamma_{\mathscr{C}}$ is given in 6.2.9. After a computation, we get for $G^m$ the next graph.

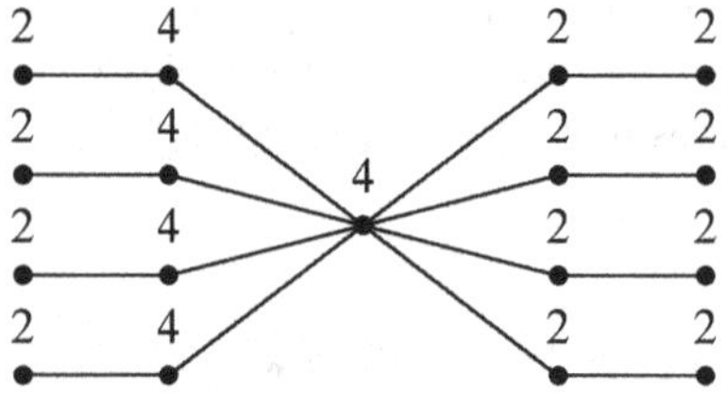


This graph $G^m$ can be transformed into its "normal form" in the sense of [33], that is, with all the Euler numbers on the legs $\leq -2$. It is the following:

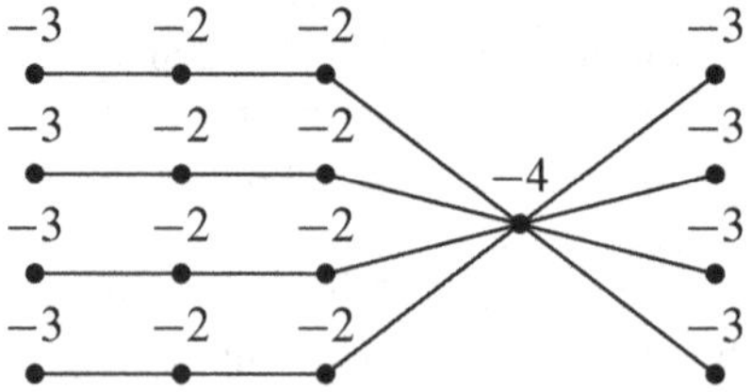


Its central vertex has Euler number $e = -4$. The eight pairs of normalized Seifert invariants $(\alpha_\ell, \omega_\ell)$, $(1 \leq \ell \leq 8)$, associated with the eight legs, are determined as Hirzebruch–Jung continued fractions associated with the entries of the corresponding legs: $\alpha_\ell/\omega_\ell$ for $1 \leq \ell \leq 4$ is $[2, 2, 3] = 7/5$, while the other four are $[3] = 3/1$; cf. 5.3.5.

Recall that the *orbifold Euler number* of the Seifert 3-manifold is defined as $e^{orb} := e + \sum_\ell \omega_\ell/\alpha_\ell$, and the normal form graph is negative definite if and only if $e^{orb} < 0$. In this case $e^{orb} = 4/21 > 0$, hence the graph is *not* negative definite.

In particular, this graph cannot be transformed into a negative definite graph by plumbing calculus.

Note that $G^m$ with opposite orientation (that is, $-G^m$) is negative definite.

**Example 10.4.6.** For $f = x^2y^2 + z^2(x + y)$ a possible graph $\Gamma_{\mathscr{C}}$ is given in 6.2.7. For a possible $G^m$ we get:

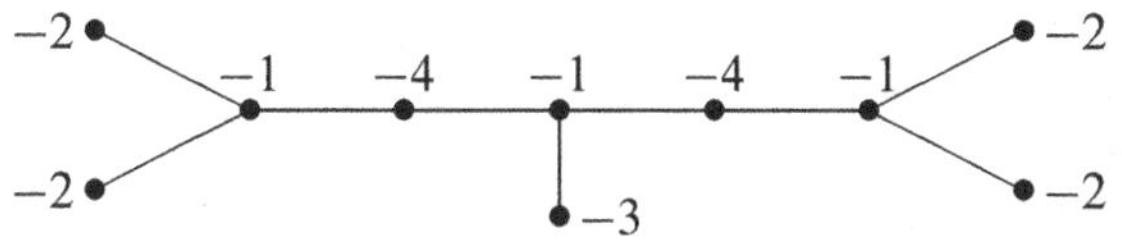


(Notice that for this graph it would be possible to use the $\mathbb{RP}^2$-absorption of the *non-orientable* calculus, but we do not do that.)

In this case $\partial F$ is a rational homology sphere, since $\det(A) \neq 0$.

**Example 10.4.7.** Assume that $f = y^3 + (x^2 - z^4)^2$. Then using 6.2.8 we get for $G^m$:

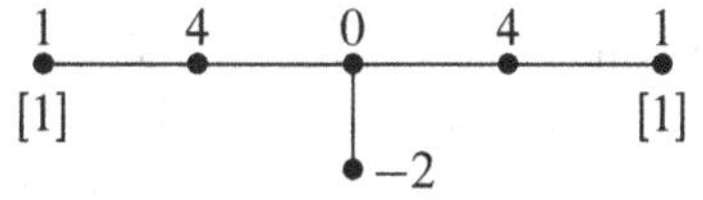


**Example 10.4.8.** Finally, the last example is the 1-parameter infinite family $f = x^a y(x^2 + y^3) + z^2$ with $a \geq 2$. The reader may consider this as a model for other infinite families.

A graph $\Gamma_{\mathscr{C}}$ with $g = x + y + z$ is given in 9.3.4.

**Case 1.** Assume that $a$ is even. We determine $G$ in several steps.

First, the graph $G_1$ can be determined easily (in particular, the normalization of $V_f$ is the $D_5$ singularity):

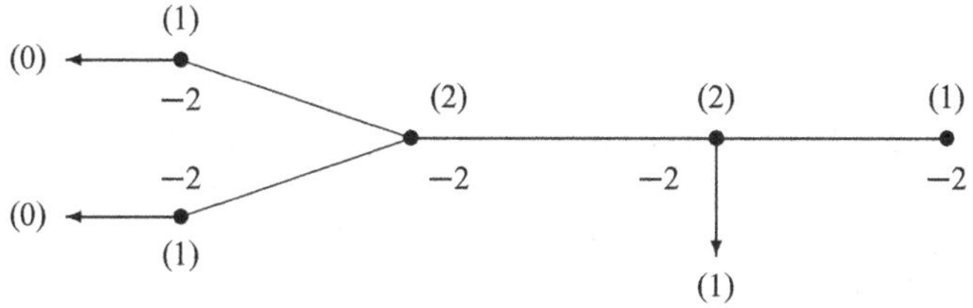


Clearly, we have two gluing tori. Let $v_1$ and $v_2$ be the vertices of $G_1$ which support the (0)-arrows. Next, we wish to determine the multiplicity $m_1$ of the vertex $v_1'$ of $G$, which is not in $G_1$ and is a neighbour of $v_1$ in $G$. For this we have to analyze the cutting edge with weights $(a; 2a+4, 1)$ and $(1; 2a+4, 1)$. By (4.3.6) (pay attention to the left-right ordering of the ends of the string), $m_1$ satisfies $a+\lambda = m_1(2a+4)$ for some $\lambda$ with $0 \leq \lambda < 2a+4$. Hence $m_1 = 1$. In $G$ the vertex $v_1'$ is glued to $v_1$ by a $\ominus$-edge, hence the Euler number of $v_1$ in $G$ is $-1$.

Finally, we analyze the graph $\Gamma^2_{\mathscr{C}}$. Its shape and the first entries of the weights of the vertices coincide with the minimal embedded resolution graph of the (transversal type) plane curve singularity $u^2+v^a$, provided that we replace $v_1$ and $v_2$ by arrowheads with multiplicity 1. Comparing the Main Algorithm and the algorithm which provides the graph of suspension singularities (cf. 5.3), we realize that the part of $G$ above $\Gamma^2_{\mathscr{C}}$ is exactly the resolution graph of $u^2+v^a+w^{2a+4}=0$ with opposite orientation. More precisely, let

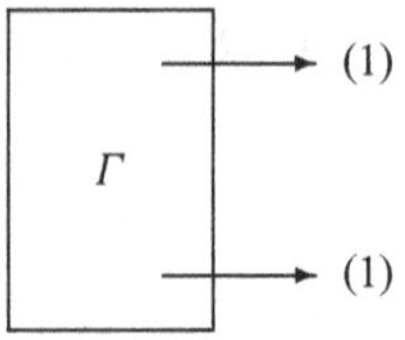


be the minimal embedded resolution graph of the germ $w: (\{u^2+v^a+w^{2a+4}=0\}, 0) \to (\mathbb{C}, 0)$, induced by the projection $(u,v,w) \mapsto w$. Let $-\Gamma$ be this graph with opposite orientation (in which one changes the signs of all Euler numbers and edge-decorations, and keeps the multiplicities). Then the graph of the open book decomposition of $(\partial F, V_g)$ is obtained by gluing $-\Gamma$ with $G_1$ such that the arrows of $-\Gamma$ are identified with $v_1$ and $v_2$ (and the Euler numbers of $v_1$ and $v_1$ are recomputed as above, or via (4.1.5) using the multiplicities):

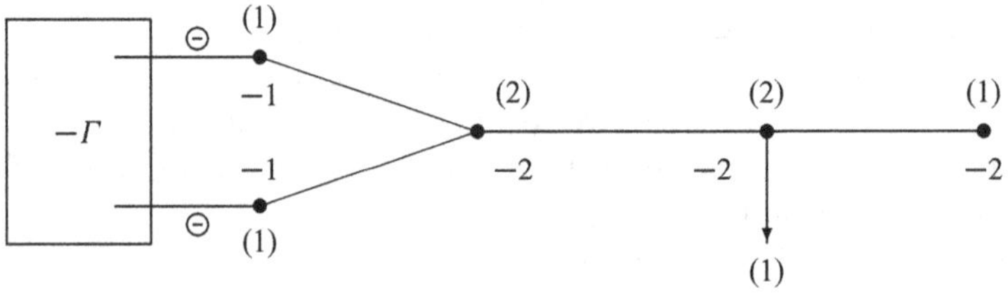


This graph has a cycle. Moreover, $\Gamma$ is a star-shaped graph whose central vertex has genus $\frac{\gcd(a,4)}{2} - 1$. Determining $\Gamma$ is standard, see 5.3, or [99].

**Case 2.** Assume that $a$ is odd, $a \geq 3$. We proceed similarly as above. The graph $G_1$ is the following:

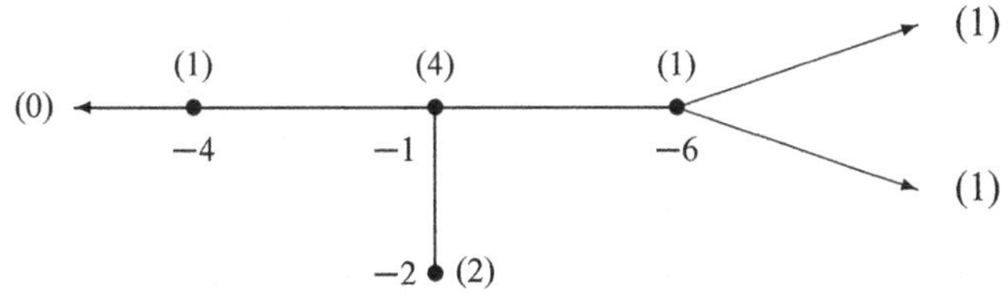


There is only one gluing torus. Let $v$ be the $(-4)$-vertex, and $v'$ its adjacent vertex of $G$ which is not in $G_1$. Then the multiplicity of $v'$ is again 1. Hence, the Euler number of $v$ in $G$ is $-3$. Therefore, the graph of $(\partial F, V_g)$ is

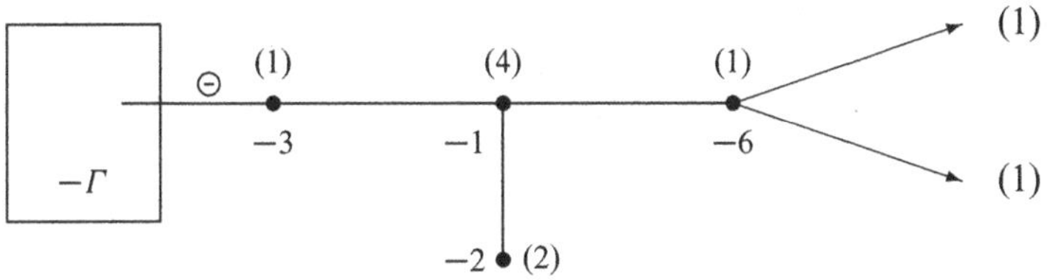


where $\Gamma$ is the minimal embedded resolution graph of the germ $w : (\{u^2 + v^a + w^{2a+4} = 0\}, 0) \to (\mathbb{C}, 0)$, and the unique arrow-head of $-\Gamma$ is identified with the $(-3)$-vertex.

# Chapter 11
# Proof of the Main Algorithm

## 11.1 Preliminary Remarks

**11.1.1.** The algorithm and its proof is a highly generalized version of the algorithm which determines the resolution graph of cyclic coverings. Its origin goes back to the case of suspensions, when one starts with an isolated plane curve singularity $f'$ and a positive integer $n$, and one determines the resolution graph of the hypersurface singularity $\{f'(x,y) + z^n = 0\}$ from the embedded resolution graph of $f'$ and the integer $n$; see 5.3.

All the geometrical constructions behind the algorithms targeting cyclic coverings are realized within the framework of complex analytic/algebraic geometry. In particular, all the graphs involved are negative definite graphs and the plumbing calculus reduces to blowing up/down $(-1)$-rational curves. Moreover, the following general principle applies: for normal surface singularities the resolution graph is a possible plumbing graph for the link, which is diffeomorphic with the boundary of the Milnor fiber of any smoothing.

**11.1.2.** The first case when a more complicated "aid-graph" was used is in [92]. The starting situation was the following: having a germ $f$ with 1-dimensional singular locus, and another germ $g$ such that the pair $(f, g)$ forms an ICIS, one wished to determine the resolution graphs of the hypersurface singularities $f + g^k$, $k \gg 0$, cf. 6.1.1. In order to find these "usual" – that is, negative definite – graphs, all the necessary information about the ICIS $(f, g)$ was stored in the "unusual" decorations of the "unusual" graph $\Gamma_{\mathscr{C}}$.

The machinery and construction developed in that article serve as a model for the present work. We start again with the very same graph $\Gamma_{\mathscr{C}}$, but rather significant differences appear. Although, in [92], the entire construction stayed within the realm of complex analytic geometry, similarly as in the case of cyclic coverings, the present case grows out of the complex analytic world. We must glue together real analytic spaces with singularities, and sometimes the gluing maps even reverse the "canonical" orientations of the regular parts. This generates additional

A. Némethi and Á. Szilárd, *Milnor Fiber Boundary of a Non-isolated Surface Singularity*, Lecture Notes in Mathematics 2037, DOI 10.1007/978-3-642-23647-1_11,

difficulties we need to be handle during the proof. The output plumbing graphs are "general plumbing graphs", which may not be definite, or not even non-degenerate). Moreover, we had to consider a larger set of moves of the smooth plumbing calculus (not standard in algebraic geometry) in order to simplify them or to reduce them to their 'normal forms'.

The explanation of the idea why the graph $\Gamma_{\mathscr{C}}$ contains all the information needed to describe $\partial F$ is given in Sect. 7.1. In fact, that is the main idea behind the whole construction. In the next section we outline the main steps of the proof.

## 11.2 The Guiding Principle and the Outline of the Proof

Consider an ICIS $\Phi = (f, g)$ as in Sect. 3.1, an embedded resolution $r : V^{emb} \to (\mathbb{C}^3, 0)$ of the divisor $V_f \cup V_g$ as in 6.1, as well as a "wedge" $\mathsf{W}_{\eta,M}$ of $\Delta_1$ for some $M \gg 0$ as in Sect. 7.1.

If one has a complex analytic isolated singularity $(\mathscr{S}, 0) \subset (\mathbb{C}^3, 0)$ for which $\Phi(\mathscr{S}) \setminus \{0\} \subset \mathsf{W}_{\eta,M}$ then one can construct a resolution of $\mathscr{S}$ in three steps.

First, consider the $r$-strict transform $\widetilde{\mathscr{S}} \subset V^{emb}$ of $\mathscr{S}$. It is contained in a tubular neighbourhood of $\mathscr{C}$, cf. (7.1.3), and its singular locus is in $\mathscr{C}$. Therefore $\widetilde{\mathscr{S}}$ can be resolved in two further steps: first taking the normalization $\mathscr{S}^{norm}$ of $\widetilde{\mathscr{S}}$, then resolving the isolated normal surface singularities of $\mathscr{S}^{norm}$. The point is that if $\mathscr{S}$ is determined by $f$ and $g$, then $\widetilde{\mathscr{S}}$ has nice local equations near any point of $\mathscr{C}$ (which can be recovered from the decorations of $\Gamma_{\mathscr{C}}$). For example, one can show that $\widetilde{\mathscr{S}}$ is an equisingular family of curves along the regular part of $\mathscr{C}$, hence the singular locus of $\mathscr{S}^{norm}$ will be situated above the double points of $\mathscr{C}$. Moreover, all these singular points will be of Hirzebruch–Jung type. In particular, the last step is the resolution of these Hirzebruch–Jung singularities, whose combinatorial data is again codified in $\Gamma_{\mathscr{C}}$.

Summing up we get the diagram:

$$\begin{array}{ccccccc} \overline{\mathscr{S}} & \xrightarrow{\text{HJ}} & \mathscr{S}^{norm} & \longrightarrow & \widetilde{\mathscr{S}} & \xrightarrow{r} & \mathscr{S} \\ & & & & \cap & & \cap \\ & & & & V^{emb} & & (\mathbb{C}^3, 0) \end{array}$$

Corresponding to the above three horizontal maps, we have the following steps at the level of graphs:

- start with the graph $\Gamma_{\mathscr{C}}$ (which stores all the local information about $\widetilde{\mathscr{S}}$);
- provide a cyclic covering graph (in the sense of Chap. 5) corresponding to the normalization step;
- modify this graph by Hirzebruch–Jung strings (see "variation" 5.1.9).

A key additional argument is a consequence of Theorem 5.1.8, which guarantees the uniqueness of the cyclic covering graph with the inserted strings.

It is exactly this guiding principle that was used in [92] to determine the resolution graph of any member of the generalized Iomdin-series $\mathscr{S} = \{f + g^k = 0\}$, $k \gg 0$.

Now, we want to obtain the plumbing-graph of the boundary $\partial F$ of the Milnor fiber of a non-isolated $f$. We show in Proposition 11.3.3 that $\partial F$ is the link of the *real* analytic germ

$$\mathscr{S}_k = \{f = |g|^k\} \subset (\mathbb{C}^3, 0),$$

and $\Phi(\mathscr{S}_k) \setminus \{0\} \subset \mathsf{W}_{\eta,M}$, provided that $k \gg 0$. Hence, we will run the same procedure as above within the world of real analytic geometry, which forces some modifications.

A final remark: the Euler number of an $S^1$-bundle over a curve is a "global object", its computation in a resolution can be rather involved (one needs more charts and gluing information connecting them). Therefore, we will determine the Euler numbers of our graphs in an indirect way: we consider the open book decomposition induced by $g$, and we determine the associated multiplicity system (this can easily be determined from local data!), then we apply (4.1.5).

## 11.3 The First Step: The Real Varieties $\mathscr{S}_k$

We fix a pair $\Phi = (f, g)$ as in Sect. 3.1, and we use all the notations and results of that part. In particular, we fix a good representative of $\Phi$ whose discriminant is $\Delta_\Phi$. Similarly as above, we write $(c, d)$ for the coordinates of $(\mathbb{C}^2, 0)$.

For any *even* integer $k$ (compare also with 11.3.2) we set

$$Z_k := \{(c, d) \in (\mathbb{C}^2, 0) \ : \ c = |d|^k\}.$$

The next lemma is elementary and its proof is left to the reader.

**Lemma 11.3.1.** *$Z_k$ is a smooth real analytic (even algebraic) surface. For $k$ sufficiently large $Z_k \cap \Delta_\Phi = \{0\}$. Moreover, $Z_k \setminus \{0\} \subset \mathsf{W}_{\eta,M}$ if $k > M$.*

**Remark 11.3.2.** As mentioned before, all the important facts regarding $Z_k$ (and the space $\mathscr{S}_k$ which will be defined next) are valid for $k$ odd as well. This is based on the additional fact that the classification of oriented 2- and 3-dimensional topological manifolds agrees with the classification of $C^\infty$ manifolds. Nevertheless, it is more convenient to use even integers $k$, since for them $|d|^k$ becomes real algebraic. In fact, later we will impose even more divisibility assumptions on $k$.

The point is that $k$ has only an auxiliary role and carries no geometric meaning, e.g., it will not appear in any "final" formula of $\partial F$. Hence its value, as soon as it is sufficiently large, is completely unimportant.

Next, define the real analytic variety $\mathscr{S}_k$ of real dimension 4 by

$$\mathscr{S}_k := \Phi^{-1}(Z_k) = \{z \in (\mathbb{C}^3, 0) \ : \ f(z) = |g(z)|^k\}.$$

**Proposition 11.3.3.** *For $k$ sufficiently large, the real variety $\mathscr{S}_k \setminus \{0\}$ is regular, hence it is a smooth oriented 4-dimensional manifold. Moreover, for sufficiently small $\epsilon > 0$, the sphere $S_\epsilon = S^5_\epsilon$ intersects $\mathscr{S}_k$ transversally. The intersection $\mathscr{S}_k \cap S_\epsilon$ is an oriented 3-manifold, which is diffeomorphic by an orientation preserving diffeomorphism to $\partial F$.*

*In particular, the link of $\mathscr{S}_k$ (i.e. $\mathscr{S}_k \cap S_\epsilon$) is independent of the choice of $k$.*

Before we start the proof let us indicate how the orientation of $\mathscr{S}_k \setminus \{0\}$ is defined. First, consider $Z_k$. It is a smooth real manifold. The projection on the $d$-axis induces a diffeomorphism; we define the orientation of $Z_k$ by the pullback of the complex orientation of the $d$-axis via this diffeomorphism. Next, all fibers of $\Phi$ are complex curves with their natural orientation. On the smooth part of $\Phi^{-1}(Z_k)$ we define the product orientation of the base and fibers.

*Proof.* The first statement follows from Lemma 11.3.1 and from the properties of the ICIS $\Phi$ (or by a direct computation). The second one is standard, using for example the "*curve selection lemma*" from [77].

Next, we prove the diffeomorphism $\mathscr{S}_k \cap S_\epsilon \simeq \partial F$.

First, recall that in certain topological arguments regarding the Milnor fiber of $f$, the sphere $S_\epsilon = \partial B_\epsilon$ is replaced by the 5-manifold with corners $\partial(\Phi^{-1}(D^2_\eta) \cap B_\epsilon)$, the Milnor fiber $F = \{f = \delta\} \cap B_\epsilon$ by $F^\square := \{f = \delta\} \cap \Phi^{-1}(D^2_\eta) \cap B_\epsilon$, and the boundary $\partial F$ by the boundary with corners $\partial F^\square$. For details, see e.g. [67], or Remark 3.1.11. By a similar argument, one shows the equivalence of $\Phi^{-1}(Z_k) \cap B_\epsilon$ with $\mathscr{S}^\square_k := \Phi^{-1}(Z_k \cap D^2_\eta) \cap B_\epsilon$, and $\Phi^{-1}(Z_k) \cap S_\epsilon$ with the 3-manifold with corners $\partial \mathscr{S}^\square_k$. Hence, we need only to show the equivalence of $\partial F^\square$ and $\partial \mathscr{S}^\square_k$.

Consider the intersection $Z_k \cap \partial D^2_\eta$, i.e., the solution of the system $\{|c|^2 + |d|^2 = \eta^2;\ c = |d|^k\}$. It is a circle along which $c$ is constant; let this value of $c$ (determined by $\eta$ and $k$) be denoted by $c_0$. Set $D_{c_0} = \{c = c_0\} \cap D^2_\eta$ as in Sect. 3.1. Then

$$\partial D_{c_0} = \partial(Z_k \cap D^2_\eta), \tag{11.3.4}$$

and $\partial F^\square = \partial(\Phi^{-1}(D_{c_0}))$ has a decomposition:

$$\partial F^\square = \Phi^{-1}(\partial D_{c_0}) \cap B_\epsilon \bigcup_{\Phi^{-1}(\partial D_{c_0}) \cap S_\epsilon} \Phi^{-1}(D_{c_0}) \cap S_\epsilon.$$

Via (11.3.4), $\partial \mathscr{S}^\square_k$ has a decomposition

$$\partial \mathscr{S}^\square_k = \Phi^{-1}(\partial D_{c_0}) \cap B_\epsilon \bigcup_{\Phi^{-1}(\partial D_{c_0}) \cap S_\epsilon} \Phi^{-1}(Z_k \cap D^2_\eta) \cap S_\epsilon.$$

Notice that there is an isotopy of $D^2_\eta$, preserving $\partial D^2_\eta$, which sends $D_{c_0}$ into $Z_k \cap D^2_\eta$. Since the restriction of $\Phi$ on $\Phi^{-1}(D^2_\eta) \cap S_\epsilon$ is a trivial fibration over $D^2_\eta$, this isotopy can be lifted. This identifies the pairs

$$(\Phi^{-1}(D_{c_0}) \cap S_\epsilon, \Phi^{-1}(\partial D_{c_0}) \cap S_\epsilon) \simeq (\Phi^{-1}(Z_k \cap D^2_\eta) \cap S_\epsilon, \Phi^{-1}(\partial D_{c_0}) \cap S_\epsilon).$$

This ends the proof. □

## 11.4 The Strict Transform $\widetilde{\mathscr{S}}_k$ of $\mathscr{S}_k$ Via $r$

Consider the resolution $r : V^{emb} \to U$ as in 6.1. Let $\widetilde{\mathscr{S}}_k$ be the strict transform of $\mathscr{S}_k$ by $r$, i.e. $\widetilde{\mathscr{S}}_k$ is the closure of $r^{-1}(\mathscr{S}_k \setminus \{0\})$ (in the euclidian topology).

**Lemma 11.4.1.**

$$\widetilde{\mathscr{S}}_k \cap r^{-1}(0) = \mathscr{C}.$$

*Proof.* The proof is similar to the proof of (7.1.3), and it is left to the reader. □

Since the restriction of $r$ induces a diffeomorphism $\widetilde{\mathscr{S}}_k \setminus \mathscr{C} \to \mathscr{S}_k \setminus \{0\}$, we get that the singular locus of $\widetilde{\mathscr{S}}_k$ satisfies

$$Sing(\widetilde{\mathscr{S}}_k) \subset \mathscr{C}.$$

Moreover, $r$ induces a diffeomorphism between $\partial\mathscr{S}_k$ (the subject of our interest) and $\partial\widetilde{\mathscr{S}}_k$. Since $\widetilde{\mathscr{S}}_k$ can be replaced by its intersection with an arbitrarily small tubular neighbourhood of $\mathscr{C}$, the boundary $\partial\widetilde{\mathscr{S}}_k$ can be localized totally near $\mathscr{C}$. In fact, this is the main advantage of the space $\mathscr{S}_k$: in this way, the wanted 3-manifold appears as a local link, or, after a resolution, as the boundary of a tubular neighbourhood of a curve configuration.

Next, we analyze the local equations of $\widetilde{\mathscr{S}}_k$ in the neighbourhood of any point of $\mathscr{C}$. For this we use the notations of Sect. 6.2. In all the cases, $U_p$ is a complex 3-ball around the point $p \in \mathscr{C}$ with three complex local coordinates $(u, v, w)$.

It is convenient to use the following notation. If $H = \{(u,v,w) \in U_p : h(u,v,w) = 0\}$ is a real analytic variety in $U_p$, then we denote by $H^+$ the closure of $H \setminus \{uvw = 0\}$. This way we neglect those components of $H$ which are included in one of the coordinate planes. Using this notation, the local equations of $\widetilde{\mathscr{S}}_k$ are as follows.

If $p$ is a generic point of a component $C$ of $\mathscr{C}$ with decoration $(m; n, \nu)$, then

$$\widetilde{\mathscr{S}}_k \cap U_p = \{(u,v,w) \, : \, u^m v^n = |v|^{\nu k}\}^+ = \{(u,v,w) \, : \, u^m = v^{\frac{\nu k}{2} - n} \bar{v}^{\frac{\nu k}{2}}\} \quad (11.4.2)$$

with $m, \nu > 0$.

If $p$ is an intersection (singular) point of $\mathscr{C}$ of type 1 (i.e. if the corresponding edge has decoration 1), then

$$\widetilde{\mathscr{S}}_k \cap U_p = \{(u,v,w) \, : \, u^m v^n w^l = |v|^{\nu k} |w|^{\lambda k}\}^+ \quad (11.4.3)$$

with $m, \nu, \lambda > 0$.

Finally, if $p$ is an intersection (singular) point of $\mathscr{C}$ of type 2, then

$$\widetilde{\mathscr{S}}_k \cap U_p = \{(u,v,w) \,:\, u^m v^{m'} w^n = |w|^{\nu k}\}^+ \tag{11.4.4}$$

with $m, m', \nu > 0$.

## 11.5 Local Complex Algebraic Models for the Points of $\widetilde{\mathscr{S}}_k$

Notice that for $k \gg 0$ and for $p$ as in (11.4.2)–(11.4.3)–(11.4.4), $\widetilde{\mathscr{S}}_k \cap U_p$ is a *real* algebraic variety. We will show that any such germ is homeomorphic with the germ of a certain *complex* algebraic hypersurface. In these computations we will assume that $k/2$ is a multiple of all the integers appearing in the decorations of $\Gamma_{\mathscr{C}}$. More precisely: whenever in the next discussion a fraction $k/l$ appears for some $l$, then we assume that $k/l$ is, in fact, an even integer.

In the next paragraphs $U$ will denote a local neighbourhood of the origin in $\mathbb{C}^3$.

**11.5.1.** Assume that $p$ **is a generic point of** $\mathscr{C}$ as in (11.4.2). Consider the map

$$\psi_p : \{(x,y,z) \in U \,:\, x^m = y^n\} \longrightarrow \{(u,v,w) \in U_p \,:\, u^m v^n = |v|^{\nu k}\}^+$$

given by the correspondences

$$\begin{cases} u = x^{-1}|y|^{\nu k/m} \\ v = y \\ w = z \end{cases} \qquad \begin{cases} x = u^{-1}|v|^{\nu k/m} \\ y = v \\ z = w. \end{cases} \tag{11.5.2}$$

Then $\psi_p$ is regular real algebraic (i.e. it extends over $x = 0$ too), it is birational and a homeomorphism. Moreover, it is a partial normalization of $\widetilde{\mathscr{S}}_k \cap U_p$, i.e. the coordinates $x, y, z$ of $\{x^m = y^n\} \cap U$ are integral over the ring of regular functions of $\widetilde{\mathscr{S}}_k \cap U_p$. Indeed, birationality follows from the fact that the second set of equations provides the inverse of the first one, and regularity follows from a limit computation, or by rewriting the first equation into $u = x^{-1}|x|^{\nu k/n}$. This formula also shows that $\psi_p$ is bijective and a homeomorphism. Moreover, since $x^m = v^n$, $x$ is integral over the ring of regular functions of $\widetilde{\mathscr{S}}_k \cap U_p$ (a similar statement for $y$ and $z$ is trivial).

In particular, the normalizations of the source and of the target of $\psi_p$ canonically coincide.

**11.5.3.** Assume that $p$ **is a singular point of** $\mathscr{C}$ **of type 1** as in (11.4.3). Consider the map

$$\psi_p : \{(\alpha,\beta,\gamma) \in U \,:\, \alpha^m = \beta^n \gamma^l\} \longrightarrow \{(u,v,w) \in U_p \,:\, u^m v^n w^l = |v|^{\nu k}|w|^{\lambda k}\}^+$$

given by

$$\begin{cases} u = \alpha^{-1}|\beta|^{\nu k/m}|\gamma|^{\lambda k/m} \\ v = \beta \\ w = \gamma \end{cases} \qquad \begin{cases} \alpha = u^{-1}|v|^{\nu k/m}|w|^{\lambda k/m} \\ \beta = v \\ \gamma = w. \end{cases} \tag{11.5.4}$$

Then, again, $\psi_p$ is regular real algebraic, birational, and additionally, it is a homeomorphism. Moreover, it is a partial normalization of $\widetilde{\mathcal{S}}_k \cap U_p$, i.e. the coordinates $\alpha, \beta, \gamma$ are integral over the ring of regular functions of $\widetilde{\mathcal{S}}_k \cap U_p$. Indeed, the regularity follows from

$$u = \alpha^{m-1}\overline{\alpha}^m|\beta|^{\nu k/m-2n}|\gamma|^{\lambda k/m-2l},$$

where $k \gg 0$ and $m > 0$. Moreover, $\alpha$ is integral over the ring of $\widetilde{\mathcal{S}}_k \cap U_p$ since $\alpha^m = v^n w^l$.

Hence again, the normalizations of the source and the target of $\psi_p$ canonically coincide.

**11.5.5.** Assume that $p$ **is a singular point of** $\mathcal{C}$ **of type 2** as in (11.4.4). In this case we can prove considerably less (from the analytic point of view). We consider the map

$$\psi_p : \{(\alpha, \beta, \gamma) \in U \ : \alpha^n = \beta^m \gamma^{m'}\} \longrightarrow \{(u, v, w) \in U_p \ : u^m v^{m'} w^n = |w|^{\nu k}\}^+$$

given by

$$\begin{cases} u = \beta^{-1}|\beta|^{\nu k/n} \\ v = \gamma^{-1}|\gamma|^{\nu k/n} \\ w = \alpha. \end{cases} \tag{11.5.6}$$

It is regular real algebraic and a homeomorphism, but it is *not* birational.

**11.5.7.** Notice also that the above maps, in all three cases, preserve the coordinate axes.

## 11.6 The Normalization $\mathcal{S}_k^{norm}$ of $\widetilde{\mathcal{S}}_k$

**11.6.1.** Let $n_{\mathcal{S}} : \mathcal{S}_k^{norm} \to \widetilde{\mathcal{S}}_k$ be the normalization of $\widetilde{\mathcal{S}}_k$; for its existence, see [14]. Since the normalization is compatible with restrictions on smaller open sets, we get the globally defined $\mathcal{S}_k^{norm}$ whose restrictions above an open set of type $\widetilde{\mathcal{S}}_k \cap U_p$ are the normalization of that $\widetilde{\mathcal{S}}_k \cap U_p$. In particular, the local behaviour of the normalization $\mathcal{S}_k^{norm}$ over the different open neighbourhoods $\widetilde{\mathcal{S}}_k \cap U_p$ can be tested in the charts considered in the previous section.

In the first case, if $p$ **is a generic point** of $\mathscr{C}$, and $\psi_p$ is the "partial normalization" from 11.5.1, then it induces an isomorphism of normalizations:

$$\psi_p^{norm} : \{(x,y,z) \in U : x^m = y^n\}^{norm} \longrightarrow \{(u,v,w) \in U_p : u^m v^n = |v|^{\nu k}\}^{+,norm}.$$

Since the left hand side is smooth, we get that $\mathscr{S}_k^{norm}$ is smooth over the regular points of $\mathscr{C}$, hence, after normalization, only finitely many singular points survive in $\mathscr{S}_k^{norm}$, and they are situated above the double points of $\mathscr{C}$.

If $p$ **is a double point of** $\mathscr{C}$ **of type 1**, then $\psi_p$ from 11.5.3 induces again an isomorphism at the level of normalizations:

$$\psi_p^{norm} : \{(\alpha,\beta,\gamma) : \alpha^m = \beta^n \gamma^l\}^{norm} \longrightarrow \{(u,v,w) : u^m v^n w^l = |v|^{\nu k} |w|^{\lambda k}\}^{+,norm}.$$

Hence, the singular points in $\mathscr{S}_k^{norm}$, situated above the double points of $\mathscr{C}$ of type 1, are equivalent with complex analytic singularities of Hirzebruch-Jung type. Recall that these singularities are determined completely combinatorially (e.g. by the integers $m,n,l$ above), and by the above chart, this combinatorial data can also be recovered from $\Gamma_{\mathscr{C}}$.

**11.6.2.** We emphasize that the two types of charts above in 11.6.1 are compatible. By this we mean the following: consider a double point $p$ of $\mathscr{C}$ of type 1, and a neighbourhood $U_p$ as above. Then $\mathscr{C} \cap U_p$ is the union of the $v$ and $w$ axis. Let $q$ be a generic point of $\mathscr{C} \cap U_p$ and consider a sufficiently small local neighbourhood $U_q \subset U_p$ (where we denote this inclusion by $j$), and consider also the chart $\psi_p$ over $\mathscr{S}_k \cap U_p$ as in 11.5.1, respectively $\psi_q$ over $\mathscr{S}_k \cap U_q$ as in 11.5.3. Then $\psi_p^{-1} \circ j \circ \psi_q$ is a complex analytic isomorphism onto its image which at the level of normalization induces an isomorphism of complex analytic smooth germs. Indeed, if $q$ is a generic point of the $w$-axis, with non-zero $w$-coordinate, and the inclusion

$$\{(u',v',w') \in U_q \,:\, (u')^m (v')^n = |v'|^{\nu k}\}^+ \stackrel{j}{\longrightarrow} \{(u,v,w) \,:\, u^m v^n w^l = |v|^{\nu k} |w|^{\lambda k}\}^+$$

is given by $u = u'(w')^{-l/m}|w'|^{\lambda k/m}$, $v = v'$ and $w = w'$, then $\psi_p^{-1} \circ j \circ \psi_q$ is given by $\alpha = xz^{l/m}$, $\beta = y$ and $\gamma = z$. Then the normalizations tautologically coincide. For example, assume $\gcd(m,n) = 1$ and take the free variables $(t,\gamma)$ normalizing $\{\alpha^m = \beta^n \gamma^l\}$ by $\alpha = t^n \gamma^{l/m}$, $\beta = t^m$ and $\gamma = \gamma$. Similarly, consider the free variables $(s,z)$ normalizing $\{x^m = y^n\}$ by $x = s^n$, $y = s^m$ and $z = z$. Then $(\psi_p^{-1} \circ j \circ \psi_q)^{norm}$ is $t = s$ and $\gamma = z$.

In particular, *the two complex charts* $\psi_p^{norm}$ *and* $\psi_q^{norm}$ *of 11.6.1 induce the same orientation on their images, they identify the inverse image of* $\mathscr{C}$ *by the same orientation and induce on a normal slice of* $\mathscr{C}$ *the same orientation.* Note that these are the key gluing-data for a plumbing construction.

**11.6.3.** On the other hand, **if** $p$ **is a singular point of** $\mathscr{C}$ **of type 2**, then $\psi_p$ from 11.5.5 *does not* induce an analytic isomorphism, since $\psi_p$ itself is not birational. In this case, 11.5.5 implies that at the level of normalizations the induced map

$$\psi_p^{norm} : \{(\alpha,\beta,\gamma) : \alpha^n = \beta^m\gamma^{m'}\}^{norm} \longrightarrow \{(u,v,w) : u^m v^{m'} w^n = |w|^{\nu k}\}^{+,norm}$$

is regular and a homeomorphism. Nevertheless, one can prove slightly more:

**Lemma 11.6.4.** *$\psi_p^{norm}$ induces a diffeomorphism over $U_p \setminus \{0\}$.*

*Proof.* Let $p$ be a double point of $\mathscr{C}$ of type 2 as in 11.5.5. Then $\mathscr{C} \cap U_p$ is the union of the $u$ and $v$ axes. Let $q$ be a generic point on the $v$ axis, – the other case is completely symmetric. Then $q$ is in the image of the following map

$$\varphi_{p,v} : \{(x,y,z) \in U' \setminus \{z=0\} : x^n = y^m z^{m'}\} \longrightarrow$$
$$\{(u,v,w) \in U_p \setminus \{v=0\} : u^m v^{m'} w^n = |w|^{\nu k}\}^+$$

given by the correspondences

$$\begin{cases} u = y^{-1}|x|^{\nu k/m} \\ v = z^{-1} \\ w = x \end{cases} \qquad \begin{cases} y = u^{-1}|w|^{\nu k/m} \\ z = v^{-1} \\ x = w. \end{cases} \tag{11.6.5}$$

Then $\varphi_{p,v}$ is regular on $U_p \setminus \{z=0\}$, since

$$u = |x|^{\nu k/m-2n} y^{m-1} \overline{y}^m |z|^{2m'},$$

it is birational (its inverse is given by the second set of equations of (11.6.5)), and it is a partial normalization, since $y^m = w^n v^{m'}$. Therefore,

$$\varphi_{p,v}^{norm} : \{(x,y,z) \in U' \setminus \{z=0\} : x^n = y^m z^{m'}\}^{norm} \longrightarrow n_{\mathscr{S}}^{-1}(\widetilde{\mathscr{S}}_k \cap U_p \setminus \{v=0\})$$

is an isomorphism. Using this isomorphism, the restriction of $\psi_p^{norm}$ from 11.6.3,

$$\psi_p^{norm} : \{(\alpha,\beta,\gamma) \in U \setminus \{\gamma=0\} : \alpha^n = \beta^m\gamma^{m'}\}^{norm} \longrightarrow n_{\mathscr{S}}^{-1}(\widetilde{\mathscr{S}}_k \cap U_p \setminus \{v=0\})$$

can be understood explicitly. Indeed, the map

$$\varphi_{p,v}^{-1} \circ \psi_p : \{(\alpha,\beta,\gamma) \in U \setminus \{\gamma=0\} : \alpha^n = \beta^m\gamma^{m'}\} \longrightarrow \tag{11.6.6}$$
$$\{(x,y,z) \in U' \setminus \{z=0\} : x^n = y^m z^{m'}\}$$

is given by

$$\begin{cases} x = \alpha \\ y = \beta|\gamma|^{\nu k m'/mn} \\ z = \gamma|\gamma|^{-\nu k/n}. \end{cases} \tag{11.6.7}$$

We claim that this induces a diffeomorphism at the level of normalization. In order to verify this, we make two reductions. First, by a cyclic covering argument, we may assume that $m' = 1$. Second, we will also assume that $\gcd(m,n) = 1$ (otherwise the normalization will have $\gcd(m,n)$ components, and the normalization maps below must be modified slightly; the details are left to the reader). We fix two integers $a$ and $b$ such that $an - bm = 1$. Then the left hand side of (11.6.6) is normalized by $(t,\gamma) \in (\mathbb{C}^2,0) \setminus \{\gamma = 0\}$, $\alpha = t^m\gamma^a$, $\beta = t^n\gamma^b$, $\gamma = \gamma$; while the right hand side is normalized by $(s,z) \in (\mathbb{C}^2,0) \setminus \{z = 0\}$, $x = s^m z^a$, $y = s^n z^b$ and $z = z$. Hence, at the normalization level

$$(\varphi_{p,\nu}^{-1} \circ \psi_p)^{norm} : (\mathbb{C}^2,0) \setminus \{\gamma = 0\} \to (\mathbb{C}^2,0) \setminus \{z = 0\}$$

is given by the diffeomorphism

$$\begin{cases} s = t|\gamma|^{a\nu k/mn} \\ z = \gamma|\gamma|^{-\nu k/n}. \end{cases} \tag{11.6.8}$$

□

**11.6.9.** Consider a singular point in $\mathscr{S}_k^{norm}$ above a double point of $\mathscr{C}$ of type 2, say $p$. By the results of 11.6.3, the type of this singularity can again be identified with a (complex analytic) Hirzebruch–Jung singularity, identified via the homeomorphism $\psi_p^{norm}$. In particular, corresponding to that point, in the plumbing graph we have to insert an appropriate Hirzebruch–Jung string. In order to do this we need to clarify orientation-compatibilities at the intersection points of this string with the inverse image of $\mathscr{C}$. More precisely, we have to clarify the compatibility of the chart $\psi_p$ with "nearby" charts of type 11.5.1.

Let $p$ be as in the previous paragraph, fix one of its neighbourhoods $\widetilde{\mathscr{S}}_k \cap U_p$ as in 11.6.3, and a generic point $q$ on the $v$-axis with small neighbourhood $\widetilde{\mathscr{S}}_k \cap U_q$ and chart $\psi_q : \{(x')^n = (y')^m\} \to \{(u')^m(w')^n = |w'|^{\nu k}\} = \widetilde{\mathscr{S}}_k \cap U_q$ given by $u' = y'^{-1}|x'|^{\nu k/m}$, $w' = x'$ and $v' = z'$, cf. Sect. 11.5.1. The inclusion $j : \widetilde{\mathscr{S}}_k \cap U_q \longrightarrow \widetilde{\mathscr{S}}_k \cap U_p$ is given by the equations $u = u'(v')^{-m'/m}$, $v = v'$ and $w = w'$. Hence $\varphi_{p,\nu}^{-1} \circ j \circ \psi_q$ is given by $x = x'$, $y = y'(z')^{m'/m}$ and $z = (z')^{-1}$. This combined with (11.6.7), the map $\psi_p^{-1} \circ j \circ \psi_q : \{x'^n = y'^m, z' \neq 0\} \to \{\alpha^n = \beta^m\gamma^{m'}, \gamma \neq 0\}$ is given by (the inverse of)

$$\begin{cases} x' = \alpha \\ y' = \beta\gamma^{m'/m} \\ z' = \gamma^{-1}|\gamma|^{\nu k/n}. \end{cases} \tag{11.6.10}$$

For simplicity assume again that $\gcd(m,n) = 1$ (the interested reader can reproduce the general case). We take integers $a$ and $b$ with $an - bm = m'$ as above. Then the free coordinates $(t,\gamma)$ normalize $\{\alpha^n = \beta^m\gamma^{m'}\}$ by $\alpha = t^m\gamma^a$, $\beta = t^n\gamma^b$ and $\gamma = \gamma$, while the free coordinates $(s,z')$ normalize $\{x'^n = y'^m\}$ by $x' = s^m$, $y' = s^n$ and $z' = z'$. In particular, the isomorphism $(\varphi_{p,\nu}^{-1} \circ j \circ \psi_q)^{norm}$ is given by the correspondence $(\mathbb{C}^2,0) \setminus \{\gamma = 0\} \leftrightarrow (\mathbb{C}^2,0) \setminus \{z' = 0\}$:

$$\begin{cases} s &= t\gamma^{a/m} \\ z' &= \gamma^{-1}|\gamma|^{\nu k/n}. \end{cases} \tag{11.6.11}$$

Notice that the inverse image of $\mathscr{C}$ (in the two charts) is given by $t = 0$, respectively by $s = 0$.

Now, consider the natural orientations of $n_{\mathscr{S}}^{-1}(\widetilde{\mathscr{S}}_k \cap U_p)$ provided via $\psi_p$ by the complex structure of the source of $\psi_p$, and also the orientation of $n_{\mathscr{S}}^{-1}(\mathscr{C} \cap U_p)$ via the same procedure. In a similar way, consider the orientations of $n_{\mathscr{S}}^{-1}(\widetilde{\mathscr{S}}_k \cap U_q)$ and $n_{\mathscr{S}}^{-1}(\mathscr{C} \cap U_q)$ induced by $\psi_q$ and the complex structure of its source. Then (11.6.11) shows the following fact:

**Lemma 11.6.12.** $(\varphi_{p,\nu}^{-1} \circ j \circ \psi_q)^{norm}$ *is a diffeomorphism which reverses the orientations. Moreover, its restriction on the inverse images of* $\mathscr{C}$ *reverses the orientation of these Riemann surfaces as well. On the other hand, the orientation of the transversal slices to the strict transforms of* $\mathscr{C}$ *are preserved.*

## 11.7 The "Resolution" $\overline{\mathscr{S}_k}$ of $\widetilde{\mathscr{S}}_k$

The singularities of $\mathscr{S}_k^{norm}$ are situated above the double points of $\mathscr{C}$. Above a double point of type 1 they are isomorphic with complex analytic Hirzebruch–Jung singularities; their resolution follows the resolution procedure of these germs, see 11.6.2.

We did not determine here the resolution and the real analytic type of the singularity situating above a double point of type 2. Nevertheless, these singularities are also identified by Lemma 11.6.12, up to an orientation reversing homeomorphism, with complex analytic Hirzebruch–Jung singularities. This is enough to determine the topology of $\mathscr{S}_k^{norm}$ and to describe the plumbing representation of its boundary. This will be done in the next section.

Although, for the purpose of the present work the above topological representation is sufficient, if we would like to handle real analytic invariants read from the structure sheaf of the resolution (like, say, the geometric genus is read from the resolution of a normal surface singularity), then an explicit description of this variety would be more than necessary. This type of analytic questions are beyond the aims of the present work, however we formulate this problem as an important goal for further research.

**11.7.1. Problem.** Find an explicit description of the real analytic/algebraic resolution of the singularity

$$\{(u,v,w) \in (\mathbb{C}^3,0) \,:\, u^m v^{m'} w^n = |w|^k\}^+,$$

where $m$, $m' > 0$ and $k$ is a sufficiently large (even) integer.

## 11.8 The Plumbing Graph: The End of the Proof

Once the geometry of the tubular neighbourhood of the divisor $\mathscr{C}$ is clarified, it is standard to describe the plumbing representation of the boundary of this neighbourhood. We follow the strategy of [92], with a modification above the double points of type 2.

**11.8.1.** Consider a component $C$ of $\mathscr{C}$ with decoration $(m;n,\nu)$. By 11.5.1, the local equation of $\widetilde{\mathscr{S}}_k$ in a neighbourhood of a generic point of $C$ is $x^m = y^n$, hence $n_{\mathscr{S}}^{-1}(C) \to C$ is a regular covering of degree $\gcd(m,n)$ over the regular part of $C$. Let $C^{norm}$ be the normalization of $C$, i.e. the curve obtained by separating the self-intersection points of $C$ (which are codified by loops of $\Gamma_{\mathscr{C}}$ attached to the vertex $v_C$ which corresponds to $C$). Then $q : n_{\mathscr{S}}^{-1}(C) \to C^{norm}$ is a cyclic branched covering whose branch points $B$ are situated above the double points of $\mathscr{C}$. They correspond bijectively to the legs of the star of $v_C$, cf. 10.2.1. Notice that if $v_C \in \mathscr{V}^1(\Gamma_{\mathscr{C}})$ (see 7.2.2 for notation) then $m = 1$, hence the covering is trivial. Otherwise $C^{norm}$ is rational by Proposition 7.4.8.

Fix a branch point $b \in B$ whose neighbourhood has a local equation of type $z^c = x^a y^b$. Then $q^{-1}(b)$ has exactly $\gcd(a,b,c)$ points; this number automatically divides $\gcd(m,n)$ for any choice of $b$. The number $\mathfrak{n}_{v_C}$ of connected components of $n_{\mathscr{S}}^{-1}C$ is the order of $\operatorname{coker}(\pi_1(C^{norm} \setminus B) \to \mathbb{Z}_{gcd(m,n)})$, where a small loop around $b$ is sent to the class of $\gcd(a,b,c)$. Hence the formula (10.2.2) for $\mathfrak{n}_{v_C}$ follows, and (10.2.4) follows too by an Euler-characteristic argument.

In the local charts of 11.5.1, $g = v^\nu = y^\nu$. Since, by normalization, $y = t^{m/gcd(m,n)}$, and $t = 0$ is the local equation of the strict transform of $C$, the vanishing order of $y^\nu$ along the strict transform of $C$ (i.e. the multiplicity of the open book decomposition of $\arg(g)$) is $m\nu/\gcd(m,n)$, proving (10.2.3). This ends the proof of Step 1 of the Main Algorithm and of Theorem 10.2.10.

**11.8.2.** Next, one has to insert the Hirzebruch–Jung strings corresponding to the singularities of $\mathscr{S}_k^{norm}$. Type 1 singular points behave similarly as those appearing in the case of cyclic coverings [86], or in the case of Iomdin series [92] (or anywhere in complex analytic geometry). In particular, the orientation compatibilities of 11.6.2 imply that these strings should be glued in with all edges decorated $+$, as usual for dual graphs of complex analytic curve configurations. One the other hand, the way how the Hirzebruch–Jung strings of type 2 should be inserted is dictated by Lemma 11.6.12. Assume that the corresponding singularity is above the intersection point $C_1 \cap C_2$ of two components of $\mathscr{C}$. Then, in the plumbing representation we have to connect their strict transforms (denoted by the same symbols) by a string $E_1, \ldots, E_s$. By 11.6.12, when $C_1$ is glued to the string, its orientation is reversed. In order to keep the ambient orientation, we have to change the orientation of its transversal slice too. But this is identified with the first curve $E_1$ of the string. If the orientation of $E_1$ is changed, then, similarly as above, we have to change the orientation of its transversal slice, which is identified with $E_2$. By iteration, we see, that all decorations of all edges of the string, inserted by Step 2, Case 2 in 10.2.5, should be $\ominus$.

The multiplicity decorations are given by the vanishing orders of $g$, and are computed by the usual procedures, see 4.3.5. This proves Step 2 of the Main Algorithm. Finally, Step 3 does not require any further explanation, cf. (4.1.5). This ends the proof of Theorem 10.2.10.

Theorems 10.3.3 and 10.3.5 are particular cases, which are obtained by forgetting some information from the graph of $\partial F$.

## 11.9 The "Extended" Monodromy Action

Usually, when one has a plumbing graph $G$, besides the 3-manifold constructed by gluing $S^1$-bundles, one can consider the plumbed 4-manifold constructed by gluing disc-bundles too. This is the case here as well; in fact, as it is clear from the constructions of this section, the plumbed 4-manifold associated with $G$ is exactly the manifold $\overline{\mathcal{S}_k}$. The point we wish to stress in this section is that there is a natural monodromy action on the pair $(\overline{\mathcal{S}_k}, \partial\overline{\mathcal{S}_k})$ such that the induced action on $\partial\overline{\mathcal{S}_k}$ coincides with the Milnor monodromy action of $\partial F$.

Indeed, instead of only defining the space $Z_k = \{c = |d|^k\}$, as in 11.3, one can take the family of spaces $Z_k(t) := \{c = |d|^k e^{it}\}$ for all values $t \in [0, 2\pi]$, and repeat the constructions of the present section. In particular, one can define in a natural way $\mathcal{S}_k(t)$, $\widetilde{\mathcal{S}}_k(t)$ and $\overline{\mathcal{S}_k}(t)$ for all $t$. This is a locally trivial bundle over the parameter $t$, hence moving $t$ from 0 to $2\pi$, we get the wished action on the pair $(\overline{\mathcal{S}_k}, \partial\overline{\mathcal{S}_k})$.

The monodromy action on the cohomology long exact sequence of $(\overline{\mathcal{S}_k}, \partial\overline{\mathcal{S}_k})$ will have important consequences, see for example the proof of Corollary 16.2.11.

**Proposition 11.9.1.** *Consider the above monodromy action on the pair* $(\overline{\mathcal{S}_k}, \partial\overline{\mathcal{S}_k})$. *Then the following facts hold:*

*(a) The action on* $\partial\overline{\mathcal{S}_k}$ *coincides with the Milnor monodromy action on* $\partial F$.
*(b) At homological level, the generalized 1-eigenspace* $H_1(\overline{\mathcal{S}_k}, \mathbb{C})_1$ *equals* $H_1(\widetilde{\mathcal{S}}_k, \mathbb{C})$. *In particular,*

$$\begin{aligned} \text{rank } H_1(\overline{\mathcal{S}_k}) &= 2g(G) + c(G), \\ \text{rank } H_1(\overline{\mathcal{S}_k})_1 &= \text{rank } H_1(\widetilde{\mathcal{S}}_k) = 2g(\Gamma_{\mathcal{C}}) + c(\Gamma_{\mathcal{C}}). \end{aligned} \tag{11.9.2}$$

*Proof.* (a) Similarly as in the proof of Proposition 11.3.3, when we compared the spaces $\Phi^{-1}(Z_k) \cap S_\epsilon$ and $\Phi^{-1}(D_{c_0}) \cap S_\epsilon$, one can identify the spaces $\Phi^{-1}(Z_k(t)) \cap S_\epsilon$ and $\Phi^{-1}(D_{c_0 e^{it}}) \cap S_\epsilon$ uniformly for any $t$. The second family enters as a building block in $\partial F$ via the decomposition from 11.3.3, and the above action is exactly the Milnor monodromy action.

(b) Let $\Gamma$ be a plumbing graph and $P(\Gamma)$ the associated plumbed 4-manifold, cf. 4.1.4. Then one has the homotopy equivalences of $P(\Gamma)$ with the core curve configuration of the plumbing. On the other hand, the first homology of this curve configuration is $2g(\Gamma)+c(\Gamma)$. Hence, rank $H_1(P(\Gamma)) = 2g(\Gamma)+c(\Gamma)$. Therefore, in the present situation, rank $H_1(\overline{\mathcal{S}_k}) = 2g(G) + c(G)$.

A similar argument shows that rank $H_1(\widetilde{\mathscr{S}}_k) = 2g(\Gamma_{\mathscr{C}}) + c(\Gamma_{\mathscr{C}})$.

Clearly, $H_1(\overline{\mathscr{S}_k}) = H_1(\mathscr{S}_k^{norm})$ too.

Next, we wish to understand the effect of the monodromy on $\widetilde{\mathscr{S}}_k$ and $\mathscr{S}_k^{norm}$ induced by the $t$-parameter family.

We will consider a generic point of the exceptional curve of $\widetilde{\mathscr{S}}_k$. Note that via the local equations $c = f = u^m v^n$ and $d = g = v^\nu$ from Sect. 6.2, the parameterized equation $c = |d|^k e^{it}$ transforms into $u^m v^n = |v|^{\nu k} e^{it}$. This equation, via the isomorphism 11.5.2, transforms into $y^n = x^m e^{it}$. In other words, $\widetilde{\mathscr{S}}_k(t)$ locally is the tubular neighbourhood of the $z$-axis in the variety given by local equation $\{(x, y, z) : y^n = x^m e^{it}\}$. Homotopically, this set is equivalent with the $z$-axis, the core curve, and the monodromy action induced on it is trivial. Analyzing all the other points too, we get that the monodromy action on $\widetilde{\mathscr{S}}_k(0)$ is homotopically trivial.

Let us analyze now the graph covering $\mathscr{S}_k^{norm}(t) \to \widetilde{\mathscr{S}}_k(t)$.

Again, let us take the same local situation as in the above discussion. For any fixed $t$, the normalization of the variety $y^n = x^m e^{it}$ has $\gcd(m,n)$ disjoint components; therefore, the $z$-axis in the normalization is covered by $\gcd(m,n)$ local discs. These components are cyclically permuted by the monodromy.

Again, analyzing all the points, we get a finite cyclic branched covering of the core curve configuration of $\widetilde{\mathscr{S}}_k(t)$ by the core curve configuration of $\mathscr{S}_k^{norm}(t)$, and the monodromy action corresponds to the cyclic action of the covering. Therefore, $H_1(\mathscr{S}_k^{norm}, \mathbb{C})_1 = H_1(\mathscr{S}_k^{norm}/\text{action}, \mathbb{C}) = H_1(\widetilde{\mathscr{S}}_k, \mathbb{C})$. $\square$

# Chapter 12
# The Collapsing Main Algorithm

## 12.1 Elimination of Assumption B

**12.1.1. Preliminary remarks.** In the formulation and the proof of the Main Algorithm 10.2 the absence of "vanishing 2-edges" in $\Gamma_{\mathscr{C}}$ is essential. If a certain $\Gamma_{\mathscr{C}}$ has such an edge, it can be modified by a blow up, which replaces the unwanted edge by three "acceptable" edges, see 10.1.3. Therefore, in any situation, it is easy to assure the condition of Assumption B, and the Main Algorithm serves as a complete algorithm for $\partial F$, $\partial F_1$, $\partial_2 F$ and for the different multiplicity systems.

Nevertheless, if the graph $\Gamma_{\mathscr{C}}$ is constructed by a canonical geometric procedure, and it has vanishing 2-cycles, the above procedure of the Main Algorithm, which starts with blowing up these edges, has some inconveniences.

First of all, in the new graph $\Gamma_{\mathscr{C}}$ we create several new vertices and edges; on the other hand, it turns out that in the output final graph $G$ all these extra vertices/edges can be eliminated, collapsed, see 12.1.3. This indicates that blowing up $\Gamma_{\mathscr{C}}$ might be unnecessary, and there should be a better procedure to eliminate the vanishing 2-edges in such a way that the new graph is not 'increasing'.

But, in fact, the main reason to search for another approach/solution is dictated by a more serious reason: we will see that "unicolored" graphs/subgraphs (that is, graphs with uniform edge decorations) have big advantages in the determination of the geometrical properties (like the structure of Jordan blocks, monodromy operators). On the other hand, by the blow up step 10.1.3 we might destroy such a property. Take for example the case of cylinders. As constructed in 9.1, and before applying the blowing up procedure, the graph $\Gamma_{\mathscr{C}}$ is unicolored: all the edge-decorations are 2. This property is not preserved after the extra blow ups of 10.1.3.

Moreover, we will see in 17.1.7, that the "twist" (local variation map) associated with a *vanishing* 2-edge is *vanishing*. Hence, the separating annulus (in the page of the open book) codified by such an edge is 'rigid', and thus it should be glued rigidly with its neighbourhood. Therefore, in the language of the graph, such an edge should be rather collapsed than blown up!

A. Némethi and Á. Szilárd, *Milnor Fiber Boundary of a Non-isolated Surface Singularity*, 131
Lecture Notes in Mathematics 2037, DOI 10.1007/978-3-642-23647-1_12,

This last statement can also be reinterpreted in the following way. We will prove in 17.1.7 that the twist of a 1-edge is negative, of a non-vanishing 2-edge is positive, while, as we already said, of a vanishing 2-edge, is zero. Since by blowing up a vanishing 2-edge we create two new 1-edges and one new 2-edge, we replace the zero-twist-contribution by two contributions of different signs. Nevertheless, when handling operators (see a concrete situation in Sects. 17.1.8–17.1.11), sometimes it is more convenient to have a semi-definite matrix rather than a non-degenerate one, which is not definite.

**12.1.2. The goal of the chapter.** In this section we present an alternative way to modify $\Gamma_{\mathscr{C}}$ and the steps of the Main Algorithm in the presence of vanishing 2-edges. This second method is also based on the algorithm just proved: we blow up such a vanishing 2-edge, we run the Main Algorithm 10.2, then we apply the reduced plumbing calculus for that part of the graph whose ancestor is that vanishing 2-edge and its adjacent vertices, and we show that this part collapses into a single orbit of vertices. Moreover, any connected subgraph whose edges are vanishing 2-edges, by this procedure collapses into a single orbit of vertices.

We keep the output of all these steps as a shortcut, which will be built in the new version of the Main Algorithm, called '*Collapsing Main Algorithm*'.

Obviously, if the original graph has no vanishing 2-edges, then the two algorithms are the same.

**12.1.3. Discussion.** Consider a vanishing 2-edge $e$ as in 10.1.3:

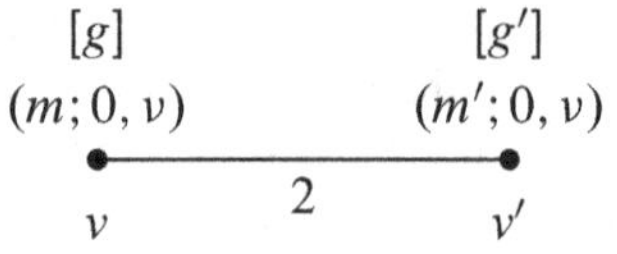


Assume that $v, v' \in \mathscr{W}$ and $v \neq v'$. Assume that the 1-legs (cf. 10.1.4) of $v$ have weights $\{(m; n_i, \nu_i)\}_{i=1}^{s}$, and the 2-legs of $v$, other than $e$, are decorated by $\{(m_j; 0, \nu)\}_{j=1}^{t}$. Set $N := \gcd(m, n_1, \dots, n_s, m_1, \dots, m_t)$. We will have similar notations $s'$, $t'$, $n'_i$, $\nu'_i$, $m'_j$, $N'$ for $v'$ too.

In 10.1.3 we have replaced $e$ by the string *Str*($e$):

*Str*($e$) :
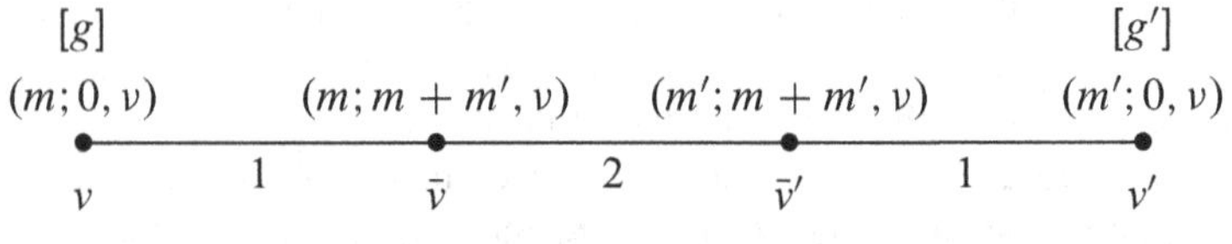


Then, let us run the Main Algorithm for this part of the graph. In the covering graph $G$ the number of vertices over $v$ is $\mathfrak{n}_v := \gcd(N, m')$, over $v'$ is $\mathfrak{n}_{v'} := \gcd(N', m)$, and over the new vertices $\bar{v}$ and $\bar{v}'$ the number of vertices is $\mathfrak{n}_e := \gcd(m, m')$. Moreover, over all the edges we have to put $\mathfrak{n}_e$ edges, they form $\mathfrak{n}_e$ strings, containing the $\mathfrak{n}_e$ vertices sitting above the new vertices, and the ends of these strings are cyclically identified with the vertices sitting over $v$ and $v'$ respectively. The vertices over $v$ have multiplicity decoration $(\nu)$ and genus decoration $\tilde{g}$ determined by (10.2.4), namely

$$\mathfrak{n}_v \cdot (2 - 2\tilde{g}) = (2 - 2g - s - t - 1) \cdot m \qquad (12.1.4)$$
$$+ \sum_{i=1}^{s} \gcd(m, n_i) + \sum_{j=1}^{t} \gcd(m, m_j) + \mathfrak{n}_e.$$

There is a similar statement for vertices over $v'$ too. The vertices of $G$ over the new vertices $v$ and $v'$ have zero genera and multiplicity decorations $(m\nu/\mathfrak{n}_e)$ and $(m'\nu/\mathfrak{n}_e)$ respectively. Furthermore, if we apply Step 2 of the algorithm from 10.2.5, then we realize that above the 1-edges the corresponding strings are degenerate (hence we insert $+$-edges only), while above the 2-edge of $Str(e)$ any inserted string $Str^{\ominus}$ has only one vertex with multiplicity $(\nu)$ and Euler decoration $(m + m')/\mathfrak{n}_e$. In particular, the $\mathfrak{n}_e$ strings above $Str(e)$ have the form

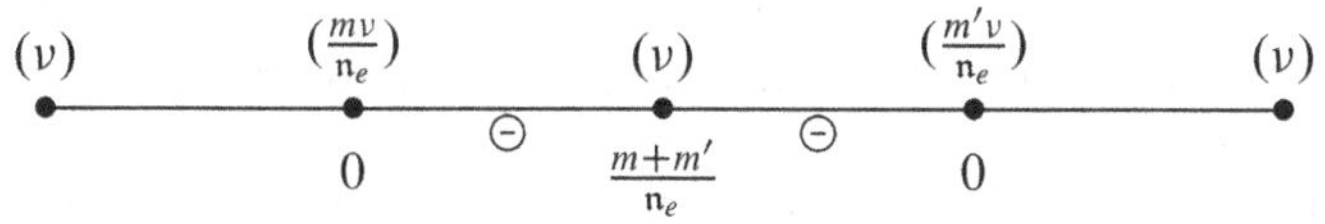

The configuration of all the vertices and edges situating above $e$ and its end-vertices form $\mathfrak{n} := \gcd(\mathfrak{n}_v, \mathfrak{n}_{v'})$ connected components.

Now, we run the plumbing calculus of oriented plumbed 3-manifolds, cf. 4.1. Notice that by two 0-chain absorptions the above string can be collapsed. Hence, after *0-chain* and *oriented handle absorptions* each connected component collapses into a single vertex. Their number will be $\mathfrak{n}$ and all of them will carry multiplicity $(\nu)$. The genus decoration $g_e$ of such a vertex can be computed as follows. First we have a contribution from 0-chain absorptions, namely the sum of all the genera of the vertices in the corresponding connected component, namely $(\mathfrak{n}_v\tilde{g} + \mathfrak{n}_{v'}\tilde{g}')/\mathfrak{n}$. Then, corresponding to oriented handle absorptions, we have to add one for each 1-cycle of that component. This, by an Euler-characteristic argument is $(\mathfrak{n}_e - \mathfrak{n}_v - \mathfrak{n}_{v'} + \mathfrak{n})/\mathfrak{n}$. Hence

$$\mathfrak{n} g_e = \mathfrak{n}_v\tilde{g} + \mathfrak{n}_{v'}\tilde{g}' + \mathfrak{n}_e - \mathfrak{n}_v - \mathfrak{n}_{v'} + \mathfrak{n},$$

that is

$$\mathfrak{n}(1 - g_e) = \mathfrak{n}_v(1 - \tilde{g}) + n_{v'}(1 - \tilde{g}') - \mathfrak{n}_e.$$

This combined with (12.1.4), gives

$$\mathfrak{n} \cdot (2 - 2g_e) = (2 - 2g - s - t - 1) \cdot m + \sum_i \gcd(m, n_i) + \sum_j \gcd(m, m_j)$$
$$+ (2 - 2g' - s' - t' - 1) \cdot m' + \sum_i \gcd(m', n'_i) + \sum_j \gcd(m', m'_j).$$

**12.1.5.** Clearly, a certain vertex of $\Gamma_{\mathscr{C}}$ may be the end-vertex of more than one vanishing 2-edge. Hence, if we run the above procedure 12.1.3 for all the vanishing

2-edges simultaneously, a bigger part of the graph will be collapsed. We make this fact more precise in the next paragraphs.

**Lemma 12.1.6.** *If $V_g$ has at most an isolated singularity then there is no cutting edge of $\Gamma_{\mathscr{C}}$ which is simultaneously a vanishing 2-edge and both its end-vertices are non-arrowheads.*

*Proof.* Assume that we have such an edge; consider the corresponding intersection point of $\mathscr{C}$ and the local equation around it as in Sect. 6.2: $f \circ r|_{U_p} = uv^{m'}$ and $g \circ r|_{U_p} = w^{\nu}$, with $m' > 1$. Then the local component $u = 0$ is in the strict transform of $V_f$. Since along the local component $w = 0$ only $g$ is vanishing, and $V_g$ has an isolated singularity, we get that $w = 0$ is situating in the strict transform of $V_g$ (otherwise $w = 0$ would be contained in an exceptional divisor which is above the origin, but along such a divisor $f \circ r$ is also vanishing). But then we have a compact curve in the intersection of the strict transforms of $V_f$ and $V_g$, which is not possible. □

Consider again the graph $\Gamma_{\mathscr{C}}$, as it is given by a certain resolution, and *unmodified* by the blowing up procedure 10.1.3. Thus it may not even satisfy Assumption B. Nevertheless, for simplicity of the discussion, we assume that it has no cutting edge which is simultaneously a vanishing 2-edge and both its end vertices are non-arrowheads. (We believe that this condition is always automatically satisfied. If $V_g$ has at most an isolated singularity this is guaranteed by 12.1.6).

Consider a maximal connected subgraph $\Gamma_{van}$ of $\Gamma_{\mathscr{C}}$ with only non-arrowhead vertices and such that all its edges are *vanishing 2-edges* connecting these vertices. In particular, $\Gamma_{van}$ has no edges supporting arrowheads. Such a subgraph is supported either by $\Gamma^1_{\mathscr{C}}$ or by $\Gamma^2_{\mathscr{C}}$. If it is a subgraph of $\Gamma^1_{\mathscr{C}}$ then it has only one vertex. In the other case it might have several vertices. Since $\Gamma_{\mathscr{C},j}$ is a tree by 7.4.12,

$$\Gamma_{van} \text{ is always a tree.} \tag{12.1.7}$$

We wish to define for each component $\Gamma_{van}$ the numbers $\mathfrak{n}_{\Gamma_{van}}$, $m_{\Gamma_{van}}$ and $g_{\Gamma_{van}}$.

Assume first that $\Gamma_{van}$ contains exactly one vertex, say $w$. Let the decoration of $w$ be $(m; n, \nu)$. Then using the star of $w$ we define $\mathfrak{n}_{\Gamma_{van}} = \mathfrak{n}_w$ as in (10.2.2), $m_{\Gamma_{van}} = m\nu/\gcd(m,n)$ as in (10.2.3), and $g_{\Gamma_{van}} = \tilde{g}_w$ as in (10.2.4), exactly as in the Main Algorithm.

Next, assume that $\Gamma_{van}$ contains several vertices. For any vertex $w$ of $\Gamma_{van}$, consider its star in $\Gamma_{\mathscr{C}}$:

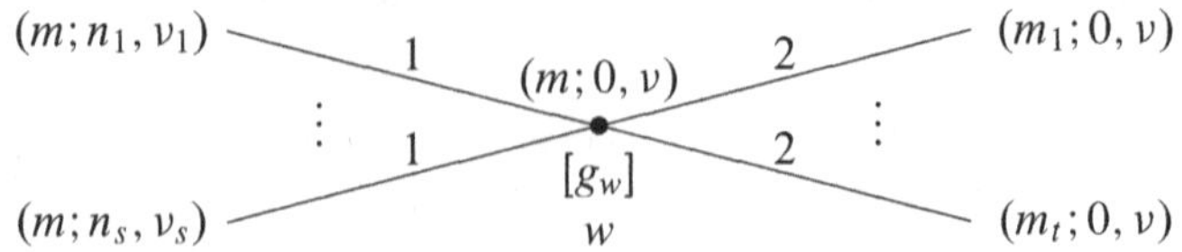

Notice that the 2-legs of this star come from two sources: either they are associated with the edges of $\Gamma_{van}$, or they are 2-edges supporting arrowheads. Let $\hat{t}_w$ be the number of this second group. The end decorations of these legs are $(1;0,1)$, hence if there is a vertex $w$ of $\Gamma_{van}$ with $\hat{t}_w > 1$, then $\nu = 1$ automatically.

Associated with the star of $w$ we consider the integers $\mathfrak{n}_w$, $\tilde{g}_w$ and $\hat{g}_w$ as follows:

- $\mathfrak{n}_w := \gcd(m, n_1, \ldots, n_s, m_1, \ldots, m_t)$, as in (10.2.2);
- $\mathfrak{n}_w(2 - 2\tilde{g}_w) = (2 - 2g_w - s - t)m + \sum_{i=1}^{s} \gcd(m, n_i) + \sum_{j=1}^{t} \gcd(m, m_j)$, as in (10.2.4);
- $\mathfrak{n}_w(2 - 2\hat{g}_w) = (2 - 2g_w - s - t)m + \sum_{i=1}^{s} \gcd(m, n_i) + \hat{t}_w$.

Furthermore, for any edge $e \in \mathscr{E}(\Gamma_{van})$, with decorations as in 12.1.3, define

- $\mathfrak{n}_e := \gcd(m, m')$, as in 12.1.3.

Then, similarly as in 12.1.3, if we eliminate the vanishing 2-edges of $\Gamma_{van}$ by the blow up procedure of 10.1.3, and run the Main Algorithm 10.2, then above $\Gamma_{van}$ the graph will have

$$\mathfrak{n}_{\Gamma_{van}} := \gcd\{\, \mathfrak{n}_w \ : \ w \in \mathscr{V}(\Gamma_{van})\} \tag{12.1.8}$$

connected components. Indeed, this follows from (12.1.7) and 5.1.6(1). Then, after 0-chain and oriented handle absorptions, the whole subgraph above $\Gamma_{van}$ will collapse into $\mathfrak{n}_{\Gamma_{van}}$ vertices, all with multiplicity

$$m_{\Gamma_{van}} = \nu, \tag{12.1.9}$$

and genus decoration $g_{\Gamma_{van}}$, which is determined similarly as in 12.1.3. More precisely, the contribution from the genera of the vertices is $\sum_w \mathfrak{n}_w \tilde{g}_w / \mathfrak{n}_{\Gamma_{van}}$, while the contribution from the cycles is

$$\Big( \sum_{e \in \mathscr{E}(\Gamma_{van})} \mathfrak{n}_e - \sum_{w \in \mathscr{V}(\Gamma_{van})} \mathfrak{n}_w + \mathfrak{n}_{\Gamma_{van}} \Big) / \mathfrak{n}_{\Gamma_{van}}. \tag{12.1.10}$$

In particular, for $g_{\Gamma_{van}}$ we get:

$$\mathfrak{n}_{\Gamma_{van}}(2 - 2g_{\Gamma_{van}}) = \sum_{w \in \mathscr{V}(\Gamma_{van})} \mathfrak{n}_w(2 - 2\hat{g}_w). \tag{12.1.11}$$

**12.1.12.** If a 2-edge supports an arrowhead then it is automatically a vanishing 2-edge. Consider such an edge $e$ of $\Gamma_{\mathscr{C}}$, whose non-arrowhead vertex $w$ has weight $(m; 0, 1)$. Let $\Gamma_{van}$ be the subgraph as in 12.1.5 which contains $w$. Since $\mathfrak{n}_w = 1$, one obtains that $\mathfrak{n}_{\Gamma_{van}} = 1$ too, hence $\Gamma_{van}$ can be collapsed by the procedure described in 12.1.5 to a unique vertex.

In $G$, above $e$, similarly as above we get exactly one string of the form

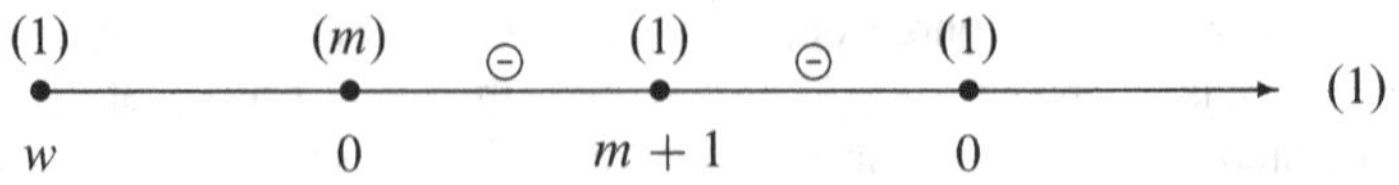


The first 0-vertex can be eliminated by 0-chain absorption. The obtained shorter string is glued to the unique vertex constructed in 12.1.5 corresponding to $\Gamma_{van}$.

Now we are able to formulate the new version of the Main Algorithm.

## 12.2 The Collapsing Main Algorithm

**12.2.1.** Start again with a graph $\Gamma_{\mathscr{C}}$, as it is given by a certain resolution, and *unmodified* by the blowing up procedure 10.1.3. Assume that it has no cutting edge which is simultaneously a vanishing 2-edge and both its end vertices are non-arrowheads. If it has some vanishing 2-edges, we will not blow them up, as in 10.1.3; instead, we will "collapse" them by the procedure described in the previous Sect. 12.1.

Denote by $\widehat{\Gamma_{\mathscr{C}}}$ the undecorated graph obtained from $\Gamma_{\mathscr{C}}$ by contracting (independently) each subgraphs of type $\Gamma_{van}$ into a unique vertex. All 1-edges, non-vanishing 2-edges, arrowhead vertices and vanishing 2-edges supporting arrowhead vertices survive inheriting the natural adjacency relations.

Then, we construct a plumbing graph $\widehat{G}$ of the open book of $\partial F$ with binding $V_g$ and fibration $\arg(g) : \partial F \setminus V_g \to S^1$ as follows. It will be determined as a covering graph of $\widehat{\Gamma_{\mathscr{C}}}$, modified with strings as in 5.1.9. In order to identify it, we have to provide the covering data of the covering $\widehat{G} \to \widehat{\Gamma_{\mathscr{C}}}$, cf. 5.1.2.

**12.2.2. Step 1. – The covering data of the vertices of $\widehat{\Gamma_{\mathscr{C}}}$.**

Over a vertex of $\widehat{\Gamma_{\mathscr{C}}}$, obtained by the contraction of the subgraph $\Gamma_{van}$ of $\Gamma_{\mathscr{C}}$, we insert $\mathfrak{n}_{\Gamma_{van}}$ vertices in $\widehat{G}$, all of them with genus decoration $g_{\Gamma_{van}}$ and multiplicity $m_{\Gamma_{van}}$. These numbers are defined in (12.1.8), (12.1.11) and (12.1.9) respectively.

Any arrowhead vertex of $\widehat{\Gamma_{\mathscr{C}}}$ is covered by one arrowhead vertex of $\widehat{G}$, decorated by multiplicity decoration (1), similarly as in the original version 10.2.1.

**12.2.3. Step 2. – The covering data of edges and the types of the inserted strings.**

**Case 1.** The case of 1-edges is the same as in the original version 10.2.5. Over such an edge $e$, which in $\Gamma_{\mathscr{C}}$ has the form

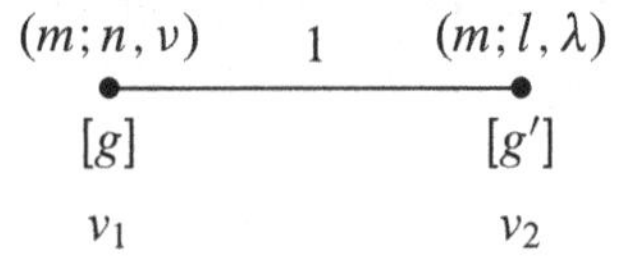


insert (cyclically) in $\widehat{G}$ exactly $\mathfrak{n}_e = \gcd(m, n, l)$ strings of type

$$Str\left(\frac{n}{\mathfrak{n}_e},\frac{l}{\mathfrak{n}_e};\frac{m}{\mathfrak{n}_e}\,\middle|\,\nu,\lambda;0\right).$$

If the edge $e$ is a loop, then the procedure is the same with the only modification that the end-vertices of the strings are identified. If the right vertex $v_2$ is an arrowhead, then complete again the same procedure with $m = 1$ and $\mathfrak{n}_e = 1$, namely: above such an edge $e$ put a single edge decorated by $+$. This edge supports that arrowhead of $\widehat{G}$ which covers the corresponding arrowhead of $\widehat{\Gamma_{\mathscr{C}}}$.

**Case 2.** The case of *non-vanishing* 2-edges is again unmodified. Above such an edge $e$, which in $\Gamma_{\mathscr{C}}$ has the form (with $n > 0$)

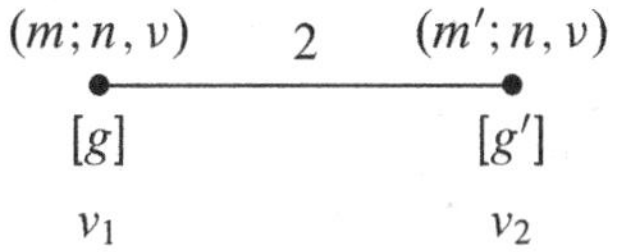


insert (cyclically) in $G$ exactly $\mathfrak{n}_e = \gcd(m, m', n)$ strings of type

$$Str^{\ominus}\left(\frac{m}{\mathfrak{n}_e},\frac{m'}{\mathfrak{n}_e};\frac{n}{\mathfrak{n}_e}\,\middle|\,0,0;\nu\right).$$

If the edge is a loop, then we modify the procedure as in the case of 1-loops.

**Case 3.** Finally, we have to consider the case of those *vanishing* 2-edges which support arrowheads (the others have been collapsed).

Above such an edge we insert in $\widehat{G}$ one string of type

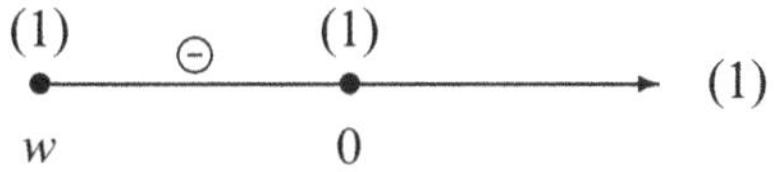


If the 2-edge is supported in $\Gamma_{\mathscr{C}}$ by a vertex which belongs to $\Gamma_{van}$, then the end-vertex $w$ of the above string should be identified with the unique vertex of $\widehat{G}$ corresponding to that subgraph $\Gamma_{van}$, keeping its $[g_{\Gamma_{van}}]$ decoration too.

**12.2.4. Step 3. – Determination of the missing Euler numbers.** The first two steps provide a graph with the next decorations: the multiplicities of all the vertices, all the genera, some of the Euler numbers and all the sign-decorations of the edges. The missing Euler numbers are determined by formula (4.1.5).

## 12.3 The Output of the Collapsing Main Algorithm

Similarly as in the first case, Theorem 5.1.8 guarantees that *there is only one cyclic graph-covering of $\Gamma_{\mathscr{C}}$ with this covering data (up to a graph-isomorphism).*

In fact, by 12.1.5, the output graph $\widehat{G}$ of the "Collapsing Algorithm" and the output $G$ of the original algorithm are connected by the reduced oriented plumbing calculus: $\widehat{G} \sim G$. Hence, Theorem 10.2.10 is valid for $\widehat{G}$ as well: *$\widehat{G}$ is a possible plumbing graph of the pair $(\partial F, \partial F \cap V_g)$, which carries the multiplicity system of the open book decomposition* $\arg(g) : \partial F \setminus V_g \to S^1$. The algorithm is again compatible with the decomposition of $\partial F$ into $\partial_1 F$ and $\partial_2 F$. The part regarding $\partial_1 F$ is unmodified (since $\Gamma^1_{\mathscr{C}}$ has no 2-edges). The graph $G_2$ transforms into $\widehat{G_2}$ similarly as $G$ transforms into $\widehat{G}$. Moreover, all the statements of 10.3 regarding $G_1$ and $G_2$ are valid for $\widehat{G_1} = G_1$ and $\widehat{G_2}$ with the natural modifications. The details are left to the reader.

**12.3.1.** On the other hand, the difference between $G$ and $\widehat{G}$, the outputs of the original and the new algorithms, both *unmodified by plumbing calculus*, can be substantial, sometimes even spectacular. See e.g. the complete computation in the case of cylinders in Chap. 20.

For example, the difference $c(G) - c(\widehat{G})$ is the sum over all subgraphs $\Gamma_{van}$ of the expression appearing in (12.1.10), which can be a rather large number. This will have crucial consequences in the discussion of the Jordan blocks of the vertical monodromies.

Similarly as in 10.2.8, we will use the notation $\widehat{G}$ for the output graph obtained by the Collapsing Main Algorithm described above *unmodified* by any operation of the plumbing calculus. We adopt similar notations for the graphs of $\partial_1 F$ and $\partial_2 F$, namely $\widehat{G_1}$ and $\widehat{G_2}$. (Recall that $G \sim \widehat{G}$. Hence there is no need to consider a "modified $\widehat{G}$", since for that we can use the already introduced notation $G^m$, cf. 10.2.8).

# Chapter 13
# Vertical/Horizontal Monodromies

## 13.1 The Monodromy Operators

Let $f : (\mathbb{C}^3, 0) \to (\mathbb{C}, 0)$ be a hypersurface singularity with a 1-dimensional singular locus. In general, it is rather difficult to determine the horizontal and vertical monodromies $\{m'_{j,hor}\}_{i=1}^{s}$ and $\{m'_{j,ver}\}_{i=1}^{s}$ of $Sing(V_f)$, especially the vertical one. (For the terminology, see Sect. 2.2.) It is even more difficult to identify the *two commuting actions simultaneously*.

This difficulty survives at the homological level too: in the literature there is no general treatment of the corresponding two commuting operators. In lack of general theory, the existing literature is limited to few sporadic examples, which are obtained by ad hoc methods.

Our goal here is to provide a general procedure to treat these homological objects and to produce (in principle without any obstruction) examples as complicated as we wish.

Similarly, if we fix another germ $g$ such that $\Phi = (f, g)$ is an ICIS, then one of the most important tasks is the computation of the algebraic monodromy representation of $\mathbb{Z}^2$ induced by $m_{\Phi,hor}$ and $m_{\Phi,ver}$ (for their definition, see Sect. 3.1). Our treatment will include the determination of these objects as well.

In fact, our primary targets are the following algebraic monodromy operators:

- the commuting pair $M'_{j,hor}$ and $M'_{j,ver}$, acting on $H_1(F'_j)$, induced by $m'_{j,hor}$ and $m'_{j,ver}$ $(1 \leq j \leq s)$,
- the commuting pair $M^{\Phi}_{j,hor}$ and $M^{\Phi}_{j,ver}$, acting on $H_1(F_\Phi \cap T_j) = H_1(F'_j)^{\oplus d_j}$ (cf. 3.3.1), induced by $m^{\Phi}_{j,hor}$ and $m^{\Phi}_{j,ver}$ $(1 \leq j \leq s)$,
- the commuting pair $M_{\Phi,hor}$ and $M_{\Phi,ver}$, acting on $H_1(F_\Phi)$, induced by $m_{\Phi,hor}$ and $m_{\phi,ver}$, cf. Sect. 3.1.

Here, usually, we considered homology with complex coefficients, but obviously, one might also consider the integral case. In fact, in some of our examples, the additional $\mathbb{Z}$-invariants will also be discussed.

A. Némethi and Á. Szilárd, *Milnor Fiber Boundary of a Non-isolated Surface Singularity*, Lecture Notes in Mathematics 2037, DOI 10.1007/978-3-642-23647-1_13, 

We separate our discussion into two parts. In this chapter we determine completely (via $\Gamma_{\mathscr{C}}$) the character decomposition (i.e. the semi-simple part) of the relevant $\mathbb{Z}^2$-representations. This includes the characteristic polynomials of all the monodromy operators. Moreover, we connect the ranks of some generalized eigenspaces with the combinatorics of the plumbing graph $G$ as well. We wish to emphasize that, although we get all our results rather automatically from the graph $\Gamma_{\mathscr{C}}$, all these results are new, and were out of reach (in this generality) with previous techniques. This shows once more the power of $\Gamma_{\mathscr{C}}$.

The second part treats the structure of the Jordan blocks. This is considerably harder. Our main motivation in this part is the computation of the homology of $\partial F$ and its algebraic monodromy action. Since the homology of $\partial F$ will be determined via the homology of $\partial F \setminus V_g$, that is, using the Wang exact sequence in which the operator $M_{\Phi,ver} - I$ appears, the determination of the 2-Jordan blocks of $M_{\Phi,ver}$ with eigenvalue one is a crucial ingredient.

Therefore, in this work, regarding the vertical monodromies, we will concentrate only on the computation of the Jordan blocks with eigenvalue one, although for some cases we will provide the complete picture. This second part (including the discussion regarding the Jordan blocks of the vertical monodromies for eigenvalue one, and the computation of homology and algebraic monodromy of $\partial F$) constitutes the next Chaps. 14–17.

**Remark 13.1.1.** Although $F_\Phi$ is not the local Milnor fiber of a hypersurface singularity, a convenient restriction of $\Phi$ provides a map over a sufficiently small disc (a transversal slice of the $d$-axis at one of its generic points) with generic fiber $F_\Phi$ such that the horizontal monodromy $M_{\Phi,hor}$ is the monodromy over the small punctured disc. In other words, the horizontal monodromy $M_{\Phi,hor}$ can be 'localized', that is, it can be represented as the monodromy of a family of curves over an arbitrarily small punctured disc. On the other hand, the vertical monodromy $M_{\Phi,ver}$ *cannot* be localized in this sense. In particular, general results about the monodromy of families over a punctured disc cannot be applied for the vertical monodromies. (This is one of the reasons why their computation is so difficult. This difficulty will be overcome here using the graph $\Gamma_{\mathscr{C}}$.)

## 13.2 General Facts

The statements of the next lemma are well-known for the horizontal monodromies by the celebrated Monodromy Theorem (see e.g. [22,59] for the global case, [15,67] for the local case, or [56] for a recent monograph). It may be known for the vertical monodromies as well, however we were not able to find a reference for it:

**Lemma 13.2.1.** *The eigenvalues of the operators $M'_{j,hor}$, $M'_{j,hor}$, $M^{\Phi}_{j,hor}$, $M^{\Phi}_{j,ver}$ $(1 \leq j \leq s)$, respectively $M_{\Phi,hor}$ and $M_{\Phi,ver}$, are roots of unity. Moreover, the size of the Jordan blocks cannot be larger than two.*

*Proof.* For the operators acting on $H_1(F_\Phi)$ use the decomposition 7.1.6 of $F_\Phi$, and the fact that the restriction of the geometric actions on each subset $\widetilde{F}_v$ is isotopic to a finite action. (As a model for the proof, see e.g. [33, §13].) The same is true for the operators acting on $H_1(F_\Phi \cap T_j)$. In this case only those subsets $\widetilde{F}_v$ appear which are indexed by the non-arrowhead vertices of $\Gamma^2_{\mathscr{C},j}$. Finally, the operators $M^\Phi_{j,*}$ and $M'_{j,*}$ are connected by a simple algebraic operation, see 3.3.1(2). Compare with the proofs of 13.4.6 and 16.2.3 as well. □

**Remark 13.2.2.** By the Monodromy Theorem valid for isolated hypersurface singularities, the Jordan blocks of $M'_{j,hor}$ with eigenvalue one must have size one, see e.g. [56, (3.5.9)]. By the correspondence 3.3.1(2), this fact is true for $M^\Phi_{j,hor}$ as well. Nevertheless, for the other four operators, such a restriction is *not* true anymore: In 19.5.2 we provide an example when all $M'_{j,ver}$, $M^\Phi_{j,ver}$ $M_{\Phi,hor}$ and $M_{\Phi,ver}$ have Jordan blocks of size 2 with eigenvalue one.

In fact, examples with $M'_{j,ver}$ having such a Jordan block can be constructed as follows: Fix a topological/equisingularity type of isolated plane curve singularity $S$ whose monodromy has a 2-Jordan block. Let $o$ be the order of the eigenvalue of this block. Then one can construct a projective plane curve $C$ of degree $d$ (sufficiently large), which is a multiple of $o$, and such that $C$ has a local singularity of type $S$. Let $V_f$ be the cone over $C$. By a result of Steenbrink [125] $M'_{j,ver} = (M'_{j,hor})^{-d}$, hence $M'_{j,ver}$ has a Jordan block with eigenvalue one and size two.

For the number of Jordan blocks we will use the following notation:

**Definition 13.2.3.** *For any operator $M$ let $\#^k_\lambda M$ denote the number of Jordan blocks of $M$ of size $k$ with eigenvalue $\lambda$.*

## 13.3 Characters: Algebraic Preliminaries

Let $H$ be a finite dimensional $\mathbb{C}$-vector space, and assume that $M \in Aut(H)$. Let $P_M(t)$ (or $P_{H,M}(t)$) be the characteristic polynomial $\det(tI - M)$ of $M$. For each eigenvalue $\lambda$, let $H_{M,\lambda} = \ker((\lambda I - M)^N)$ (for $N$ large) be the generalized $\lambda$-eigenspace of $M$. Obviously, the multiplicity of $t-\lambda$ in $P_M(t)$ is exactly $\dim H_{M,\lambda}$. When $M$ is clear from the context, we simply write $H_\lambda = H_{M,\lambda}$.

Sometimes it is more convenient to replace $P_M(t)$ by its *divisor*

$$Div(H; M) := \sum_\lambda \dim H_{M,\lambda} \cdot (\lambda) \in \mathbb{Z}[\mathbb{C}^*]. \tag{13.3.1}$$

More generally, assume that two commuting automorphisms $M_1$ and $M_2$ act on $H$. Then, for each pair $(\lambda, \xi) \in \mathbb{C}^* \times \mathbb{C}^*$, set $H_{(\lambda,\xi)} := H_{M_1,\lambda} \cap H_{M_2,\xi}$ and

$$Div(H; M_1, M_2) := \sum_{(\lambda,\xi)} \dim(H_{(\lambda,\xi)}) \cdot (\lambda, \xi) \in \mathbb{Z}[\mathbb{C}^* \times \mathbb{C}^*]. \tag{13.3.2}$$

Above $\mathbb{Z}[\mathbb{C}^*]$ and $\mathbb{Z}[\mathbb{C}^* \times \mathbb{C}^*]$ are the group rings of $\mathbb{C}^*$ and $\mathbb{C}^* \times \mathbb{C}^*$ over $\mathbb{Z}$.

**13.3.3. Key Example.** Fix a triple of integers $(m;n,\nu)$ with $m,\ \nu>0$ and $n\geq 0$. Let $\mathscr{P}$ be the set of points

$$\mathscr{P}:=\{(u,v)\in\mathbb{C}^*\times\mathbb{C}^*\mid v^\nu=1,\ u^m v^n=1\}.$$

In fact, $\mathscr{P}$ is a finite subgroup of $\mathbb{C}^*\times\mathbb{C}^*$ of order $m\nu$. On this set of points we define two commuting permutations: For each pair of real numbers $(t_{hor},t_{ver})$, consider the set of points

$$\mathscr{P}(t_{hor},t_{ver}):=\{(u,v)\in\mathbb{C}^*\times\mathbb{C}^*\mid v^\nu=e^{it_{ver}},\ u^m v^n=e^{it_{hor}}\}.$$

Fixing $t_{ver}=0$ and moving $t_{hor}$ from 0 to $2\pi$, we get a locally trivial family of $m\nu$ points, which defines a permutation $\sigma_{hor}$ of $\mathscr{P}$. Similarly, fixing $t_{hor}=0$ and moving $t_{ver}$ from 0 to $2\pi$, we get the permutation $\sigma_{ver}$ of $\mathscr{P}$. One can verify that the two permutations commute, based for example on the fact that the torus $\{|u|=|v|=1\}$ has an abelian fundamental group.

Let $H:=H_0(\mathscr{P},\mathbb{C})$ be the vector space with base elements indexed by the points from $\mathscr{P}$, i.e. the vector space of elements of type $\sum_{p\in\mathscr{P}}c_p\cdot p$, where $c_p\in\mathbb{C}$. For any permutation $\sigma$ of $\mathscr{P}$, define $\sigma_*\in Aut(H)$ by

$$\sigma_*(\textstyle\sum_p c_p\cdot p):=\sum_p c_p\cdot\sigma(p).$$

Our goal is to determine

$$\Lambda(m;n,\nu):=Div(H;\sigma_{hor,*},\sigma_{ver,*})\in\mathbb{Z}[\mathbb{C}^*\times\mathbb{C}^*].$$

Consider the following two elements of $\mathscr{P}$:

$$\mathfrak{h}:=(e^{2\pi i/m},1)\quad\text{and}\quad\mathfrak{v}:=(e^{-2\pi in/m\nu},e^{2\pi i/\nu}).$$

By a computation one can verify that $\sigma_{hor}$ and $\sigma_{ver}$ can be obtained by multiplication in $\mathscr{P}$ by $\mathfrak{h}$ and $\mathfrak{v}$ respectively. Let $\widetilde{\mathscr{P}}$ be the subgroup of the permutation group of $\mathscr{P}$ generated by $\sigma_{hor}$ and $\sigma_{ver}$. Having the forms of $\mathfrak{h}$ and $\mathfrak{v}$, one can easily verify that $\mathfrak{h}$ and $\mathfrak{v}$ (hence $\sigma_{hor}$ and $\sigma_{ver}$ in $\widetilde{\mathscr{P}}$ too) satisfy the relations

$$\mathfrak{h}^m=\mathfrak{v}^{m\nu/(m,n)}=\mathfrak{h}^n\mathfrak{v}^\nu=1,$$

and that $\widetilde{\mathscr{P}}$ acts transitively on $\mathscr{P}$. (Here $(m,n)=\gcd(m,n)$.) Define the group

$$G(m;n,\nu):=\{(\lambda,\xi)\in\mathbb{C}^*\times\mathbb{C}^*\ :\ \lambda^m=\lambda^n\xi^\nu=1\}.\tag{13.3.4}$$

Note that for $(\lambda,\xi)\in G(m;n,\nu)$, one automatically has $\xi^{m\nu/(m,n)}=1$.

Then $G(m;n,\nu)$ is isomorphic to $\widetilde{\mathscr{P}}$, and both have order $m\nu$. Since $\widetilde{\mathscr{P}}$ acts transitively on $\mathscr{P}$ (which has the same order) both are isomorphic to $\mathscr{P}$ too. An isomorphism $\widetilde{\mathscr{P}}\longrightarrow\mathscr{P}$ can be generated by $\sigma_{hor}\mapsto\mathfrak{h}$ and $\sigma_{ver}\mapsto\mathfrak{v}$.

Therefore, for any $(\lambda,\xi)\in G(m;n,\nu)$ there is a $(\lambda,\xi)$-eigenvector of the action $(\sigma_{hor},\sigma_{ver})$ which has the form:

$$\sum_{k,l}\lambda^k\xi^l\cdot\sigma_{hor}^{-k}\sigma_{ver}^{-l}(p_0),$$

where $p_0=(1,1)\in\mathscr{P}$, and the index set $k,l$ is taken in such a way that the set $\{\sigma_{hor}^k\sigma_{ver}^l\}_{k,l}$ is exactly $\widetilde{\mathscr{P}}$, and each element is represented once. In particular,

$$\Lambda(m;n,\nu)=\sum_{(\lambda,\xi)\in G(m;n,\nu)}(\lambda,\xi). \tag{13.3.5}$$

As a particular example, assume that above one has $\nu=1$. Then $G(m;n,1)=\{\lambda:\lambda^m=1\}$ and $\xi=\lambda^{-n}$. Hence

$$\Lambda(m;n,1)=\sum_{\lambda^m=1}(\lambda,\lambda^{-n}). \tag{13.3.6}$$

From (13.3.5) it also follows that the characteristic polynomials of $\sigma_{hor,*}$, respectively of $\sigma_{ver,*}$ acting on $H$ are the following:

$$P_{\sigma_{hor,*}}(t)=(t^m-1)^\nu,\quad P_{\sigma_{ver,*}}(t)=(t^{m\nu/(m,n)}-1)^{(m,n)}. \tag{13.3.7}$$

Moreover, the characteristic polynomial of $\sigma_{hor,*}$ restricted on the generalized 1-eigenspace $H_{\sigma_{ver,*},1}$ of $\sigma_{ver,*}$ is

$$P_{\sigma_{hor,*}|H_{\sigma_{ver,*},1}}(t)=t^{(m,n)}-1. \tag{13.3.8}$$

**13.3.9. The '$d$-covering'.** Let $H$ be a finite dimensional vector space with two commuting automorphisms $M_1$ and $M_2$. Furthermore, fix a positive integer $d$. We define $H^{(d)}:=H^{\oplus d}$ and its automorphisms $M_1^{(d)}$ and $M_2^{(d)}$ by

$$M_1^{(d)}(x_1,\dots,x_d)=(M_1(x_1),\dots,M_1(x_d)),$$

and

$$M_2^{(d)}(x_1,\dots,x_d)=(M_2(x_d),x_1,\dots,x_{d-1}).$$

It is not hard to see that $Div(H^{(d)};M_1^{(d)},M_2^{(d)})$ can be recovered from $Div(H;M_1,M_2)$ and the integer $d$. Indeed, consider the morphism

$$\Xi^{(d)}:\mathbb{Z}[\mathbb{C}^*\times\mathbb{C}^*]\to\mathbb{Z}[\mathbb{C}^*\times\mathbb{C}^*],\ \text{ where }\ \Xi^{(d)}((\lambda,\xi)):=\sum_{\alpha^d=\xi}(\lambda,\alpha).$$

Then, one shows that

$$\Xi^{(d)}(Div(H;M_1,M_2))=Div(H^{(d)};M_1^{(d)},M_2^{(d)}).$$

**Lemma 13.3.10.** *For any fixed positive integer $d$ one has:*

*(a)* $\Xi^{(d)}$ *is injective.*
*(b)* $\Xi^{(d)}(\Lambda(m;n,\nu)) = \Lambda(m;n,d\nu)$.

*Proof.* (a) If one takes $\Omega^{(d)} : \mathbb{Z}[\mathbb{C}^* \times \mathbb{C}^*] \to \mathbb{Q}[\mathbb{C}^* \times \mathbb{C}^*]$, defined by $\Omega^{(d)}(\lambda, \alpha) = \frac{1}{d}(\lambda, \alpha^d)$, then $\Omega^{(d)} \circ \Xi^{(d)}$ is the inclusion $\mathbb{Z}[\mathbb{C}^* \times \mathbb{C}^*] \hookrightarrow \mathbb{Q}[\mathbb{C}^* \times \mathbb{C}^*]$. For (b) notice that the system $\{\lambda^m = \lambda^n \xi^\nu = 1,\ \alpha^d = \xi\}$ is equivalent to $\{\lambda^m = \lambda^n \alpha^{d\nu} = 1\}$. □

We will also need the following property of $d$-coverings provided by elementary linear algebra. Consider the vector space $H$ and the two commuting automorphisms $M_1$ and $M_2$ as above. Let $H_{M_2,1}$ be the generalized 1-eigenspace associated with $M_2$, and consider the restrictions of $M_1$ and $M_2$ (denoted by the same symbols $M_1$ and $M_2$) to this subspace. In this way we get the triple $(H_{M_2,1}; M_1, M_2)$. Similarly, for any positive integer $d$, we can consider the triple $(H^{(d)}_{M_2^{(d)},1}; M_1^{(d)}, M_2^{(d)})$.

**Lemma 13.3.11.** *For any triple $(H; M_1, M_2)$ and for any $d$, one has an isomorphism of triples*

$$(H_{M_2,1}; M_1, M_2) \approx (H^{(d)}_{M_2^{(d)},1}; M_1^{(d)}, M_2^{(d)}).$$

*Proof.* First note that we may assume that all the eigenvalues of $M_2$ are equal to 1. Then, it is convenient to write $M_2$ as $\widetilde{M}_2^d$ for some $\widetilde{M}_2$ which commutes with $M_1$. This can be done as follows: if $M_2 = I + N$, where $I$ is the identity and $N$ is a nilpotent operator, then

$$\widetilde{M}_2 = (I + N)^{1/d} := I + \frac{1}{d} N + \frac{1}{2!d}\Big(\frac{1}{d} - 1\Big) N^2 + \cdots .$$

Consider the matrix identities

$$\begin{bmatrix} 0 & 0 & \cdots & I \\ \widetilde{M}_2^{d-1} & 0 & \cdots & 0 \\ \cdots & \cdots & \cdots & \cdots \\ 0 & \cdots & \widetilde{M}_2 & 0 \end{bmatrix} \begin{bmatrix} 0 & I & 0 & \cdots \\ 0 & 0 & I & \cdots \\ \cdots & \cdots & \cdots & \cdots \\ \widetilde{M}_2^d & 0 & 0 & \cdots \end{bmatrix} \begin{bmatrix} 0 & \widetilde{M}_2^{-(d-1)} & 0 & \cdots \\ 0 & 0 & \widetilde{M}_2^{-(d-2)} & \cdots \\ \cdots & \cdots & \cdots & \cdots \\ I & 0 & 0 & \cdots \end{bmatrix} = \begin{bmatrix} 0 & \widetilde{M}_2 & 0 & \cdots \\ 0 & 0 & \widetilde{M}_2 & \cdots \\ \cdots & \cdots & \cdots & \cdots \\ \widetilde{M}_2 & 0 & 0 & \cdots \end{bmatrix}$$

and

$$\frac{1}{d} \begin{bmatrix} 1 & 1 & \cdots & 1 \\ 1 & \bar{\xi}_2 & \cdots & \bar{\xi}_2^{d-1} \\ \cdots & \cdots & \cdots & \cdots \\ 1 & \bar{\xi}_d & \cdots & \bar{\xi}_d^{d-1} \end{bmatrix} \begin{bmatrix} 0 & 1 & 0 & \cdots \\ 0 & 0 & 1 & \cdots \\ \cdots & \cdots & \cdots & \cdots \\ 1 & 0 & 0 & \cdots \end{bmatrix} \begin{bmatrix} 1 & 1 & \cdots & 1 \\ 1 & \xi_2 & \cdots & \xi_d \\ \cdots & \cdots & \cdots & \cdots \\ 1 & \xi_2^{d-1} & \cdots & \xi_d^{d-1} \end{bmatrix} = \begin{bmatrix} 1 & & & \\ & \xi_2 & & \\ & & \cdots & \\ & & & \xi_d \end{bmatrix}$$

where $1 = \xi_1, \xi_2, \cdots, \xi_d$ are the $d$-roots of unity, and $\bar{\xi}_i$ their conjugates. They show that $(H^{(d)}; M_1^{(d)}, M_2^{(d)})$ is isomorphic to $\oplus_{i=1}^d (H; M_1, \xi_i \widetilde{M}_2)$.

On the other hand, $(H; M_1, \widetilde{M}_2)$ and $(H; M_1, M_2)$ are isomorphic. □

## 13.4 The Divisors $Div_\Phi$, $Div_j^\Phi$ and $Div'_j$ in Terms of $\Gamma_{\mathscr{C}}$

The three pairs of operators listed in Sect. 13.1 define three divisors. These are the following.

**Definition 13.4.1.** *We set*

$$Div'_j := Div(H_1(F'_j); M'_{j,hor}, M'_{j,ver}) \quad (1 \le j \le s), \tag{13.4.2}$$

$$Div_j^\Phi := Div(H_1(F'_j)^{\oplus d_j}; M^\Phi_{j,hor}, M^\Phi_{j,ver}), \ (1 \le j \le s), \tag{13.4.3}$$

*and*

$$Div_\Phi := Div(H_1(F_\Phi); M_{\Phi,hor}, M_{\Phi,ver}). \tag{13.4.4}$$

**13.4.5. Some old/new notations.** Recall that $\mathscr{W}(\Gamma_{\mathscr{C}})$ (respectively $\mathscr{W}(\Gamma^2_{\mathscr{C},j})$) denote the set of non-arrowhead vertices of $\Gamma_{\mathscr{C}}$ (respectively of $\Gamma^2_{\mathscr{C},j}$, for any $j = 1, \dots, s$). For each $w \in \mathscr{W}(\Gamma_{\mathscr{C}})$, let $C_w$ be the corresponding irreducible curve in $\mathscr{C}$, $g_w$ its genus, and $\delta_w$ the number of legs associated with the star of $v$, i.e. the number of edges in $\Gamma_{\mathscr{C}}$ adjacent to $w$, where each loop contributes twice (cf. 10.1.4). Moreover, assume that the decoration of $w$ in $\Gamma_{\mathscr{C}}$ is $(m_w; n_w, \nu_w)$.

With these notations, one has the following *A'Campo type identities*, generalizations of the identity (5.2.8) proved in [4]:

**Theorem 13.4.6.**

$$Div_\Phi - (1,1) = \sum_{w \in \mathscr{W}(\Gamma_{\mathscr{C}})} (2g_w + \delta_w - 2) \cdot \Lambda(m_w; n_w, \nu_w). \tag{13.4.7}$$

*Moreover, for any* $j = 1, \dots, s$,

$$Div_j^\Phi - \sum_{\xi^{d_j}=1} (1, \xi) = \sum_{w \in \mathscr{W}(\Gamma^2_{\mathscr{C},j})} (\delta_w - 2) \cdot \Lambda(m_w; n_w, \nu_w), \tag{13.4.8}$$

$$Div'_j - (1,1) = \sum_{w \in \mathscr{W}(\Gamma^2_{\mathscr{C},j})} (\delta_w - 2) \cdot \Lambda(m_w; n_w, \nu_w/d_j), \tag{13.4.9}$$

*where* $d_j = \deg(g|\Sigma_j)$, *or* $d_j = \gcd(\nu_w)_{w \in \mathscr{W}(\Gamma^2_{\mathscr{C},j})}$ *by 7.4.12.*

*In particular, the above formulae provide the ranks of* $H_1(F_\Phi)$ *and* $H_1(F'_j)$ *as well:*

$$\begin{aligned} \operatorname{rank} H_1(F_\Phi) &= 1 + \sum_{w\in\mathscr{W}(\Gamma_{\mathscr{C}})} (2g_w + \delta_w - 2)\cdot m_w\nu_w, \\ \mu'_j = \operatorname{rank} H_1(F'_j) &= 1 + \sum_{w\in\mathscr{W}(\Gamma^2_{\mathscr{C},j})} (\delta_w - 2)\cdot m_w\nu_w/d_j. \end{aligned} \tag{13.4.10}$$

*Proof.* As in any "A'Campo type" formula (cf. [4] or (5.2.8) above), it is more convenient to work with a zeta-function of an action instead of its characteristic polynomial. In the present case also, we will determine first the element

$$D(F_\Phi) := Div\,(H_0(F_\Phi); M^0_{\Phi,hor}, M^0_{\Phi,ver}) - Div\,(H_1(F_\Phi); M_{\Phi,hor}, M_{\Phi,ver})$$

in $\mathbb{Z}[\mathbb{C}^*\times\mathbb{C}^*]$. Above, $M^0_{\Phi,hor}$ and $M^0_{\Phi,ver}$ are the horizontal and the vertical monodromies acting on $H_0(F_\Phi)$. Since $F_\Phi$ is connected, this space is $\mathbb{C}$, and $M^0_{\Phi,hor} = M^0_{\Phi,ver} = Id_{\mathbb{C}}$. Hence $Div(H_0(F_\Phi); M^0_{\Phi,hor}, M^0_{\Phi,ver}) = (1,1)$, and thus the left hand side of (13.4.7) is $-D(F_\Phi)$.

The point is that $D(F_\Phi)$ is additive with respect to a Mayer–Vietoris exact sequence. More precisely, if we consider the decomposition 7.1.6, then

$$D(F_\Phi) = \sum_{w\in\mathscr{W}(\Gamma_{\mathscr{C}})} D(\widetilde{F}_w) - D(B),$$

where $B$ is the union of "cutting circles". Since $H_0(B) = H_1(B)$ and the monodromy actions on them can also be identified, $D(B) = 0$. On the other hand, $\widetilde{F}_w$ is a regular covering over the regular part $C_w^{reg}$ of the curve $C_w$ with a finite fiber which can be identified with $\mathscr{P}$ in the Key Example 13.3.3. Moreover, the horizontal and vertical actions on $\widetilde{F}_w$ are induced by the corresponding actions on $\mathscr{P}$. Hence $D(\widetilde{F}_w) = \chi(C_w^{reg})\cdot D(\mathscr{P}) = \chi(C_w^{reg})\cdot\Lambda(m_w; n_w, \nu_w)$, where $\chi(C_w^{reg})$ stands for the Euler-characteristic of $C_w^{reg}$, and equals $2 - 2g_w - \delta_w$.

The proof of (13.4.8) is similar. Using the results and notations of 3.3.1, one gets that $F_\Phi \cap T_j$ is cut by "cutting circles" into the pieces $\{\widetilde{F}_w\}_{w\in\Gamma^2_{\mathscr{C},j}}$. But $F_\Phi\cap T_j$ consists of $d_j$ copies of $F'_j$. Hence, with the additional fact that $g_w = 0$ for all $w\in\mathscr{W}(\Gamma^2_{\mathscr{C},j})$, we get

$$-D(\cup_{d_j} F'_j; m^\Phi_{j,hor}, m^\Phi_{j,hor}) = \sum_{w\in\mathscr{W}(\Gamma^2_{\mathscr{C},j})} (\delta_w - 2)\cdot\Lambda(m_w; n_w, \nu_w).$$

On the other hand, in this case, the 0-homology is different: $H_0(\cup_{d_j} F'_j, \mathbb{C}) = \mathbb{C}^{d_j}$ on which the horizontal monodromy acts by identity and the vertical one by cyclic permutation –, therefore its contribution is $\sum_{\xi^{d_j}=1}(1,\xi)$.

Now, using the special forms of $m^\Phi_{j,hor}$ and $m^\Phi_{j,hor}$ from 3.3.1(2), by 13.3.9 and 13.3.10 one gets that $\Omega^{(d_j)}(Div^\Phi_j) = Div'_j$, thus (13.4.9) follows too. □

The results 13.4.6, (13.3.7) and (13.3.8) imply:

**Corollary 13.4.11.** *(a) The characteristic polynomial of $M_{\Phi,hor}$ and $M_{\Phi,ver}$, acting on $H = H_1(F_\Phi, \mathbb{C})$, are*

$$P_{M_{\Phi,hor}}(t) = (t-1)\cdot \prod_{w\in\mathscr{W}(\Gamma_{\mathscr{C}})} (t^{m_w}-1)^{\nu_w(2g_w+\delta_w-2)},$$

$$P_{M_{\Phi,ver}}(t) = (t-1)\cdot \prod_{w\in\mathscr{W}(\Gamma_{\mathscr{C}})} (t^{m_w\nu_w/(m_w,n_w)}-1)^{(m_w,n_w)(2g_w+\delta_w-2)}.$$

*The characteristic polynomial of the restriction of $M_{\Phi,hor}$ on the generalized eigenspace $H_1(F_\Phi, \mathbb{C})_{M_{\Phi,ver},1}$ is*

$$P_{M_{\Phi,hor}|H_{M_{\Phi,ver},1}}(t) = (t-1)\cdot \prod_{w\in\mathscr{W}(\Gamma_{\mathscr{C}})} (t^{(m_w,n_w)}-1)^{2g_w+\delta_w-2}.$$

*(b) There are similar formulae for the operators acting on $H = H_1(F_\Phi \cap T_j)$:*

$$P_{M^\Phi_{j,hor}}(t) = (t-1)^{d_j}\cdot \prod_{w\in\mathscr{W}(\Gamma^2_{\mathscr{C},j})} (t^{m_w}-1)^{\nu_w(\delta_w-2)},$$

$$P_{M^\Phi_{j,ver}}(t) = (t^{d_j}-1)\cdot \prod_{w\in\mathscr{W}(\Gamma^2_{\mathscr{C},j})} (t^{m_w\nu_w/(m_w,n_w)}-1)^{(m_w,n_w)(\delta_w-2)},$$

$$P_{M^\Phi_{j,hor}|H_{M^\Phi_{j,ver},1}}(t) = (t-1)\cdot \prod_{w\in\mathscr{W}(\Gamma^2_{\mathscr{C},j})} (t^{(m_w,n_w)}-1)^{\delta_w-2}.$$

*(c) Finally, the characteristic polynomials of the local horizontal/vertical monodromies acting on $H = H_1(F'_j, \mathbb{C})$ are*

$$P_{M'_{j,hor}}(t) = (t-1)\cdot \prod_{w\in\mathscr{W}(\Gamma^2_{\mathscr{C},j})} (t^{m_w}-1)^{\nu_w(\delta_w-2)/d_j},$$

$$P_{M'_{j,ver}}(t) = (t-1)\cdot \prod_{w\in\mathscr{W}(\Gamma^2_{\mathscr{C},j})} (t^{m_w\nu_w/d_j(m_w,n_w)}-1)^{(m_w,n_w)(\delta_w-2)},$$

$$P_{M'_{j,hor}|H_{M'_{j,ver},1}}(t) = (t-1)\cdot \prod_{w\in\mathscr{W}(\Gamma^2_{\mathscr{C},j})} (t^{(m_w,n_w)}-1)^{\delta_w-2}.$$

*Notice that the output of the right hand sides of the formulas in (c) (a posteriori) should be independent of the choice of the germ g, since the left hand sides depend only on the germ f.*

*Notice also that* $P_{M^{\Phi}_{j,hor}|H_{M^{\Phi}_{j,ver},1}}(t) = P_{M'_{j,hor}|H_{M'_{j,ver},1}}(t)$.

**Remark 13.4.12.** The above formulas from 13.4.6 and 13.4.11 are valid even if $\Gamma_{\mathscr{C}}$ does not satisfy Assumption A. In this case, it might happen that $\mathscr{W}(\Gamma^2_{\mathscr{C},j}) = \emptyset$, see e.g. 8.1.5. In such a situation, by convention, $\sum_{\mathscr{W}(\Gamma^2_{\mathscr{C},j})} = 0$ and $\prod_{\mathscr{W}(\Gamma^2_{\mathscr{C},j})} = 1$.

## 13.5 Examples

**13.5.1.** Assume that $f = x^3y^7 - z^4$; see 6.2.9 for a graph $\Gamma_{\mathscr{C}}$ with $g = x + y + z$. Then, by 13.4.11(c), the characteristic polynomials of the two vertical monodromies $M'_{j,ver}$ $(j = 1, 2)$ are $(t^7-1)^3/(t-1)^3$ corresponding to the transversal type $y^7 - z^4$, and $(t^3 - 1)^3/(t - 1)^3$ corresponding to the transversal type $x^3 - z^4$. This can also be verified geometrically in an elementary way: by the Thom–Sebastiani theorem (see [115]), in the first case $F'_j$ homotopically is the join of 7 points with 4 points. Analyzing the equation of $f$ we get that the vertical monodromy is the join of the cyclic permutation of the 7 points with the trivial permutation of the 4 points. A similar geometric description is valid for the second case as well.

In particular, in this case, these vertical operators have no eigenvalue 1.

**13.5.2.** If $f = y^3 + (x^2 - z^4)^2$ (see 6.2.8 for $\Gamma_{\mathscr{C}}$), then the transversal type is $A_2$ and $M'_{1,ver}$ has characteristic polynomial $(t - 1)^2$. Since the eigenvalues of the commuting operator $M'_{1,hor}$ are distinct, $M'_{1,ver}$ is the identity.

**13.5.3.** If $f = x^a + y^2 + xyz$ $(a = 3$ or $5)$, cf. 9.4.8 and 9.4.9, then $s = 1$, the transversal type is $A_1$, and $Div'_1 = (1, 1)$.

**13.5.4.** If $f = x^2y^2 + z^2(x + y)$, or $f = x^2y + z^2$, cf. 6.2.7 and 9.3.1, then again each transversal type is $A_1$, but $Div'_j = (1, -1)$.

**Remark 13.5.5.** Assume that $f$ is homogeneous of degree $d$. For $g$ a generic linear function, $\Gamma_{\mathscr{C}}$ was constructed in Chap. 8. This says that any vertex has a decoration of type $(m; d, 1)$. Moreover, by (13.3.6),

$$\Lambda(m; d, 1) = \sum_{\lambda^m=1} (\lambda, \lambda^{-d}).$$

Therefore, the statement of 13.4.6 is compatible with $M_{\Phi,ver} = (M_{\Phi,hor})^{-d}$, or $M^{\Phi}_{j,ver} = (M^{\Phi}_{j,hor})^{-d}$ already mentioned. See also 19.1 for more details.

## 13.6 Vertical Monodromies and the Graph $G$

**13.6.1.** As we already explained, we are primarily interested in the generalized eigenspaces of the vertical monodromies corresponding to eigenvalue one. Section 13.4 provides their ranks in terms of $\Gamma_{\mathscr{C}}$. In this section we compute them in terms of the *combinatorics of the plumbing graph* $G$.

The reader is invited to recall the definition of the graphs $G$ and $G_{2,j}$, the plumbing graphs of $(\partial F, V_g)$ and $\partial_{2,j} F$, which were introduced in 10.2.10 and 10.3.4, and are kept *unmodified by plumbing calculus*.

For any graph $Gr$ with arrowheads $\mathscr{A}(Gr)$ and non-arrowheads $\mathscr{W}(Gr)$, and where the arrowheads are supported by usual or dash-edges, we also define $\mathscr{E}_{\mathscr{W}}(Gr)$ as the set of edges connecting non-arrowhead vertices. Recall that $c(Gr)$ denotes the number of independent cycles in $Gr$ and $g(Gr)$ the sum of the genus decorations of $Gr$. These numbers, clearly, are not independent. For example, if $Gr$ is connected, then by an Euler-characteristic argument:

$$1 - c(Gr) = |\mathscr{W}(Gr)| - |\mathscr{E}_{\mathscr{W}}(Gr)|. \tag{13.6.2}$$

**Example 13.6.3.** Let $G$ be one of the output graphs of the Main Algorithm 10.2. In order to determine $c(G)$ and $g(G)$, for each $w \in \mathscr{W}(\Gamma_{\mathscr{C}})$ we will rewrite the decorations $m, n, n_1, \dots, n_s, m_1, \dots, m_t$ used in the Main Algorithm as $m_w, n_w, n_{w,1}, \dots, n_{w,s}, m_{w,1}, \dots, m_{w,t}$. Then, using the formulae of 10.2, we have the following expressions in terms of $\Gamma_{\mathscr{C}}$ for the cardinalities $|\mathscr{A}(G)|$, $|\mathscr{W}(G)|$, and $|\mathscr{E}_{\mathscr{W}}(G)|$ of the corresponding sets associated with $G$:

$$\begin{aligned}
|\mathscr{A}(G)| &= |\mathscr{A}(\Gamma_{\mathscr{C}})|,\\
|\mathscr{W}(G)| &= \sum_{w \in \mathscr{W}(\Gamma_{\mathscr{C}})} \gcd(m_w, n_w, n_{w,1}, \dots, n_{w,s}, m_{w,1}, \dots, m_{w,t}),\\
2|\mathscr{E}_{\mathscr{W}}(G)| + |\mathscr{A}(G)| &= \sum_{w \in \mathscr{W}(\Gamma_{\mathscr{C}})} \Big( \sum_i \gcd(m_w, n_w, n_{w,i}) + \sum_j \gcd(m_w, n_w, m_{w,j}) \Big).
\end{aligned}$$

Moreover,

$$2g(G) = \sum_{w \in \mathscr{W}(\Gamma_{\mathscr{C}})} (2g_w + \delta_w - 2)\gcd(m_w, n_w) + 2|\mathscr{W}(G)| - 2|\mathscr{E}_w(G)| - |\mathscr{A}(G)|;$$

and $c(G)$ also follows via (13.6.2).

**Remark 13.6.4.** One can verify that, in general,

$$g(G) \geq g(\Gamma_{\mathscr{C}}) \quad \text{and} \quad c(G) \geq c(\Gamma_{\mathscr{C}}).$$

Compare with Remark 5.1.7 above or with [86, (3.11)].

**Proposition 13.6.5.** *The ranks of the generalized eigenspaces* $H_1(F_\Phi,\mathbb{C})_{M_{\Phi,ver},1}$ *and* $H_1(F'_j,\mathbb{C})_{M'_{j,ver},1}$ *satisfy the following identities:*

$$\dim\ H_1(F_\Phi,\mathbb{C})_{M_{\Phi,ver},1} = 2g(G) + 2c(G) + |\mathscr{A}(G)| - 1. \tag{13.6.6}$$

$$\dim\ H_1(F'_j,\mathbb{C})_{M'_{j,ver},1} = 2g(G_{2,j}) + 2c(G_{2,j}) + |\mathscr{A}(G_{2,j})| - 1. \tag{13.6.7}$$

*Notice also that by 13.4.11, one also has the identity*

$$\dim H_1(F_\Phi \cap T_j,\mathbb{C})_{M^\Phi_{j,ver},1} = \dim H_1(F'_j,\mathbb{C})_{M'_{j,ver},1}. \tag{13.6.8}$$

Above, $|\mathscr{A}(G_{2,j})|$ is the number of arrowheads of $G_{2,j}$, which, in fact, are all dash-arrows. Their number is equal to $|\mathscr{E}_{cut,j}|$, the number of "cutting edges" adjacent to $\Gamma^2_{\mathscr{C},j}$.

*Proof.* By the third formula of 13.4.11(a) one has

$$\dim\ H_1(F_\Phi)_{M_{\Phi,ver},1} = 1 + \sum_{w\in\mathscr{W}(\Gamma_\mathscr{C})} \gcd(m_w,n_w)\cdot(2g_w+\delta_w-2).$$

Then use the identities of 13.6.3 and (13.6.2). The proof of the second identity is similar. □

**Remark 13.6.9.** Although, the graph $\Gamma^2_{\mathscr{C},j}$ is a tree whose vertices have zero genus-decorations (cf. 7.4.12), in general, both $g(G_{2,j})$ and $c(G_{2,j})$ can be non-zero.

Consider for example a line arrangement with $d$ lines and its graph $\Gamma_\mathscr{C}$ as in 8.2. Fix an intersection point $j \in \Pi$ contained in $m_j$ lines, and consider its corresponding vertex $v_j$ in $\Gamma_\mathscr{C}$. Above $v_j$ there is only one vertex in $G$, whose genus $\widetilde{g}_j$ via (10.2.4) is

$$2-2\widetilde{g}_j = (2-m_j)\cdot\gcd(m_j,d)+m_j.$$

Therefore, $\widetilde{g}_j = 0$ if and only if $(m_j-2)\cdot(\gcd(m_j,d)-1) = 0$; hence $\widetilde{g}_j$ typically is not zero.

An example when $c(G_{2,j}) \neq 0$ is provided in 19.4.5 for $d$ even.

It might happen that $G_1$ and $G_2$ have no cycles, while $G$ does; for such examples see 9.4.8 or 9.4.9.

**13.6.10.** Proposition 13.6.5 points out an important fact. Although the graph $\Gamma_\mathscr{C}$ depends on the choice of the resolution $r$ in 6.1, hence the graphs $G$ inherit this dependency as well, certain numerical invariants of $G$, describing geometrical invariants of the original germ $f$ or of the pair $(f,g)$, are independent of this ambiguity. Sometimes, even more surprisingly, the role of $g$ is irrelevant too. Here is the start of the list of such numerical invariants:

- $|\mathscr{A}(G)|$= the number of irreducible components of $V_f \cap V_g$;
- $|\mathscr{A}(G_{2,j})|$= the number of gluing tori of $\partial_{2,j}F$, cf. 10.3.6 (independent of $g$);

- $g(G) + c(G)$ is an invariant of $(f, g)$ by (13.6.6);
- $g(G_{2,j}) + c(G_{2,j})$ is an invariant of $f$ (independent of $g$) by (13.6.7).

For the continuation of the list and more comments on $g(G)$ and $c(G)$, see 15.1.7 and 15.2.

Note that all these invariants are stable with respect to the *reduced* oriented plumbing calculus (in fact, the definition of the reduced set of operations relies exactly on this observation). In particular, we also have:

**Corollary 13.6.11.** *The statements of Proposition 13.6.5 are valid for any graph $G^m$ with $G^m \sim G$, and for any $G^m_{2,j}$ with $G^m_{2,j} \sim G_{2,j}$ respectively. In particular, for $\widehat{G}$ and $\widehat{G}_{2,j}$ too.*

# Chapter 14
# The Algebraic Monodromy of $H_1(\partial F)$: Starting Point

Let us fix again an ICIS $(f, g)$.

In order to determine the characteristic polynomial of the Milnor monodromy acting on $H_1(\partial F)$ we need to understand two key geometrical objects:

$$\text{the pair } (\partial F, \partial F \setminus V_g) \text{ and the fibration } \arg(g) : \partial F \setminus V_g \to S^1.$$

The first pair compares the homology of $\partial F$ and $\partial F \setminus V_g$. Then, from the fibration we can try to determine the cohomology of $\partial F \setminus V_g$. This discussion will run through several chapters. It ends with the complete description in the most significant cases, however, the program will be obstructed in the general case.

The main reason for the lack of a complete general description lies in the fact that for the variation map involved, the "uniform twist property", usually valid in *complex* geometry, is not valid in the present *real analytic* situation. Here for edges with different decorations we glue together pieces with different orientations; for technical details see 17.1.

## 14.1 The Pair $(\partial F, \partial F \setminus V_g)$

From the long homological exact sequence of the pair $(\partial F, \partial F \setminus V_g)$ we get

$$\longrightarrow H_2(\partial F, \partial F \setminus V_g) \xrightarrow{\partial} H_1(\partial F \setminus V_g) \longrightarrow H_1(\partial F) \to 0. \tag{14.1.1}$$

By excision,

$$H_2(\partial F, \partial F \setminus V_g) = H_0(\partial F \cap V_g) \otimes H_2(D, \partial D),$$

where $D$ is a real 2-disc, hence it is free of rank $|\mathscr{A}(G)| = |\mathscr{A}(\Gamma_{\mathscr{C}})|$, the number of components of $\partial F \cap V_g$. Since the monodromy is trivial in a neighbourhood of $\partial F \cap V_g$, cf. Theorem 3.2.2 and Proposition 3.3.1, we get:

A. Némethi and Á. Szilárd, *Milnor Fiber Boundary of a Non-isolated Surface Singularity*, Lecture Notes in Mathematics 2037, DOI 10.1007/978-3-642-23647-1_14,

**Lemma 14.1.2.** *The characteristic polynomials of the restrictions of the Milnor monodromy action of $F$ on $\partial F$ and on $\partial F \setminus V_g$ satisfy*

$$P_{H_1(\partial F),M}(t) = P_{H_1(\partial F \setminus V_g),M}(t) \cdot (t-1)^{-\text{rank im }\partial},$$

*where*

$$1 \leq \text{rank im }\partial \leq |\mathscr{A}(G)|. \qquad (14.1.3)$$

Here, the first inequality follows from the fact that $H_1(\partial F) \neq H_1(\partial F \setminus V_g)$. Indeed, consider a component of $\partial F \cap V_g$ and a small loop around it in $\partial F$. Then its homology class in $H_1(\partial F)$ is zero, but it is sent into $\pm 1 \in H_1(S^1)$ by $\arg(g)_*$, and hence it is non-zero in $H_1(\partial F \setminus V_g)$.

At this generality, it is impossible to say more about rank im $\partial$: both bounds in (14.1.3) are sharp. For example, one can prove (see Chap. 19), that if $f$ is homogeneous of degree $d$, and the projective curve $C = \{f = 0\}$ has $|\Lambda|$ irreducible components, then rank im $\partial = d - |\Lambda| + 1$ and $d = |\mathscr{A}(G)|$. Therefore, in the case of arrangements rank im $\partial = 1$, and if $C$ is irreducible then rank im $\partial = d = |\mathscr{A}(G)|$.

## 14.2 The Fibrations arg($g$)

By Theorem 3.2.2, the fibration $\arg(g) : \partial F \setminus V_g \to S^1$ is equivalent to the fibration $\Phi^{-1}(\partial D_\delta) \to \partial D_\delta$ with fiber $F_\Phi$ and monodromy $m_{\Phi,ver}$. Furthermore, the Milnor monodromy on $\partial F \setminus V_g$ is identified with the induced monodromy by $m_{\Phi,hor}$. Therefore, from the Wang exact sequence of this second fibration

$$H_1(F_\Phi) \xrightarrow{M_{\Phi,ver}-I} H_1(F_\Phi) \longrightarrow H_1(\partial F \setminus V_g) \longrightarrow H_0(F_\Phi) = \mathbb{Z} \to 0 \quad (14.2.1)$$

we get

$$P_{H_1(\partial F \setminus V_g),M}(t) = (t-1) \cdot P_{M_{\Phi,hor}|\text{coker}\,(M_{\Phi,ver}-I)}(t). \qquad (14.2.2)$$

This formula can be "localized" around the singular locus: the Wang exact sequence of the fibration $\partial_{2,j} F \to S^1$ provides

$$P_{H_1(\partial_{2,j} F),M}(t) = (t-1) \cdot P_{M^\Phi_{j,hor}|\text{coker}\,(M^\Phi_{j,ver}-I)}(t). \qquad (14.2.3)$$

Since the horizontal/Milnor monodromy on $\partial_1 F$ is trivial, the left hand side of (14.2.2) differs from the product (over $j$) of the left hand side of (14.2.3) only by a factor of type $(t-1)^N$. This, of course, is true for the right hand sides too: for some $N \in \mathbb{Z}$ one has

$$P_{M_{\Phi,hor}|\text{coker}\,(M_{\Phi,ver}-I)}(t) = (t-1)^N \cdot \prod_j P_{M^\Phi_{j,hor}|\text{coker}\,(M^\Phi_{j,ver}-I)}(t). \qquad (14.2.4)$$

On the other hand, by Proposition 3.3.1, $(H_1(\partial_{2,j}F), M^{\Phi}_{j,hor}, M^{\Phi}_{j,ver})$ is the "$d$-covering" of $(H_1(F'_j), M'_{j,hor}, M'_{j,ver})$. Therefore, by Lemma 13.3.11, the generalized 1-eigenspaces of their vertical monodromies can be identified:

$$(H_1(\partial_{2,j}F)_{M^{\Phi}_{j,ver},1}, M^{\Phi}_{j,hor}, M^{\Phi}_{j,ver}) = (H_1(F'_j)_{M'_{j,ver},1}, M'_{j,hor}, M'_{j,ver}). \quad (14.2.5)$$

In particular,

$$P_{M^{\Phi}_{j,hor}|\mathrm{coker}\,(M^{\Phi}_{j,ver}-I)}(t) = P_{M'_{j,hor}|\mathrm{coker}\,(M'_{j,ver}-I)}(t), \quad (14.2.6)$$

the second polynomial being computed at the level of the homology of the local transversal fiber $H_1(F'_j)$. Summing up, we get that

$$P_{H_1(\partial F),M}(t) = (t-1)^N \cdot \prod_j P_{M'_{j,hor}|\mathrm{coker}\,(M'_{j,ver}-I)}(t), \quad (14.2.7)$$

for some integer $N$. This shows clearly, that in order to determine the characteristic polynomial $P_{H_1(\partial F),M}$, we need to clarify the triplet $(H_1(F'_j)_{M'_{j,ver},1}, M'_{j,hor}, M'_{j,ver})$, and the rank of $H_1(\partial F)$ which will take care of the integer $N$ in (14.2.7).

**14.2.8.** It is more convenient to replace in the above expressions the coker of the operators by their corresponding images. Moreover, similarly as above, it is useful to study in parallel both "local" (i.e. the right hand side of (14.2.7)) and "global" (the right hand side of (14.2.2)) expressions.

Accordingly, we introduce the following polynomials:

**Definition 14.2.9.** *Let $P^{\#}(t)$ be the characteristic polynomial of $M_{\Phi,hor}$ induced on the image of $(M_{\Phi,ver} - I)$ on the generalized 1-eigenspace $H_1(F_\Phi)_{M_{\Phi,ver},1}$.*

*Similarly, for any $1 \le j \le s$, let $P^{\#}_j(t)$ be the characteristic polynomial of $M'_{j,hor}$ induced on the image of $(M'_{j,ver}-I)$ on the generalized 1-eigenspace $H_1(F'_j)_{M'_{j,ver},1}$.*

Clearly, $P^{\#}(t)$ has degree $\#^2_1 M_{\Phi,ver}$ while the degree of $P^{\#}_j(t)$ is $\#^2_1 M'_{j,ver}$.

Since the characteristic polynomials of the horizontal monodromies acting on $H_1(F_\Phi)_{M_{\Phi,ver},1}$ and $H_1(F'_j)_{M'_{j,ver},1}$ are determined in Corollary 13.4.11, the above facts give

**Lemma 14.2.10.**

$$P_{M_{\Phi,hor}|H_1(\partial F\setminus V_g)}(t) = \frac{(t-1)^2}{P^{\#}(t)} \cdot \prod_{w\in\mathscr{W}(\Gamma_{\mathscr{C}})} (t^{(m_w,n_w)}-1)^{2g_w+\delta_w-2}.$$

*Moreover, for any $j$, and for the horizontal monodromy of $\Phi$ induced on $H_1(\partial_{2,j}F)$*

$$P_{M_{\Phi,hor}|H_1(\partial_{2,j}F)}(t) = \frac{(t-1)^2}{P^{\#}_j(t)} \cdot \prod_{w\in\mathscr{W}(\Gamma^2_{\mathscr{C},j})} (t^{(m_w,n_w)}-1)^{\delta_w-2}.$$

Again, since the monodormy on $\partial_1 F$ is trivial, we have

$$P_{M_{\Phi,hor}|H_1(\partial F\setminus V_g)}(t) = \prod_j P_{M_{\Phi,hor}|H_1(\partial_{2,j}F)}(t) \text{ up to a factor } (t-1)^N,$$

and

$$m_w = 1 \text{ for } w \notin \bigcup_j \mathscr{W}(\Gamma^2_{\mathscr{C},j}).$$

Thus Lemma 14.2.10 implies

$$P^\#(t) = \prod_j P_j^\#(t) \text{ up to a factor } (t-1)^N. \quad (14.2.11)$$

Therefore, 14.2.10 and 14.1.2 combined gives that in order to determine $P_{H_1(\partial F),M}(t)$ one needs to find $P^\#(t)$ (or, equivalently, all $P_j^\#(t)$) and the rank of $H_1(\partial F)$.

In the next chapters we will treat these missing terms by different geometric methods.

**14.2.12. The size of the Jordan blocks.** By the above discussion we can now easily prove an addendum of Lemma 13.2.1 regarding the size of the Jordan blocks of different operators.

**Proposition 14.2.13.** *All the Jordan blocks of the monodromy operators acting on $H_1(\partial F \setminus V_g)$ and $H_1(\partial F)$ have size at most two. The number of Jordan blocks of size two of the monodromy acting on $H_1(\partial F \setminus V_g)$ agrees for any fixed eigenvalue with the number of size two Jordan blocks of $M_{\Phi,hor}$ acting on* coker $(M_{\Phi,ver} - I)$. *Moreover, this number is an upper bound for the number of Jordan blocks of size two of the monodromy acting on $H_1(\partial F)$ for any fixed eigenvalue.*

*Proof.* By the exact sequence (14.1.1) it is enough to prove the statement for $H_1(\partial F \setminus V_g)$. This homology group can be inserted in the Wang exact sequence (14.2.1). Since the size of the Jordan blocks of the monodromy acting on $H_1(F_\Phi)$ is at most two by 13.2.1, it is enough to show that the sequence (14.2.1) has an equivariant splitting. Note that the last surjection of (14.2.1) is the same as $\arg_* : H_1(\partial F \setminus V_g) \to \mathbb{Z}$, and the monodromy on $\mathbb{Z}$ acts trivially. Consider a component of $\partial F \cap V_g$, let $\gamma$ be a small oriented meridian around it in $\partial F$. Then the class of $\gamma$ is preserved by the Milnor monodromy (as being part of $\partial_1 F$) and $\arg_*([\gamma]) = 1$; hence such a splitting exists. The last two statements also follow from this discussion. □

**Remark 14.2.14.** In order to understand the homology of $\partial F$, one does not need any information regarding the generalized ($\lambda \neq 1$)-eigenspaces of $M_{\Phi,ver}$ and $\oplus_j M^\Phi_{j,ver}$, although they codify important information about the ICIS $\Phi$. This will be the subject of forthcoming research. Nevertheless, for $f$ homogeneous, we will determine the complete Jordan-block structure via the identities 19.1, see 19.5.

# Chapter 15
# The Ranks of $H_1(\partial F)$ and $H_1(\partial F \setminus V_g)$ via Plumbing

## 15.1 Plumbing Homology and Jordan Blocks

We start with general facts regarding the rank of the first homology group of plumbed 3-manifolds. The statements are known, at least for negative definite graphs; see Propositions 4.4.2 and 4.4.5, which serve as models for the next discussion. For simplicity, we will state the results for the 3-manifolds $\partial F$, $\partial F \setminus V_g$ and $\partial_{2,j} F$, that is for the graphs $G^m \sim G$ and $G^m_{2,j} \sim G_{2,j}$ (cf. 10.2.10 and 10.3.4), although they are valid for any plumbed 3-manifold.

In the next definition, $Gr$ denotes either the graph $G^m$ or $G^m_{2,j}$ (or any other graph with similar decorations, and with two types of vertices: non-arrowheads $\mathscr{W}$ and arrowheads $\mathscr{A}$).

Recall from 4.1.7 that $A$ denotes the *intersection matrix* of $Gr$, $\mathfrak{I}$ the *incidence matrix* of the arrows of $Gr$, and $(A, \mathfrak{I})$ is the block matrix of size $|\mathscr{W}| \times (|\mathscr{W}| + |\mathscr{A}|)$.

**Definition 15.1.1.** *Set*

$$\operatorname{corank} A_{Gr} := |\mathscr{W}| - \operatorname{rank} A \quad \textit{and} \quad \operatorname{corank}(A, \mathfrak{I})_{Gr} := |\mathscr{W}| + |\mathscr{A}| - \operatorname{rank}(A, \mathfrak{I}).$$

Note that if a graph $Gr$ has some dash-arrows (like $G_{2,j}$), then the Euler number of the non-arrowhead supporting such dash-arrows is not well-defined; hence $\operatorname{rank} A_{Gr}$ is not well-defined either. Nevertheless, $\operatorname{rank}(A, \mathfrak{I})_{Gr}$ is well-defined even for such graphs.

**Lemma 15.1.2.** *For any $G^m \sim G$ and $G^m_{2,j} \sim G_{2,j}$ one has*

$$\operatorname{rank} H_1(\partial F) = 2g(G^m) + c(G^m) + \operatorname{corank} A_{G^m},$$

$$\operatorname{rank} H_1(\partial F \setminus V_g) = 2g(G^m) + c(G^m) + \operatorname{corank}(A, \mathfrak{I})_{G^m},$$

$$\operatorname{rank} H_1(\partial_{2,j} F) = 2g(G^m_{2,j}) + c(G^m_{2,j}) + \operatorname{corank}(A, \mathfrak{I})_{G^m_{2,j}} \quad (1 \le j \le s).$$

A. Némethi and Á. Szilárd, *Milnor Fiber Boundary of a Non-isolated Surface Singularity*, Lecture Notes in Mathematics 2037, DOI 10.1007/978-3-642-23647-1_15,

*In particular, for* rank im $\partial$ *from Lemma 14.1.2 one has*

$$\operatorname{rank}\operatorname{im}\partial = \operatorname{corank}(A,\mathfrak{I})_{G^m} - \operatorname{corank} A_{G^m}.$$

*Proof.* Let $P$ be the plumbed 4-manifold associated with a plumbing graph $Gr$ obtained by plumbing disc-bundles, cf. 4.1.4. Assume that the arrowheads of $Gr$ represent the link $K \subset \partial P$. Consider the homology exact sequence of the pair $(P, \partial P \setminus K)$:

$$H_2(P) \xrightarrow{i} H_2(P, \partial P \setminus K) \longrightarrow H_1(\partial P \setminus K) \longrightarrow H_1(P) \longrightarrow H_1(P, \partial P \setminus K)$$

Notice that $H_1(P, \partial P \setminus K) = H^3(P, K) = 0$, while $H_2(P, \partial P \setminus K) = H^2(P, K) = H^2(P) \oplus H^1(K)$ (since $K \hookrightarrow P$ is homotopically trivial). Moreover, the morphism $i$ can be identified with $(A, \mathfrak{I})$. Since the rank of $H_1(P)$ is $2g(Gr) + c(Gr)$, the second identity follows. Taking $K = \mathscr{A} = \emptyset$ in this argument, we get the first identity. The last identity follows similarly. □

**Remark 15.1.3.** Since $H_1(P, \mathbb{Z})$ is free of rank $2g + c$, the same argument over $\mathbb{Z}$ shows that $H_1(\partial P, \mathbb{Z}) = \mathbb{Z}^{2g+c} \oplus \operatorname{coker}(A)$. The point is that coker $(A)$ usually has a $\mathbb{Z}$-torsion summand, as it is shown by many examples of the present work, see for example 19.2.1.

Lemma 15.1.2 via the identities (14.2.2), (14.2.3) and (14.2.6) reads as

**Corollary 15.1.4.**

$$\dim\operatorname{coker}(M_{\Phi,ver} - I) = 2g(G^m) + c(G^m) + \operatorname{corank}(A,\mathfrak{I})_{G^m} - 1,$$
$$\dim\operatorname{coker}(M'_{j,ver} - I) = 2g(G^m_{2,j}) + c(G^m_{2,j}) + \operatorname{corank}(A,\mathfrak{I})_{G^m_{2,j}} - 1 \ (1 \le j \le s).$$

This combined with 13.6.5 and 13.6.11 gives

**Corollary 15.1.5.**

$$\#_1^2 M_{\Phi,ver} = c(G^m) - \operatorname{corank}(A,\mathfrak{I})_{G^m} + |\mathscr{A}(G)|,$$
$$\#_1^2 M'_{j,ver} = c(G^m_{2,j}) - \operatorname{corank}(A,\mathfrak{I})_{G^m_{2,j}} + |\mathscr{E}_{cut,j}| \quad (1 \le j \le s).$$

**Remark 15.1.6.** Although $c(G)$ and $g(G)$ can be computed easily from the graphs $\Gamma_{\mathscr{C}}$ or $G$, for the ranks of the matrices $A$ and $(A, \mathfrak{I})$ the authors found no "easy" formula. Even in case of concrete examples their direct computation can be a challenge. These "global data" of the graphs resonates with the "global information" codified in the Jordan block structure of the vertical monodromies.

**Remark 15.1.7.** Clearly, the integers $g(Gr)$, $c(Gr)$ and corank $A_{Gr}$ might change under the reduced calculus (namely, under R4), see e.g. the construction of the "collapsing" algorithm 12.2, or the next typical example realized in Sect. 19.7(3c):

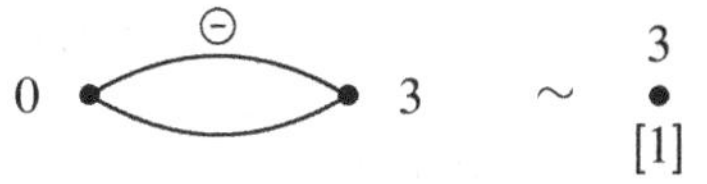


(oriented handle absorption)

For the first graph $A = \begin{bmatrix} 0 & 0 \\ 0 & 3 \end{bmatrix}$, hence corank $A = c = 1$ and $g = 0$, while for the second one corank $A = c = 0$ and $g = 1$.

On the other hand, the expressions $2g(G) + c(G) + \operatorname{corank} A_G$ and $2g(G) + c(G) + \operatorname{corank}(A, \mathfrak{J})_G$ are stable under the reduced plumbing calculus; in fact, they are stable even under all the operations of the oriented plumbing calculus. This fact together with 13.6.10 show that the following graph-expressions are independent of the construction of $G$ and are also *stable under the reduced calculus*:

- $|\mathscr{A}(G)|$, $g(G) + c(G)$, cf. 13.6.10;
- $g(G) + \operatorname{corank} A_G$ and $g(G) + \operatorname{corank}(A, \mathfrak{J})_G$ (from 15.1.2); in particular, $\operatorname{corank}(A, \mathfrak{J})_G - \operatorname{corank} A_G$ and $c(G) - \operatorname{corank} A_G$ as well;
- and all the corresponding expressions for $G_{2,j}$: $|\mathscr{A}(G_{2,j})|$, $g(G_{2,j}) + c(G_{2,j})$, $g(G_{2,j}) + \operatorname{corank}(A, \mathfrak{J})_{G_{2,j}}$.

## 15.2 Bounds for corank $A$ and corank $(A, \mathfrak{J})$

**15.2.1. Bounds for** $\operatorname{corank}(A, \mathfrak{J})_{G^m}$**.** From Corollary 15.1.5 and $|\mathscr{W}| \geq rank(A, \mathfrak{J})$, we get

$$|\mathscr{A}(G)| \leq \operatorname{corank}(A, \mathfrak{J})_{G^m} \leq c(G^m) + |\mathscr{A}(G)|. \tag{15.2.2}$$

These inequalities are sharp: the lower bound is realized for example in the case of homogeneous singularities (see 19.2.3), while the upper bound is realized in the case of cylinders, see (20.1.5). Decreasing $c(G^m)$ by (reduced) calculus, we decrease the difference between the two bounds as well.

**15.2.3. Bounds for** $\operatorname{corank} A_{G^m}$**.** By the last identity of Lemma 15.1.2 and the left inequality of (14.1.3) we get

$$\operatorname{corank} A_{G^m} \leq \operatorname{corank}(A, \mathfrak{J})_{G^m} - 1. \tag{15.2.4}$$

This and (15.2.2) imply

$$0 \leq \operatorname{corank} A_{G^m} \leq c(G^m) + |\mathscr{A}(G)| - 1. \tag{15.2.5}$$

Here, again, both bounds can be realized: the lower bound for $f$ irreducible homogeneous, cf. 19.3.7, while the upper bound for cylinders, see (20.1.5).

**Remark 15.2.6.** Finally, since corank $(A, \mathfrak{I})_{G^m} \geq |\mathscr{A}(G)|$, 15.1.5 implies

$$\#_1^2 M_{\Phi,ver} \leq c(G^m), \tag{15.2.7}$$

and, similarly, for any $j$

$$\#_1^2 M'_{j,ver} \leq c(G^m_{2,j}). \tag{15.2.8}$$

In particular, if we succeed to decrease $c(G^m)$ by reduced calculus, we get a better estimate for the number of Jordan blocks. In particular, $c(\widehat{G})$, in general, is a much better estimate than $c(G)$.

In the light of Theorem 3.2.2, the global inequality (15.2.7) reads as follows: Fix a germ $f$ with 1-dimensional singular locus, and choose $g$ such that $\Phi = (f, g)$ forms an ICIS. Consider the fibration $\arg(g) : \partial F \setminus V_g \to S^1$. Then (15.2.7) says that the number of 2-Jordan blocks with eigenvalue 1 of the algebraic monodromy of the fibration $\arg(g)$, *for any germ* $g$, is dominated by $c(G^m)$. The surprising factor here is that $G^m$ is a possible plumbing graph of $\partial F$, and $\partial F$ is definitely independent of the germ $g$.

It is instructive to compare the above identities and inequalities with similar statements valid in the world of complex geometry, see e.g. Remark 5.2.7(2).

# Chapter 16
# The Characteristic Polynomial of $\partial F$ Via $P^{\#}$ and $P^{\#}_j$

## 16.1 The Characteristic Polynomial of $G \to \Gamma_{\mathscr{C}}$ and $\widehat{G} \to \widehat{\Gamma_{\mathscr{C}}}$

**16.1.1.** In order to continue our discussion regarding the polynomials $P^{\#}$ and $P^{\#}_j$, $1 \le j \le s$ (cf. 14.2.9 and 14.2.10), we have to consider some natural "combinatorial" characteristic polynomials associated with the graph coverings involved.

Consider the cyclic graph covering $G \to \Gamma_{\mathscr{C}}$, cf. 10.2. Recall that above a vertex $w \in \mathscr{W}(\Gamma_{\mathscr{C}})$ there are $\mathfrak{n}_w$ vertices, while above an edge $e \in \mathscr{E}_w(\Gamma_{\mathscr{C}})$ there are exactly $\mathfrak{n}_e$ edges of $G$, where the integers $\mathfrak{n}_w$ and $\mathfrak{n}_e$ are given in (10.2.2) and (10.2.6) of the Main Algorithm. In particular, they can easily be read from the decorations of $\Gamma_{\mathscr{C}}$. The cyclic action on $G$ cyclically permutes the vertices situating above a fixed $w$ and the edges situating above a fixed $e$. Let $|G|$ be the topological realization (as a topological connected 1-complex) of the graph $G$. Then the action induces an operator, say $\mathfrak{h}(|G|)$, on $H_1(|G|, \mathbb{C})$.

Similarly, we consider the covering $\widehat{G} \to \widehat{\Gamma_{\mathscr{C}}}$ from 12.2. The vertices of $\widehat{\Gamma_{\mathscr{C}}}$ are the contracted subtrees $[\Gamma_{van}]$; and above $[\Gamma_{van}]$, in $\widehat{G}$ there are exactly $\mathfrak{n}_{\Gamma_{van}}$ vertices. Those edges of $\widehat{\Gamma_{\mathscr{C}}}$ which do not support arrowheads are inherited from those edges $\mathscr{E}^*_{\mathscr{W}}$ of $\Gamma_{\mathscr{C}}$ which connect non-arrowheads and are not vanishing 2-edges. Each of them is covered by $\mathfrak{n}_e$ edges. The cyclic action induces the operator $\mathfrak{h}(|\widehat{G}|)$ on $H_1(|\widehat{G}|, \mathbb{C})$.

**Definition 16.1.2.** *We denote the characteristic polynomial of* $\mathfrak{h}(|G|)$ *and of* $\mathfrak{h}(|\widehat{G}|)$ *by* $P_{\mathfrak{h}(|G|)}(t)$ *and* $P_{\mathfrak{h}(|\widehat{G}|)}(t)$ *respectively.*

By the connectivity of $G$ and $\widehat{G}$, and by the fact that the cyclic action acts trivially on $H_0(|G|) = H_0(|\widehat{G}|)$, we get

**Lemma 16.1.3.**

$$P_{\mathfrak{h}(|G|)}(t) = (t-1) \cdot \frac{\prod_{e \in \mathscr{E}_{\mathscr{W}}(\Gamma_{\mathscr{C}})} (t^{\mathfrak{n}_e} - 1)}{\prod_{w \in \mathscr{W}(\Gamma_{\mathscr{C}})} (t^{\mathfrak{n}_w} - 1)},$$

A. Némethi and Á. Szilárd, *Milnor Fiber Boundary of a Non-isolated Surface Singularity*, Lecture Notes in Mathematics 2037, DOI 10.1007/978-3-642-23647-1_16, 

$$P_{\mathfrak{h}(|\widehat{G}|)}(t) = (t-1)\cdot \frac{\prod_{e\in\mathscr{E}^*_{\mathscr{W}}(\Gamma_{\mathscr{C}})}(t^{\mathfrak{n}_e}-1)}{\prod_{\Gamma_{van}}(t^{\mathfrak{n}_{\Gamma_{van}}}-1)}.$$

**16.1.4.** We list some additional properties of these polynomials:

(a) Their degrees are rank $H_1(|G|) = c(G)$ and rank $H_1(|\widehat{G}|) = c(\widehat{G})$.
(b) Analyzing 12.1.5, one verifies the divisibility $P_{\mathfrak{h}(|\widehat{G}|)} \mid P_{\mathfrak{h}(|G|)}$.
(c) By Lemma 16.1.3, the multiplicity of the factor $(t-1)$ in $P_{\mathfrak{h}(|G|)}(t)$ is exactly $c(\Gamma_{\mathscr{C}})$. This fact remains true for $P_{\mathfrak{h}(|\widehat{G}|)}$ too, since each $\Gamma_{van}$ is a tree and $c(\Gamma_{\mathscr{C}}) = c(\widehat{\Gamma_{\mathscr{C}}})$. Hence 1 is not a root of the polynomial $P_{\mathfrak{h}(|G|)}/P_{\mathfrak{h}(|\widehat{G}|)}$.
(d) It might happen that $c(\widehat{G}) > c(\Gamma_{\mathscr{C}})$ (see for example 19.5.2). Hence $\mathfrak{h}(|\widehat{G}|)$ might have non-trivial eigenvalues.

**16.1.5.** Obviously, the above discussion can be "localized" above the graph $\Gamma^2_{\mathscr{C}}$. With the natural notations, we set

$$P_{\mathfrak{h}(|G_{2,j}|)}(t) = (t-1)\cdot \frac{\prod_{e\in\mathscr{E}_{\mathscr{W}}(\Gamma^2_{\mathscr{C},j})}(t^{\mathfrak{n}_e}-1)}{\prod_{w\in\mathscr{W}(\Gamma^2_{\mathscr{C},j})}(t^{\mathfrak{n}_w}-1)},$$

$$P_{\mathfrak{h}(|\widehat{G_{2,j}}|)}(t) = (t-1)\cdot \frac{\prod_{e\in\mathscr{E}^*_{\mathscr{W}}(\Gamma^2_{2,j})}(t^{\mathfrak{n}_e}-1)}{\prod_{\Gamma_{van}\subset\Gamma^2_{2,j}}(t^{\mathfrak{n}_{\Gamma_{van}}}-1)}.$$

In fact, the product over $j$ of these localized polynomials contains *all the non-trivial eigenvalues* of $P_{\mathfrak{h}(|G|)}$ and $P_{\mathfrak{h}(|\widehat{G}|)}$ respectively. Indeed, in the formulae of 16.1.3, if $w$ is a vertex of $\Gamma^1_{\mathscr{C}}$ then $\mathfrak{n}_w = 1$. Similarly, if $e$ is an edge with at least one of its end-vertices in $\Gamma^1_{\mathscr{C}}$, then $\mathfrak{n}_e = 1$ as well.

## 16.2 The Characteristic Polynomial of $\partial F$

**16.2.1.** In this section we compute the polynomials $P^\#$ and $P^\#_j$ under certain additional assumptions. Via the identities of 14.2.10 and 14.1.2 this is sufficient (together with the results of Chap. 15 regarding the rank $H_1(\partial F)$) to determine the characteristic polynomial of the Milnor monodromy of $H_1(\partial F)$.

Let $\Gamma_{\mathscr{C}}$ be the graph read from a resolution as in 6.1.2, which might have some vanishing 2-edges that are not yet eliminated by blow ups considered in 10.1.3.

**Definition 16.2.2.** *Let $Gr$ be either the graph $\Gamma_{\mathscr{C}}$, or $\Gamma^2_{\mathscr{C},j}$ for some $j$. We say that $Gr$ is* "unicolored", *if all its edges connecting non-arrowheads have the same sign-decoration and there are no vanishing 2-edges among them. We say that $Gr$ is* almost unicolored, *if those edges which connect non-arrowheads and are not vanishing 2-edges, have the same sign-decoration.*

Consider the polynomials $P_{\mathfrak{h}(|G|)}(t)$ and $P_{\mathfrak{h}(|G_{2,j}|)}(t)$ introduced in 16.1. Recall that their degrees are $c(G)$ and $c(G_{2,j})$ respectively.

**Theorem 16.2.3.** *(I) For any fixed $j$, the polynomial $P_j^{\#}(t)$ divides the polynomial $P_{\mathfrak{h}(|G_{2,j}|)}(t)$.*

*Moreover, the following statements are equivalent:*

$$\begin{aligned} &(a)\;\; P_j^{\#}(t) = P_{\mathfrak{h}(|G_{2,j}|)}(t)\\ &(b)\;\; \#_1^2 M'_{j,ver} = c(G_{2,j}) \\ &(c)\;\; \text{corank}\,(A,\mathfrak{I})_{G_{2,j}} = |\mathscr{E}_{cut,j}|. \end{aligned} \tag{16.2.4}$$

*These equalities hold in the following situations: either (i) $\Gamma^2_{\mathscr{C},j}$ is unicolored, or after determining $G_{2,j}$ via the Main Algorithm 10.2, the graph $G_{2,j}$ satisfies either (ii) $c(G_{2,j}) = 0$, or (iii)* corank $(A,\mathfrak{I})_{G_{2,j}} = |\mathscr{E}_{cut,j}|$.

*(II) The polynomial $P^{\#}(t)$ divides the polynomial $P_{\mathfrak{h}(|G|)}(t)$.*

*Moreover, the following statements are equivalent:*

$$\begin{aligned} &(a)\;\; P^{\#}(t) = P_{\mathfrak{h}(|G|)}(t)\\ &(b)\;\; \#_1^2 M_{\Phi,ver} = c(G) \\ &(c)\;\; corank(A,\mathfrak{I})_G = |\mathscr{A}(G)|. \end{aligned} \tag{16.2.5}$$

*These equalities hold in the following situations: either (i) $\Gamma_{\mathscr{C}}$ is unicolored, or after finding the graph $G$, it either satisfies (ii) $c(G) = 0$ or (iii)* corank $(A,\mathfrak{I}) = |\mathscr{A}(G)|$.

*But, even if (16.2.5) does not hold, one has*

$$P^{\#}(t) = P_{\mathfrak{h}(|G|)}(t) \;\; \textit{up to a multiplicative factor of type } (t-1)^N \tag{16.2.6}$$

*whenever (16.2.4) holds for all $j$.*

**Remark 16.2.7.** The equivalent statements (16.2.5) are satisfied e.g. by all homogeneous singularities (see 19.2.3), as well as by all cylinders, provided that the algebraic monodromy of the corresponding plane curve singularity is finite, see (20.1.5). On the other hand, they are not satisfied by those cylinders, which do not satisfy the above monodromy restriction. Nevertheless, their case will be covered by the "collapsing" version 16.2.13.

The proof of Theorem 16.2.3 is given in Chap. 17. The major application targets the characteristic polynomials of the monodromy acting on $H_1(\partial F)$:

**Theorem 16.2.8.** *Assume that (16.2.4) holds for all $j$, or (16.2.5) holds. Then*

$$P_{H_1(\partial F),M}(t) = \frac{(t-1)^{2+\text{corank}\,A_G - |\mathscr{A}(G)|}}{P_{\mathfrak{h}(|G|)}(t)} \cdot \prod_{w\in\mathscr{W}(\Gamma_{\mathscr{C}})} (t^{(m_w,n_w)}-1)^{2g_w+\delta_w-2}.$$

*In particular, one has the following formulae for the ranks of eigenspaces:*

$$\text{rank } H_1(\partial F)_1 = 2g(\Gamma_{\mathscr{C}}) + c(\Gamma_{\mathscr{C}}) + \operatorname{corank} A_G, \tag{16.2.9}$$

$$\text{rank } H_1(\partial F)_{\neq 1} = 2g(G) + c(G) - 2g(\Gamma_{\mathscr{C}}) - c(\Gamma_{\mathscr{C}}). \tag{16.2.10}$$

*More generally, in any situation (i.e. even if the above assumptions are not satisfied), there exists a polynomial $Q$ with $Q(1) \neq 0$, which divides both $P_{\mathfrak{h}(|G|)}$ and $\prod_{w \in \mathscr{W}(\Gamma_{\mathscr{C}})} (t^{(m_w, n_w)} - 1)^{2g_w + \delta_w - 2}$, such that*

$$P_{H_1(\partial F), M}(t) = \frac{(t-1)^N}{Q(t)} \cdot \prod_{w \in \mathscr{W}(\Gamma_{\mathscr{C}})} (t^{(m_w, n_w)} - 1)^{2g_w + \delta_w - 2},$$

*where* $N = 2 + \operatorname{corank} A_G - |\mathscr{A}(G)| - c(G) + \deg(Q)$.

*Proof.* Use 13.4.11, 15.1.2, 15.1.4 and 16.2.3. □

**Corollary 16.2.11.** *Assume that (16.2.4) holds for all $j$, or (16.2.5) holds. Then the following facts hold over coefficients in $\mathbb{C}$:*

*(a) The intersection matrix $A_G$ has a generalized eigen-decomposition $(A_G)_{\lambda=1} \oplus (A_G)_{\lambda\neq 1}$ induced by the Milnor monodromy, and $(A_G)_{\lambda \neq 1}$ is non-degenerate;*
*(b) There exist subspaces $K^i \subset H^i(\partial F)_1$ for $i = 1, 2$ with*

$$\operatorname{codim} K^1 = \dim K^2 = \operatorname{corank} A_G$$

*such that the cup-product $H^1(\partial F)_\lambda \cup H^1(\partial F)_\mu \to H^2(\partial F)_{\lambda\mu}$ has the following properties:*

*(1) $H^1(\partial F)_\lambda \cup H^1(\partial F)_\mu = 0$ for $1 \neq \lambda \neq \overline{\mu} \neq 1$;*
*(2) $\oplus_{\lambda \neq 1} H^1(\partial F)_\lambda \cup H^1(\partial F)_{\overline{\lambda}} \subset K^2$;*
*(3) $(\oplus_{\lambda \neq 1} H^1(\partial F)_\lambda) \cup K^1 = 0$;*
*(4) $K^1 \cup K^1 \subset K^2$;*
*(5) $K^1 \cup K^2 = 0$.*

*Proof.* We combine the proof of 15.1.2 with 11.9. Set $P_G := \overline{\mathscr{S}_k}$, and consider the cohomological long exact sequence associated with the pair $(P_G, \partial P_G)$. The map $H^2(P_G, \partial P_G) \to H^2(P_G)$ can be identified with $A_G$. The sequence has a generalized eigenspace decomposition. Define $K^i := \operatorname{im}[H^i(P_G)_1 \to H^i(\partial F)_1] \subset H^i(\partial F)_1$ for $i = 1, 2$. For $\lambda \neq 1$, via (16.2.10) and (11.9.2), we get that the inclusion $H^1(P_G)_{\lambda\neq 1} \to H^1(\partial P_G)_{\lambda \neq 1}$ is an isomorphism. Hence $(A_G)_{\lambda\neq 1}$ is non-degenerate.

For the second part, lift the classes of $H^1(\partial X)$ to $H^1(X)$ before multiplying them. (5) follows from $H^3(X) = 0$. (Cf. also with [129].) □

Such a result, in general, is the by-product of a structure-theorem regarding the mixed Hodge structure of the cohomology ring (that is, it is the consequence of the fact that $H^2$ has no such weight where the product would sit). Compare also with the comments from 18.1.9 and with some of the open problems from 24.4.

**16.2.12.** The next theorem, based on the "Collapsing Main Algorithm" and the corresponding improved version of the proof of Theorem 16.2.3, provides a better "estimate" in the general case, and in the special cases of *almost unicolored* graphs handles the presence of vanishing 2-edges as well.

Consider the polynomials $P_{\mathfrak{h}(|\widehat{G}|)}(t)$ and $P_{\mathfrak{h}(|\widehat{G_{2,j}}|)}(t)$ introduced in 16.1. Recall that their degrees are $c(\widehat{G})$ and $c(\widehat{G_{2,j}})$ respectively.

**Theorem 16.2.13.** *(I) For any fixed $j$, the polynomial $P_j^{\#}(t)$ divides the polynomial $P_{\mathfrak{h}(|\widehat{G_{2,j}}|)}(t)$.*

*Moreover, the following statements are equivalent:*

$$\begin{aligned}&(a)\;\; P_j^{\#}(t) = P_{\mathfrak{h}(|\widehat{G_{2,j}}|)}(t)\\ &(b)\;\; \#_1^2 M'_{j,ver} = c(\widehat{G_{2,j}})\\ &(c)\;\; \operatorname{corank}(A,\mathfrak{I})_{\widehat{G_{2,j}}} = |\mathscr{E}_{cut,j}|.\end{aligned} \tag{16.2.14}$$

*These equalities hold in the following situations: either (i) $\Gamma^2_{\mathscr{C},j}$ is almost unicolored, or after determining $\widehat{G}_{2,j}$ via the Collapsing Main Algorithm the graph $\widehat{G}_{2,j}$ satisfies either (ii) $c(\widehat{G_{2,j}})=0$, or (iii) $\operatorname{corank}(A,\mathfrak{I})_{\widehat{G_{2,j}}} = |\mathscr{E}_{cut,j}|$.*

*(II) The polynomial $P^{\#}(t)$ divides the polynomial $P_{\mathfrak{h}(|\widehat{G}|)}(t)$.*

*Moreover, the following statements are equivalent:*

$$\begin{aligned}&(a)\;\; P^{\#}(t) = P_{\mathfrak{h}(|\widehat{G}|)}(t)\\ &(b)\;\; \#_1^2 M_{\Phi,ver} = c(\widehat{G})\\ &(c)\;\; \operatorname{corank}(A,\mathfrak{I})_{\widehat{G}} = |\mathscr{A}(G)|.\end{aligned} \tag{16.2.15}$$

*These equalities hold in the following situations: either (i) $\Gamma_{\mathscr{C}}$ is almost unicolored, or after determining the graph $\widehat{G}$, it either satisfies (ii) $c(\widehat{G})=0$ or (iii) $\operatorname{corank}(A,\mathfrak{I})_{\widehat{G}} = |\mathscr{A}(G)|$.*

*But, even if (16.2.15) does not hold, one has*

$$P^{\#}(t) = P_{\mathfrak{h}(|\widehat{G}|)}(t) \text{ up to a multiplicative factor of type } (t-1)^N \tag{16.2.16}$$

*whenever (16.2.14) holds for all $j$.*

This implies:

**Theorem 16.2.17.** *Assume that (16.2.14) holds for all $j$, or (16.2.15) holds. Then*

$$P_{H_1(\partial F),M}(t) = \frac{(t-1)^{2+\operatorname{corank} A_{\widehat{G}} - |\mathscr{A}(G)|}}{P_{\mathfrak{h}(|\widehat{G}|)}(t)} \cdot \prod_{w\in\mathscr{W}(\Gamma_{\mathscr{C}})} \left(t^{(m_w,n_w)}-1\right)^{2g_w+\delta_w-2}$$

*In particular,*

$$\text{rank } H_1(\partial F)_1 = 2g(\Gamma_{\mathscr{C}}) + c(\Gamma_{\mathscr{C}}) + \operatorname{corank} A_{\widehat{G}}, \tag{16.2.18}$$

$$\text{rank } H_1(\partial F)_{\neq 1} = 2g(\widehat{G}) + c(\widehat{G}) - 2g(\Gamma_{\mathscr{C}}) - c(\Gamma_{\mathscr{C}}). \tag{16.2.19}$$

*More generally, in any situation (even if the above assumptions are not satisfied), there exists a polynomial $Q$ with $Q(1) \neq 0$, which divides both $P_{\mathfrak{h}(|\widehat{G}|)}$ and $\prod_{w\in\mathscr{W}(\Gamma_{\mathscr{C}})}(t^{(m_w,n_w)}-1)^{2g_w+\delta_w-2}$, such that*

$$P_{H_1(\partial F),M}(t) = \frac{(t-1)^N}{Q(t)} \cdot \prod_{w\in\mathscr{W}(\Gamma_{\mathscr{C}})} (t^{(m_w,n_w)}-1)^{2g_w+\delta_w-2},$$

*where* $N = 2 + \operatorname{corank} A_{\widehat{G}} - |\mathscr{A}(G)| - c(\widehat{G}) + \deg(Q)$.

**Corollary 16.2.20.** *Assume that (16.2.14) holds for all $j$, or (16.2.15) holds. Then the statement of Corollary 16.2.11 is valid provided that we replace G by $\widehat{G}$.*

# Chapter 17
# The Proof of the Characteristic Polynomial Formulae

## 17.1 Counting Jordan Blocks of Size 2

Our goal is to prove Theorem 16.2.3. The proof is based on a specific construction. The presentation is written for the graph $G$ (later adapted to $\widehat{G}$ as well), but it can be reformulated for $G_{2,j}$ as well.

**17.1.1. The vertical monodromy $m_{\Phi,ver}$ as a quasi-periodic action.** First, we wish to understand the geometric monodromy $m_{\Phi,ver} : F_\Phi \to F_\Phi$. For this the topology of fibered links, as it is described in [33, §13], will be our model. (Although, in [loc.cit.] the machinery is based on splice diagrams, Sect. 23 of [33] gives the necessary hints for plumbing graphs as well.)

Nevertheless, our situation is more complicated. First, in the present situation we will have *three* local types/contributions; two of them do not appear in the classical complex analytic case of [33]. Secondly, the basic property which is satisfied by analytic germs defined on normal surface singularities, namely that a "variation operator" has a uniform twist, in our situation is not true, it is ruined by the new local contributions.

Consider the graph $\Gamma_\mathscr{C}$ from Chap. 6. For simplicity we will write $\mathscr{E}$, $\mathscr{E}_\mathscr{W}$, $\mathscr{W}$, etc. for $\mathscr{E}(\Gamma_\mathscr{C})$, $\mathscr{E}_\mathscr{W}(\Gamma_\mathscr{C})$, $\mathscr{W}(\Gamma_\mathscr{C})$, etc. It is the dual graph of the curve configuration $\mathscr{C} \subset V^{emb}$, where $r : V^{emb} \to U$ is a representative of an embedded resolution of $(V_f \cup V_g, 0) \subset (\mathbb{C}^3, 0)$. Then, by 7.1.4, for any tubular neighbourhood $T(\mathscr{C}) \subset V^{emb}$ of $\mathscr{C}$, one has an inclusion $(\Phi \circ r)^{-1}(c_0, d_0)$ for $(c_0, d_0) \in \mathsf{W}_{\eta,M}$. For simplicity, we write $\widetilde{F}_\Phi$ for the diffeomorphically lifted fiber $(\Phi \circ r)^{-1}(c_0, d_0)$ of $\Phi$.

Next, consider the decomposition 7.1.6 of $\widetilde{F}_\Phi$. More precisely, for any intersection point $p \in C_v \cap C_u$ (or, self-intersection point of $C_v$ if $v = u$), which corresponds to the edge $e \in \mathscr{E}$ of $\Gamma_\mathscr{C}$, let $T(e)$ be a small closed ball centered at $p$. Let $T^\circ(e)$ be its interior. Then, for $(c_0, d_0)$ sufficiently close to the origin, $\widetilde{F}_\Phi \cap T(e)$ is a union of annuli. Moreover, $\widetilde{F}_\Phi \setminus \cup_e T^\circ(e)$ is a union $\cup_{v\in\mathscr{V}} \widetilde{F}_v$, where $\widetilde{F}_v$ is in a small tubular neighbourhood of $C_v$ $(v \in \mathscr{V})$.

A. Némethi and Á. Szilárd, *Milnor Fiber Boundary of a Non-isolated Surface Singularity*, Lecture Notes in Mathematics 2037, DOI 10.1007/978-3-642-23647-1_17, 

If $v$ is an arrowhead supported by an edge $e$, then the inclusion $\widetilde{F}_v \cap \partial T(e) \subset \widetilde{F}_v$ admits a strong deformation retract. Hence the pieces $\{\widetilde{F}_v\}_{v\in\mathscr{A}}$, and the separating annuli $\{\widetilde{F}_\Phi \cap T(e)\}_{e\in\mathscr{E}\setminus\mathscr{E}_\mathscr{W}}$ can be neglected. Thus, instead of $\widetilde{F}_\Phi$, we will consider only

$$\widetilde{F}^*_\Phi := \widetilde{F}_\Phi \setminus \Big( \bigcup_{e\in\mathscr{E}\setminus\mathscr{E}_\mathscr{W}} T(e) \cup \bigcup_{v\in\mathscr{A}} \widetilde{F}_v \Big).$$

In particular, $\widetilde{F}^*_\Phi$ is separated by the annuli $\{\widetilde{F}_\Phi \cap T(e)\}_{e\in\mathscr{E}_\mathscr{W}}$ in surfaces $\{\widetilde{F}_w\}_{w\in\mathscr{W}}$, and each $\widetilde{F}_w$ is the total space of a covering, where the base space is $C_w \setminus \cup_e T(e)$ and the fiber is isomorphic to $\mathscr{P}_w := \mathscr{P}$ from the Key Example 13.3.3.

Moreover, one might choose the horizontal/vertical monodromies of $\widetilde{F}^*_\Phi$ in such a way that they will preserve this decomposition, and their action on $\widetilde{F}_w$ will be induced by the permutations $\sigma_{w,hor} := \sigma_{hor}$, respectively $\sigma_{w,ver} := \sigma_{ver}$ acting on $\mathscr{P}_w$ (cf. 13.3.3). This shows that $m_{\Phi,ver}$ is isotopic to an action $\widetilde{m}_{\Phi,ver}$, which preserves the above decomposition, and its restriction on each $\widetilde{F}_w$ is finite. Let $q$ be a common multiple of the orders of $\{\sigma_{w,ver}\}_{w\in\mathscr{W}}$. Then $\widetilde{m}^q_{\Phi,ver}$ is the identity on each $\widetilde{F}_w$, and acts as a "twist map" on each separating annulus. In the topological characterization of $\widetilde{m}_{\Phi,ver}$, this twist is crucial.

**Definition 17.1.2.** *[33, §13] Let $h : A \to A$ be a homeomorphism of the oriented annulus $A = S^1 \times [0,1]$ with $h|\partial A = id$. The (algebraic) twist of $h$ is defined as the intersection number*

$$\text{twist}(h) := (x, \text{var}_h(x)),$$

*where $x \in H_1(A, \partial A, \mathbb{Z})$ is a generator, and the variation map* $\text{var}_h : H_1(A, \partial A, \mathbb{Z}) \to H_1(A, \mathbb{Z})$ *is defined by* $\text{var}_h([c]) = [h(c) - c]$, *for any relative cycle $c$.*

*More generally, if $B$ is a disjoint union of annuli and $h : B \to B$ is a homeomorphism with $h^q|\partial B = id$ for some integer $q > 0$, for any component $A$ of $B$ define*

$$\text{twist}(h; A) := \frac{1}{q}\, \text{twist}(h^q|A).$$

Notice that twist$(h; A)$, defined in this way, is independent of the choice of $q$.

**Example 17.1.3.** In the "classical" situation one considers an analytic family of curves (over a small disc), where the central fiber is a normal crossing divisor and the generic fiber is smooth. The generic fiber is cut by separating annuli, which are situated in the neighbourhood of the normal crossing intersection point of the central fiber. Around such a point $p$, in convenient local coordinates $(u, v) \in (\mathbb{C}^2, p)$, the family is given by the fibers of $f(u, v) = u^a v^b$ for two positive integers $a$ and $b$. Hence, the union of annuli is the fiber $f^{-1}(\epsilon)$ intersected with a small ball centered at $p$ (and $\epsilon$ is small with respect to the radius of this ball). It consists of gcd$(a, b)$ annuli. The Milnor monodromy action $h$ is induced by $[0, 2\pi] \ni t \mapsto f^{-1}(\epsilon e^{it})$. In order to make the computation, one must choose $h$ in such a way that its restriction on the boundary "near" the $x$-axis is finite of order $a$, and similarly, on the boundary components "near" the $y$-axis is finite of order $b$. Then, one can show (see e.g. [33, page 164]) that the twist, for each connected component $A$, is

$$\operatorname{twist}(h; A) = -\frac{\gcd(a,b)}{ab}.$$

**17.1.4.** In fact, in most of the forthcoming arguments, what is really important is not the *value* of the twist itself, but its *sign*.

**Definition 17.1.5.** *In a geometric situation as in 17.1.1, we say that we have a uniform twist, if for all the annuli the signs of all the twists are the same.*

**Example 17.1.6.** In our present situation of 17.1.1, we are interested in the twist of the separating annuli associated with the edges $e \in \mathcal{E}_{\mathcal{W}}$. Depending on whether the edge is of type 1 or 2, we have to consider two different situations. For both cases we will consider the local equations from Sect. 6.2.

If $e$ is a 1-edge, then the fiber (union of annuli) and the vertical monodromy action are given by

$$u^m v^n w^l = c, \quad v^\nu w^\lambda = de^{it} \text{ (with } (c,d) \text{ constant, and } t \in [0,2\pi]).$$

If $e$ is a 2-edge, then the fiber (union of annuli) and the vertical monodromy action are given by

$$u^m v^{m'} w^n = c, \quad w^\nu = de^{it} \text{ (with } (c,d) \text{ constant, and } t \in [0,2\pi]).$$

A 2-edge $e$ is a vanishing 2-edge if $n = 0$.

We invite the reader to compute the corresponding twists exactly, in both cases. In the present proof we need only the following statement.

**Lemma 17.1.7.** *Fix an edge $e \in \mathcal{E}_{\mathcal{W}}$, set $B := \widetilde{F}_\Phi \cap T(e)$, and let $A$ be one of the connected components of $B$.*

*(a) If $e$ is a 1-edge, then* $\operatorname{twist}(h; A) < 0$.
*(b) If $e$ is a non-vanishing 2-edge, then* $\operatorname{twist}(h; A) > 0$.
*(c) If $e$ is a vanishing 2-edge, then* $\operatorname{twist}(h; A) = 0$.

*Proof.* The first case behaves as a "covering of degree $m$" of the classical case $v^\nu w^\lambda = e^{it}$ ($t \in [0,2\pi]$), which was exemplified in 17.1.3. The monodromy of the second case behaves as the inverse of the monodromy of the classical monodromy operator: $u^m v^{m'} = e^{-nit/\nu}$ ($t \in [0,2\pi]$). Finally, assume that $n = 0$. Then the restriction of the vertical monodromy to $\partial B$ has order $\nu$, hence one can take $q = \nu$. But $h^\nu$ extends as the identity on the whole $B$. The details are left to the reader. □

**17.1.8.** Now we continue the proof of Theorem 16.2.3. The structure of the 2-Jordan blocks of $M_{\Phi,ver}$ is codified in the following commutative diagram, as given in [33, (14.2)]:

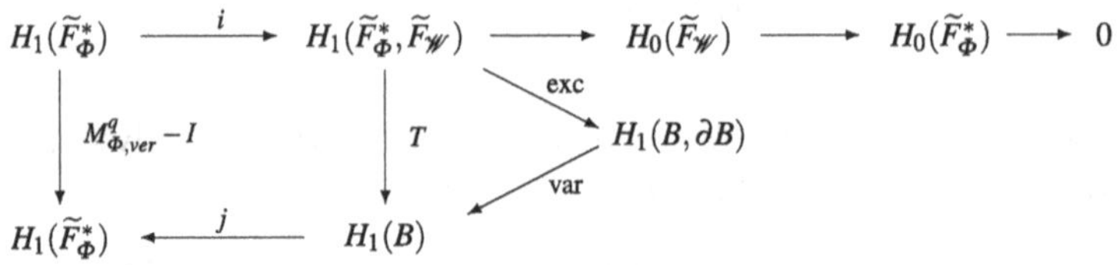
$H_1(\widetilde{F}^*_\Phi)$
$i$
$H_1(\widetilde{F}^*_\Phi, \widetilde{F}_\mathscr{W})$
$H_0(\widetilde{F}_\mathscr{W})$
$H_0(\widetilde{F}^*_\Phi)$
0
exc
$M^q_{\Phi,ver} - I$
$T$
$H_1(B, \partial B)$
var
$H_1(\widetilde{F}^*_\Phi)$
$j$
$H_1(B)$

Above we use the following notations:

$$\widetilde{F}_\mathscr{W} := \cup_{w \in \mathscr{W}} \widetilde{F}_w, \quad B := \cup_{e \in \mathscr{E}_\mathscr{W}} (\widetilde{F}^*_\Phi \cap T(e)),$$

- $q$ is a positive integer as in 17.1.1, hence $M^q_{\Phi,ver}$ is unipotent,
- var is the variation map associated with $\tilde{m}^q_{\Phi,ver}$,
- exc is the excision *isomorphism*, and $T$ is the composite var $\circ$ exc,
- the horizontal line is a homological exact sequence.

Since $B$ is the disjoint union of the separating annuli, var is a diagonal map. On the diagonal, each entry corresponds to an annulus $A$, and equals the integer

$$q \cdot \operatorname{twist}(m_{\Phi,ver}; A)$$

determined in 17.1.7.

Obviously, the number of *all* 2-Jordan blocks of $M_{\Phi,ver}$ is $\operatorname{rank} \operatorname{im}(M^q_{\Phi,ver} - I)$. On the other hand, by the commutativity of the diagram,

$$\operatorname{im}(M^q_{\Phi,ver} - I) \simeq j \circ T(\operatorname{im}(i)) \simeq \operatorname{im}(i)/\ker(j \circ T|\operatorname{im}(i)). \qquad (17.1.9)$$

*The point in (17.1.9) is that* $\operatorname{im}(M^q_{\Phi,ver} - I)$ *appears as a factor space of* $\operatorname{im}(i)$.

**17.1.10.** In some cases $\operatorname{im}(M^q_{\Phi,ver} - I)$ can be determined exactly.

Assume that $\Gamma_\mathscr{C}$ has no vanishing 2-edges, i.e. the case 17.1.7(c) does not occur. Then all diagonal entries of var are non-zero, hence both var and $T$ are isomorphisms. The next lemma is a direct consequence of [33, page 113]:

**Lemma 17.1.11.** *Assume that $\Gamma_\mathscr{C}$ is unicolored (cf. 16.2.2). Then the restriction of $j \circ T$ on* $\operatorname{im}(i)$ *is injective. In particular, if G is unicolored, then the number of all 2-Jordan blocks of $M_{\Phi,ver}$ is* $\operatorname{rank} \operatorname{im}(i)$.

*Proof.* For any $y, z \in H_1(\widetilde{F}_\Phi, \widetilde{F}_\mathscr{W})$, we consider the intersection number $(y, Tz)$, denoted by $\langle y, z\rangle$. Since $T$ is diagonal, and all entries on the diagonal have the same sign, the form $\langle y, z\rangle$ is definite. Assume that $j\,Ti(x) = 0$. Then

$$0 = (x, j\,Ti(x))_{H_1(\widetilde{F}^*_\Phi)} = (i(x), Ti(x)) = \langle i(x), i(x)\rangle,$$

hence $i(x) = 0$. □

## 17.2 Characters

The corresponding $\mathbb{Z}^2$-characters of $\mathrm{im}(i)$ can be determined by the following exact sequence (as part of diagram 17.1.8):

$$0 \longrightarrow \mathrm{im}(i) \longrightarrow H_1(\widetilde{F}^*_\Phi, \widetilde{F}_{\mathscr{W}}) \xrightarrow{\tilde{\partial}} H_0(\widetilde{F}_{\mathscr{W}}) \longrightarrow H_0(\widetilde{F}^*_\Phi) \longrightarrow 0,$$

and the action of the vertical/horizontal monodromies on this sequence. In this proof we only need those blocks of $M_{\Phi,ver}$ which have eigenvalue 1.

The action of the algebraic *vertical monodromy* on each term of this sequence is finite: it is induced by a permutation of the connected components of the spaces $\widetilde{F}_{\mathscr{W}}$ and $(B, \partial B)$. The corresponding 1-eigenspaces form the following exact sequence:

$$0 \longrightarrow \mathrm{im}(i)_{ver,1} \longrightarrow H_1(\widetilde{F}^*_\Phi, \widetilde{F}_{\mathscr{W}})_{ver,1} \xrightarrow{\tilde{\partial}} H_0(\widetilde{F}_{\mathscr{W}})_{ver,1} \longrightarrow H_0(\widetilde{F}^*_\Phi)_{ver,1} \longrightarrow 0. \tag{17.2.1}$$

This sequence will be compared with another sequence which computes the simplicial homology of the connected 1-complex $|G|$. Namely, one considers the free vector spaces $\mathbb{C}^{|\mathscr{E}_{\mathscr{W}}(G)|}$ and $\mathbb{C}^{|\mathscr{W}(G)|}$, generated by the edges $\mathscr{E}_{\mathscr{W}}(G)$ and vertices $\mathscr{W}(G)$ of $G$, as well as the boundary operator $\partial'$. Then one has the exact sequence

$$0 \longrightarrow H_1(|G|) \longrightarrow \mathbb{C}^{|\mathscr{E}_{\mathscr{W}}(G)|} \xrightarrow{\partial'} \mathbb{C}^{|\mathscr{W}(G)|} \longrightarrow H_0(|G|) \longrightarrow 0. \tag{17.2.2}$$

**Lemma 17.2.3.** *The two exact sequences (17.2.1) and (17.2.2) are isomorphic. Moreover, the horizontal monodromy acting on (17.2.1) can be identified with the action on (17.2.2) induced by the cyclic action of the covering $G \to \Gamma_{\mathscr{C}}$. (The cyclic action is induced by the positive generator of $\mathbb{Z}$, cf. 5.1.2.)*

*Proof.* It suffices to identify the second and the third terms together with the connecting morphisms, and their compatibility with the monodromy action. (In fact, the isomorphism of the last terms is trivial: $H_0(\widetilde{F}^*_\Phi)_{ver,1} = H_0(\widetilde{F}_\Phi)_{ver,1} = H_0(|G|) = \mathbb{C}$ is clear since $\widetilde{F}_\Phi$ and $|G|$ are connected.)

The identification follows from the proof of the Main Algorithm. Recall, that in the previous sections, we constructed a decomposition of $\widetilde{F}^*_\Phi$, the lifted fiber $(\Phi \circ r)^{-1}(c_0, d_0)$. Let us repeat the very same construction for $(\Phi \circ r)^{-1}(\partial D_{c_0})$, where $D_{c_0}$ is the disc $\{(c,d) : c = c_0\}$ as in Chap. 3, or in the proof of 11.3.3. That is, we decompose the total space of the fibration over $\partial D_{c_0} = S^1$ instead of only one fiber. Let $Tot(\widetilde{F}_{\mathscr{W}})$ be the space we get instead of $\widetilde{F}_{\mathscr{W}}$, which, in fact, is the total space of a fibration over $S^1$ with fiber $\widetilde{F}_{\mathscr{W}}$ and monodromy the vertical monodromy. Since the geometric vertical monodromy permutes the components of $\widetilde{F}_{\mathscr{W}}$, the algebraic vertical monodromy $M_{H_0(\widetilde{F}_{\mathscr{W}}),ver}$ acts finitely on $H_0(\widetilde{F}_{\mathscr{W}})$, and the $coker(M_{H_0(\widetilde{F}_{\mathscr{W}}),ver} - I)$ can be identified with $H_0(\widetilde{F}_{\mathscr{W}})_{ver,1}$. Hence, by the Wang exact sequence of the above fibration, we get that $H_0(\widetilde{F}_{\mathscr{W}})_{ver,1} = H_0(Tot(\widetilde{F}_{\mathscr{W}}))$. On the other hand, in the proof of 11.3.3 (complemented also with the second part of 3.2.2), $\Phi^{-1}(\partial D_{c0}) \cap B_\epsilon$ appears as a part of $\partial \widetilde{\mathscr{S}}_k$. Therefore, $Tot(\widetilde{F}_{\mathscr{W}})$ appears as a part of $\partial \widetilde{\mathscr{S}}_k$: that part which is situated in the neighbourhood of the regular part of the compact components of $\mathscr{C}$. Hence, by 11.6.1 and 11.8.1, $H_0(Tot(\widetilde{F}_{\mathscr{W}}))$ is freely generated by the collection of curves situating in the normalization $\mathscr{S}_k^{norm}$ of $\widetilde{\mathscr{S}}_k$ located above the compact components of $\mathscr{C}$. This is codified exactly by the non-arrowhead vertices of $G$.

For the second terms, first notice that by the isomorphism exc : $H_1(\widetilde{F}^*_\Phi, \widetilde{F}_{\mathscr{W}}) \to H_1(B, \partial B)$, the rank of $H_1(\widetilde{F}^*_\Phi, \widetilde{F}_{\mathscr{W}})$ counts the separating annuli of $\widetilde{F}^*_\Phi$. Now, we can repeat the entire construction and argument above for $\mathscr{E}_{\mathscr{W}}$ instead of $\mathscr{W}$. Indeed, $H_1(\widetilde{F}^*_\Phi, \widetilde{F}_{\mathscr{W}})_{ver,1}$ counts the separating tori (not considering those corresponding to the binding of the open book) of $(\Phi \circ r)^{-1}(\partial D_{c0})$, and this, by the proof of the Main Algorithm, is codified exactly by $\mathscr{E}_{\mathscr{W}}$. The details are left to the reader.

Since all the maps and identifications are natural and compatible with the action of the corresponding horizontal monodromies, the morphisms $\tilde{\partial}$ and $\partial'$ and the actions of the horizontal monodromies are all identified. □

Now, we are ready to finish the proof of 16.2.3. By the above discussion, $\#_1^2 M_{\Phi,ver}$ is equal to the dimension of $I := \mathrm{im}(i)_{ver,1}/\ker(j\, T|\mathrm{im}(i))_{ver,1}$, which is smaller than $\dim(i)_1 = \dim H_1(|G|) = c(G)$. Moreover, $P^\#$ is the characteristic polynomial of the horizontal monodromy acting on $I$, which clearly divides the characteristic polynomial of the horizontal monodromy acting on $H_1(|G|)$, which is $P_{\mathfrak{h}(|G|)}$. If $c(G) = 0$, or if $c(G) = \#_1^2 M_{\Phi,ver}$ (i.e., if corank$(A, \mathfrak{I}) = |\mathscr{A}|$, cf. 15.1.5), or if $\Gamma_{\mathscr{C}}$ is unicolored (cf. 17.1.11), then $P^\# = P_{\mathfrak{h}(|G|)}$.

**17.2.4.** The local case, valid for any $j$, follows by similar arguments if one replaces $\Gamma_{\mathscr{C}}$ by $\Gamma^2_{\mathscr{C},j}$. The last statement follows from Lemma 14.2.10, or from the sentences following it.

**17.2.5. The proof of Theorem 16.2.13.** Assume that $\Gamma_{\mathscr{C}}$ has a vanishing 2-edge $e \in \mathscr{E}_{\mathscr{W}}$, that is, the situation of 17.1.7(c) occurs. Let $h := \tilde{m}_{\Phi,ver}$ be as in 17.1.1, and fix $q$ such that the restriction of $h^q$ on $\widetilde{F}_{\mathscr{W}}$ is the identity. The proof of 17.1.7(c) shows that $h^q$ can be extended to $\widetilde{F}^*_\Phi \cap T(e)$ by the identity map. In particular, in such a situation it is better to replace the space $\widetilde{F}_{\mathscr{W}}$ from the diagram 17.1.8 by $\widetilde{F}'_{\mathscr{W}}$, defined as the union of $\widetilde{F}_{\mathscr{W}}$ with *all* separating annuli $\widetilde{F}^*_\Phi \cap T(e)$ corresponding to the vanishing 2-edges from $\mathscr{E}_{\mathscr{W}}$. Moreover, we define $B'$ as the union of the other separating annuli. Then one gets a new diagram (involving the spaces $\widetilde{F}^*_\Phi$, $\widetilde{F}'_{\mathscr{W}}$, $B'$, and morphisms $i'$ and $T'$) such that in the new "collapsing" situation $T'$ becomes *an isomorphism.* Then all the arguments above, complemented with the corresponding facts from Chap. 12 about the "Collapsing Main Algorithm", can be repeated, and the second version 16.2.13 follows as well.

**Example 17.2.6.** Assume that $f(x, y, z) = f'(x, y)$ and $g = z$ as in 9.1. Then we have only vanishing 2-edges, hence $\widetilde{F}'_{\mathscr{W}} = \widetilde{F}_\Phi$. Therefore, in the new diagram $H_1(\widetilde{F}^*_\Phi, \widetilde{F}'_{\mathscr{W}}) = 0$. In fact, since $\nu = 1$ for all vertices, we can even take $q = 1$ (use e.g. 13.3.3), hence in this case $m_{\Phi,ver}$ is isotopic to the identity. (This can also be proved by a direct argument, see Chap. 20).

# Chapter 18
# The Mixed Hodge Structure of $H_1(\partial F)$

## 18.1 Generalities: Conjectures

**18.1.1.** We believe that a substantial part of the numerical identities and inequalities obtained in the previous chapters are closely related with general properties of mixed Hodge structures (in the sequel abbreviated by MHS) supported by different (co)homology groups involved in the constructions.

Although the detailed study of the mixed Hodge structure on $H_1(\partial F)$ and related properties exceeds the aims of the present work, we decided to dedicate a few paragraphs to this subject too: we wish to formulate some of the expectations and to shed light on the results of the book from this point of view as well.

For general results, terminology and properties of MHS, see for example the articles of Deligne [26]. For the MHS on the cohomology of the Milnor fiber of local singularities see the articles of Steenbrink [121, 122, 124], or consult the monographs of Dimca, Kulikov, or Peters and Steenbrink [27, 56, 102]. On the link of an isolated singularity Durfee defined a MHS [30], for different versions and generalizations see [31, 32, 35, 122, 124].

For the convenience of the reader we recall the basic definition of MHS.

**Definition 18.1.2.** *(a) A pure Hodge structure of weight $m$ is a pair $(H, F^\bullet)$, where $H$ is a finite dimensional $\mathbb{R}$-vector space and $F^\bullet$ is a decreasing finite filtration on $H_\mathbb{C} = H \otimes \mathbb{C}$ (called the Hodge filtration) such that*

$$H_\mathbb{C} = F^p \oplus \overline{F^{m-p+1} H_\mathbb{C}}$$

*for all $p \in \mathbb{Z}$, where the conjugation $\bar{\cdot}$ on $H_\mathbb{C}$ is induced by the conjugation of $\mathbb{C}$.*

*(b) A mixed Hodge structure is a triple $(H, W_\bullet, F^\bullet)$, where $H$ is a finite dimensional $\mathbb{R}$-vector space, $W_\bullet$ is a finite increasing filtration on $H$ (called the weight filtration), and $F^\bullet$ is a finite decreasing filtration on $H_\mathbb{C}$ such that $F^\bullet$*

A. Némethi and Á. Szilárd, *Milnor Fiber Boundary of a Non-isolated Surface Singularity*, Lecture Notes in Mathematics 2037, DOI 10.1007/978-3-642-23647-1_18,

*on $Gr_m^W H$ induces a Hodge structure of weight m for all m. (Here $Gr_m^W H := W_m H/W_{m-1}H$.)*

**18.1.3. MHS on the cohomology of $\partial F$.** Without entering in the theory of derived categories and mixed Hodge modules, we outline a possible way to define a MHS on the cohomology of $\partial F$.

If $f : (\mathbb{C}^3, 0) \to (\mathbb{C}, 0)$ is a hypersurface singularity, the cohomology of its Milnor fiber $F$ carries a MHS. This can be defined via Deligne's nearby cycle functor $\psi_f$ which produces the mixed Hodge module $\psi_f \mathbb{R}_{\mathbb{C}^3}$ supported on $V_f$. If $i$ denotes the inclusion of the origin into $V_f$, then $H^k(i^*\psi_f \mathbb{R}_{\mathbb{C}^3}) = H^k(F)$, defining a MHS on $H^k(F)$.

Usually, if $V_f$ has an isolated singularity, then the MHS of $H^*(\partial F)$ is defined via the isomorphism of $H^*(\partial F)$ with the cohomology $H^*(K_f)$ of the link of $V_f$, which is identified with a local cohomology $H^{*+1}_{\{0\}}(V_f)$. The MHS on the link of a normal surface singularity can be defined in a similar way through local cohomology (see [30–32, 35, 122, 124]).

If $V_f$ has a non-isolated singularity, this procedure does not work: the link and $\partial F$ have different cohomologies. Therefore, for such a case we propose the following definition.

Consider $i^!\psi_f \mathbb{R}_{\mathbb{C}^3}$ too; this has the property that $H^k(i^!\psi_f \mathbb{R}_{\mathbb{C}^3}) = H^k_c(F)$.

**Definition 18.1.4.** *Define the MHS on the cohomology of the boundary $\partial F$ via*

$$\mathrm{cone}(i^!\psi_f \mathbb{R}_{\mathbb{C}^3} \to i^*\psi_f \mathbb{R}_{\mathbb{C}^3}). \tag{18.1.5}$$

This definition automatically implies the following fact.

**Corollary 18.1.6.** *One has an exact sequence of mixed Hodge structures:*

$$0 \to H^1(F) \to H^1(\partial F) \to H^2_c(F) \to H^2(F) \to H^2(\partial F) \to H^3_c(F) \to 0. \tag{18.1.7}$$

**Remark 18.1.8.** The above definition agrees with the "link functor" $\mathrm{cone}(i^! \to i^*)$ of Durfee and Saito [32], which can be considered in any category where the basic functors $i^!$, $i^*$ and cone are defined. The link functor can be applied to any mixed Hodge module. If one applies it to the constant sheaf on $V_f$, one gets a MHS supported by the cohomology of the link $L$ (this case was discussed in [32]). When one applies it to $\psi_f \mathbb{R}_{\mathbb{C}^3}$, one gets the MHS of the cohomology of $\partial F$.

The following cases are relevant from the point of view of the results of the present work.

**Example 18.1.9.** (a) Assume that $(X, x)$ is a normal surface singularity. Let $G_X$ be (one of its) dual resolution graphs. It is known that its intersection matrix $A$ is negative definite. We wish to give geometric meaning to the integers $g(G_X)$ and $c(G_X)$. Recall that in this case we only use the blowing up/down operations of $(-1)$ rational vertices, which keep the integers $g(G_X)$ and $c(G_X)$ stable. Therefore, they can be recovered from any negative definite plumbing graph, and thus these numbers depend only on the link $K_X$ of $(X,x)$. Recall also that $\dim H_1(K_X) = 2g(G_X) + c(G_X)$, cf. 15.1.2.

The integers $g(G_X)$ and $c(G_X)$ have the following Hodge theoretical interpretation. If $W_\bullet$ denotes the weight filtration of $H_1(\partial X, \mathbb{R})$, then $\dim Gr^W_{-1} H_1(K_X) = 2g(G_X)$ and $\dim Gr^W_0 H_1(K_X) = c(G_X)$ (and all the other graded components are zero). Hence, in this situation, $Gr^W_\bullet H_1(\partial X)$ is topological.

(b) Let us analyze how the facts from (a) are modified if we consider a more general situation. Assume that $V$ is a smooth complex surface and $C \subset V$ a normal crossing curve with all irreducible components compact. We will denote by the same $V$ a small tubular neighbourhood of $C$. Then the oriented 3-manifold $\partial V$ can be represented by a plumbing graph – the dual graph of the configuration $C$. Let this be denoted by $G_C$. Similarly as above, $A$ denotes the associated intersection matrix of $G_C$, or, equivalently, the intersection matrix of the curve-configuration of the irreducible components of $C$. Again (as in 15.1.2), $\dim H_1(\partial V) = 2g(G_C) + c(G_C) + \operatorname{corank} A$. Moreover, $H_1(\partial V, \mathbb{R})$ admits a mixed Hodge structure such that $\dim Gr^W_{-2} H_1(\partial V) = \operatorname{corank} A$, $\dim Gr^W_{-1} H_1(\partial V) = 2g(G_C)$ and $\dim Gr^W_0 H_1(\partial V) = c(G_C)$ (and all the other graded components are zero), see e.g. [36, (6.9)].

The point is that now this decomposition depends essentially *on $C$ and the analytic embedding of $C$ into $V$*, and, in general, cannot be deduced from the *topology* of $\partial V$ *alone*. (To see this, compare the following two cases: the union of three generic lines in the projective plane and a smooth elliptic curve with self-intersection zero). This discussion shows that if we wish to keep the information regarding the weight filtration of $\partial V$, then we are only allowed to use those calculus-operations which preserve $c(G_C)$, $g(G_C)$ and $\operatorname{corank} A$. In particular, the oriented handle absorption should not be allowed.

We believe that the following properties hold for the MHS defined above:

**18.1.10. Conjecture.** The weight filtration of the mixed Hodge structure of $H_1(\partial F)$, defined in (18.1.5), satisfies

$$0 = W_{-3} \subset W_{-2} \subset W_{-1} \subset W_0 = H_1(\partial F).$$

Moreover, for any graph $G$ provided by the Main Algorithm one has:

$$\dim Gr^W_i H_1(\partial F) = \begin{cases} \operatorname{corank} A_G & \text{if } i = -2, \\ 2g(G) & \text{if } i = -1, \\ c(G) & \text{if } i = 0. \end{cases} \tag{18.1.11}$$

Obviously, there is a corresponding dual cohomological statement:

**18.1.12. Conjecture.** The weight filtration of the mixed Hodge structure of $H^1(\partial F)$ satisfies

$$0 = W_{-1} \subset W_0 \subset W_1 \subset W_2 = H^1(\partial F),$$

and for any graph $G$ provided by the Main Algorithm one has:

$$\dim Gr_i^W H^1(\partial F) = \begin{cases} \operatorname{corank} A_G & \text{if } i = 2, \\ 2g(G) & \text{if } i = 1, \\ c(G) & \text{if } i = 0. \end{cases} \tag{18.1.13}$$

Similarly, the weight filtration of $H^2(\partial F)$ satisfies

$$0 = W_1 \subset W_2 \subset W_3 \subset W_4 = H^2(\partial F),$$

and for any graph $G$ provided by the Main Algorithm one has:

$$\dim Gr_i^W H^2(\partial F) = \begin{cases} \operatorname{corank} A_G & \text{if } i = 2, \\ 2g(G) & \text{if } i = 3, \\ c(G) & \text{if } i = 4. \end{cases} \tag{18.1.14}$$

**Remark 18.1.15.** (1) The monomorphism $H^1(F) \hookrightarrow H^1(\partial F)$ from (18.1.7) is *strictly compatible* with both weight and Hodge filtrations; namely,

$$W_i(H^1(F)) = H^1(F) \cap W_i(H^1(\partial F)) \quad \text{for any } i.$$

There is a similar statement for the Hodge filtration too.

Moreover, we claim that the cup product $H^1(\partial F) \otimes H^1(\partial F) \mapsto H^2(\partial F)$ is a morphism of mixed Hodge structures, in particular, it preserves the weight filtration

$$W_i(H^1(\partial F)) \cup W_j(H^1(\partial F)) \subset W_{i+j}(H^2(\partial F)). \tag{18.1.16}$$

For example, if $A_G$ is non-degenerate, then (18.1.13) and (18.1.14) would imply that the cup-product $H^1(\partial F) \otimes H^1(\partial F) \to H^2(\partial F)$ was trivial. This is compatible with the fact that the cup-product is indeed trivial, whenever $A_G$ is non-degenerate: this is a result of Sullivan [129], see also [31].

Moreover, (18.1.16) can be compared with the list of inclusions of 16.2.11(b) where the fact that $(A_G)_{\lambda \neq 1}$ is non-degenerate was also used (see also 24.2).

(2) If true, the above Conjectures would have the following consequence: the numerical invariants $\operatorname{corank} A_G$, $g(G)$ and $c(G)$ associated with the graph $G$ obtained from the Main Algorithm are independent of the choice of the resolution $r$ (or, of the graph $\Gamma_{\mathscr{C}}$ used in the algorithm). Therefore, we emphasize again, if one wishes to keep in $G^m$ all the information regarding the MHS of $\partial F$, one has to use only those operations of the reduced plumbing calculus which preserve these numbers, that is, one has to exclude the oriented handle absorption R5.

(3) Conjecture 18.1.12 is compatible with 18.1.9(a): the weights of $H^1(K_X)$ are $< 2$, while the weighs of $H^2(K_X)$ are $> 2$.

(4) In 19.9 (treating homogeneous singularities) and in Sect. 20.3 (cylinders) we provide evidences for the above Conjecture.

# Part III
# Examples

# Chapter 19
# Homogeneous Singularities

Assume that $f$ is homogeneous of degree $d \geq 2$. In order to determine a possible $\Gamma_{\mathscr{C}}$, we take for $g$ a generic linear function. We adopt the notations of Chap. 8, where the graph $\Gamma_{\mathscr{C}}$ is constructed.

We start our discussion with the list of some specific properties of the geometry of homogeneous singularities regarding the graphs $\Gamma_{\mathscr{C}}$, $G$, or the ICIS $\Phi$. Then we combine this new information with the general properties established in the previous chapters.

## 19.1 The First Specific Feature: $M_{ver} = (M_{hor})^{-d}$

Clearly

$$|\mathscr{A}(\Gamma_{\mathscr{C}})| = |\mathscr{A}(G)| = d.$$

Moreover, $d_j = 1$ for any $j$, hence

$$M'_{j,hor} = M^{\Phi}_{j,hor} \text{ and } M'_{j,ver} = M^{\Phi}_{j,ver}. \tag{19.1.1}$$

Since $f$ and $g$ are homogeneous, by [125], we have

$$\begin{cases} M'_{j,ver} = (M'_{j,hor})^{-d} \\ M_{\Phi,ver} = (M_{\Phi,hor})^{-d}. \end{cases} \tag{19.1.2}$$

In particular, for any of the pairs $(M'_{j,ver}, M'_{j,hor})$ or $(M_{\Phi,ver}, M_{\Phi,hor})$, the number of 2-Jordan blocks of the vertical operator coincides with the number of 2-Jordan blocks of the horizontal operator. Moreover, the horizontal monodromies determine the corresponding $\mathbb{Z}^2$-representations completely.

In fact, the identities (19.1.2) are true at the level of the geometric monodromies as well. Let us check this for the pair $(m_{\Phi,ver}, m_{\Phi,hor})$; the local version is similar.

A. Némethi and Á. Szilárd, *Milnor Fiber Boundary of a Non-isolated Surface Singularity*, Lecture Notes in Mathematics 2037, DOI 10.1007/978-3-642-23647-1_19, 

Consider the homogeneous action on $\mathbb{C}^3$ given by $\lambda * (x, y, z) = (\lambda x, \lambda y, \lambda z)$. This is projected via $\Phi$ to the action $\lambda * (c, d) = (\lambda^d c, \lambda d)$ of $\mathbb{C}^2$. Hence, the monodromy along the loop $(e^{itd}c, e^{it}d)$, $t \in [0, 2\pi]$, is lifted to a trivial action on the Milnor fiber, that is

$$m^d_{\Phi,hor} \cdot m_{\Phi,ver} = I \text{ and } (m'_{j,hor})^d \cdot m'_{j,ver} = I. \tag{19.1.3}$$

## 19.2 The Second Specific Feature: The Graphs $\overline{G_{2,j}}$

Consider the graph $\Gamma_{\mathscr{C}}$ of Chap. 8. In the Main Algorithm 10.2 applied for this $\Gamma_{\mathscr{C}}$, one puts exactly one vertex in $G$, say $\widetilde{v}_\lambda$, above a vertex $v_\lambda$ of $\Gamma_{\mathscr{C}}$. If we delete all these vertices from $G$, we get the union of the graphs $\overline{G_{2,j}}$, cf. 10.3.4 and Remark 10.3.8. The point is that

> $\overline{G_{2,j}}$, *with opposite orientation, is a possible embedded resolution graph associated with the $d$-suspension of the transversal singularity $T\Sigma_j$ and the germ induced by the projection on the suspension coordinate.*

More precisely, if $f'_j(u, v) = f|_{(Sl_q, q)}$ ($q \in \Sigma_j \setminus \{0\}$, cf. Sect. 2.2) is the local equation of the transversal type plane curve singularity, then its $d$-suspension is the isolated hypersurface singularity $X := (\{w^d = f'_j(u, v)\}, 0)$. Then $G_{2,j} = -\Gamma(X, w)$.

This follows from a comparison of the Main Algorithm 10.2 and the algorithm which provides the graph of suspension singularities 5.3. Independently, it can also be proved by combining Proposition 3.3.1(2) and the local identity of (19.1.3), or by the covering procedure which will be described in 19.3.

In particular, the graph $G$ consists of the disjoint union of graphs of these suspensions with opposite orientation, and $|\Lambda|$ new vertices $\widetilde{v}_\lambda$ decorated by $[g_\lambda]$, altogether supporting $d$ arrowheads. The edges connecting these new vertices to the graphs $\overline{G_{2,j}}$ are $\ominus$-edges and reflect the combinatorics of the "identification map" $c$ (i.e. the incidence of the singular points on the components of $C$, see 8.1). The arrowheads are distributed as follows: each $\widetilde{v}_\lambda$ supports $d_\lambda$ arrowheads, all of them connected by $+$ edges. The gluing property of the two pieces (i.e. of $\partial_1 F$ and $\partial_2 F$) is codified in the Euler numbers of the vertices $\widetilde{v}_\lambda$ in $G$. This is determined by (4.1.5), once we carry the multiplicity system of $g$ in the construction: in the suspension graphs one has the multiplicity system of the germ $w : (\{w^d = f'_j(u, v)\}, 0) \to (\mathbb{C}, 0)$, and each $\widetilde{v}_\lambda$ and arrowhead has multiplicity (1). This provides a very convenient "short-cut" to obtain $G$.

**Example 19.2.1.** A projective curve of the projective plane is called cuspidal if all its singularities are locally analytically irreducible.

It is well-known that there exists a rational cuspidal projective curve $C$ of degree $d = 5$ with two local irreducible singularities with local equations $u^3 + v^4 = 0$

and $u^2 + v^7 = 0$ (see e.g. [80]). Then the above procedure provides two suspension singularities, both of Brieskorn type. Their equations are $u^3 + v^4 + w^5 = 0$ and $u^2 + v^7 + w^5 = 0$. (We made the choice of $C$ carefully: we wished to get pairwise relative prime exponents in each singularity, in order to "minimize" $|H_1(\partial F, \mathbb{Z})|$, cf. 19.6.) The embedded resolution graphs of the coordinate function $w$ restricted on these suspension singularities are the following:

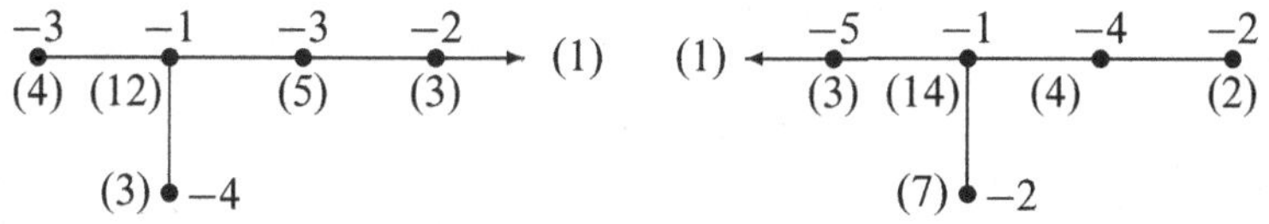

Reversing the orientations we get the following graphs, which coincide with the graphs $\overline{G_{2,j}}$ $(j = 1, 2)$:

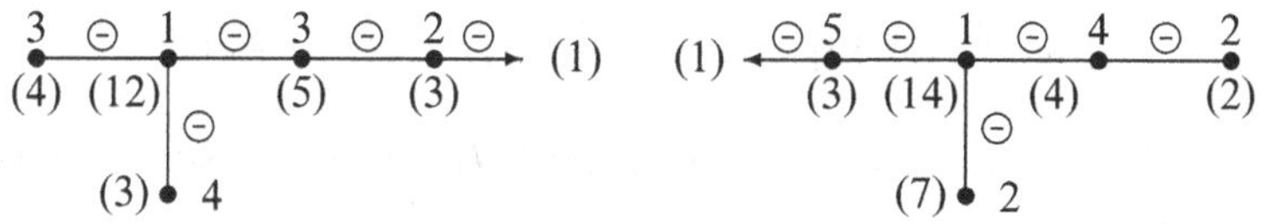

In order to get $G$, we have to insert one more vertex $\widetilde{v}$ (corresponding to $C$) with multiplicity (1) and genus decoration zero, and $d = 5$ arrowheads, all with multiplicity (1), connected with $+$ edges to $\widetilde{v}$:

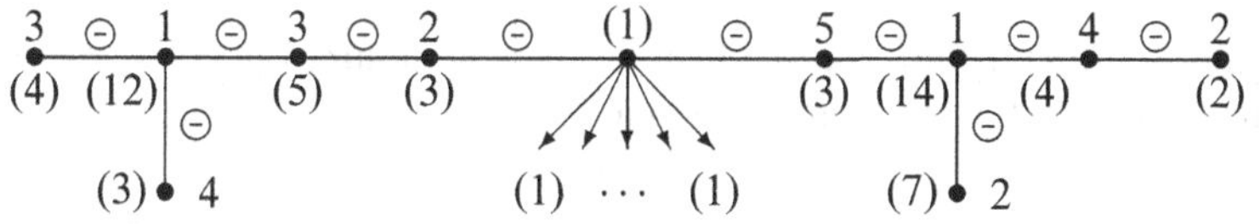

The missing Euler number $e$ of $\widetilde{v}$ is computed via (4.1.5), namely $e + 5 - 3 - 3 = 0$, hence $e = 1$. Deleting all the multiplicities we obtain the graph of $\partial F$:

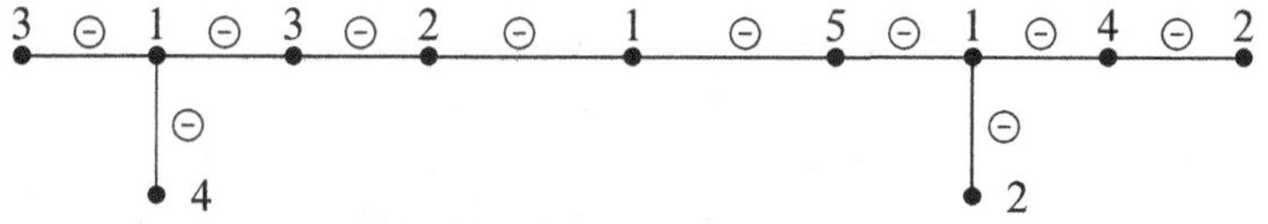

By plumbing calculus, one can blow down twice 1-curves, and also one may delete the $\ominus$'s. This possible graph of $\partial F$ is not the "normal form" of [94], the interested reader may transform it easily to get a graph with all Euler numbers negative. But even if we pass to the normal form, the intersection matrix will not be negative definite, its index will be $(-12, +1)$.

Notice that $\partial F$ is a rational homology sphere. In fact, it is easy to verify that $H_1(\partial F, \mathbb{Z}) = \mathbb{Z}_5$.

As a consequence of the above discussion, we get:

**Corollary 19.2.2.** *If $C$ is a rational unicuspidal projective plane curve (i.e. $C$ has only one singular point at which $C$ is locally analytically irreducible), and its local singularity has only one Puiseux pair, then $\partial F$ can be represented by a plumbing graph which is either star-shaped or a string.*

This will be exemplified more in Sect. 19.8.

Another consequence is the following

**Corollary 19.2.3.**

$$\operatorname{corank}(A,\mathfrak{I})_G = |\mathscr{A}(G)|.$$

*Therefore, Theorem 16.2.3 applies and $\#_1^2 M_{\Phi,ver} = c(G)$ and $P^{\#}(t) = P_{(h)(|G|)}(t)$ too.*

*In particular, the characteristic polynomial of the monodromy acting on $H_1(\partial F,\mathbb{Z})$ is determined by Theorem 16.2.8, and*

$$rank\, H_1(\partial F)_{\neq 1} = 2g(G) + c(G) - 2g(\Gamma_{\mathscr{C}}) - c(\Gamma_{\mathscr{C}}).$$

*Moreover, the equivalent statements of (16.2.4) are also satisfied, in particular* $\operatorname{corank}(A,\mathfrak{I})_{G_{2,j}} = |\mathscr{E}_{cut,j}|$ *and* $\#_j^2 M'_{j,ver} = c(G_{2,j})$ *for any* $j$.

*Proof.* Since each $\widetilde{v}_\lambda$ supports at least one arrowhead, the rank of $(A,\mathfrak{I})$ is maximal whenever all the ranks of the intersection matrices associated with $\overline{G_{2,j}}$ are maximal. But these matrices are definite, hence non-degenerate, cf. 4.3.4(3). For the other statements see (16.2.5) and (16.2.10). □

After stating the third specific feature, and determining corank $A$, we will return to the characteristic polynomial formula.

## 19.3 The Third Specific Feature: The $d$-Covering

Let $C = \{f = 0\} \subset \mathbb{P}^2$ as before. It is well-known that there is a regular $d$-covering $F = \{f = 1\} \to \mathbb{P}^2 \setminus C$ given by $(x,y,z) \mapsto [x:y:z]$. Let $T$ be a small tubular neighbourhood of $C$, and let $\partial T$ be its boundary. In the next discussion it is more convenient to orient $\partial T$ not as the boundary of $T$, but as the boundary of $\mathbb{P}^2 \setminus T^\circ$. Then the above covering induces a regular $d$-covering, which is compatible with the orientations:

$$\partial F \to \partial T.$$

Moreover, the $\mathbb{Z}_d$ covering transformation on $\partial F$ coincides with a representative of the Milnor monodromy action on $\partial F$. Using either this, or just the homogeneity of $f$, we get that

*the geometric monodromy action is finite of order $d$.*

**Example 19.3.1.** Let us exemplify this covering procedure on a simple case. Assume that $C$ has degree 3 and has a cusp singularity. Then a possible plumbing representation for $\partial T$ (oriented as the boundary of $T$) is the graph $(i)$ below.

$(i)$

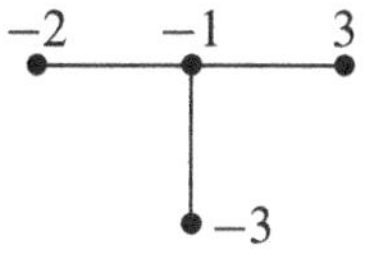


$(ii)$

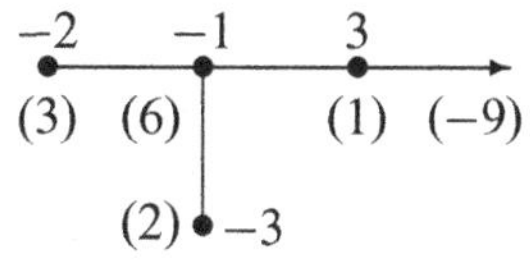


Then the $\mathbb{Z}_3$-cyclic covering of $\partial T$ can be computed as the $\mathbb{Z}_3$-covering of the divisor marked in $(ii)$ by a similar algorithm as described in 5.3 for cyclic coverings. This provides the graph $(iii)$ below (as the graph covering of the graph $(i)$). It is a possible plumbing graph of $\partial F$ with opposite orientation. Changing the orientation and after reduced calculus, we get the graph $(iv)$. It is the graph that we get by the Main Algorithm as well (after modified by reduced calculus).

The monodromy action permutes the three $-2$-curves.

$(iii)$

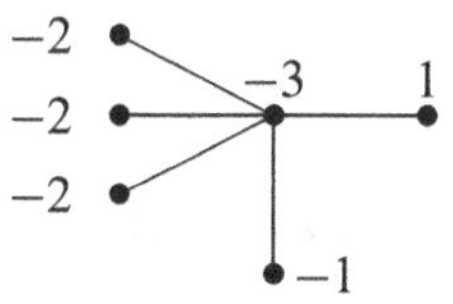


$(iv)$

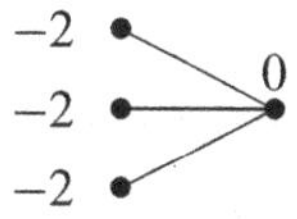


**19.3.2.** The 3-manifold $\partial T$ was studied extensively by several authors (see for example the articles [23, 24, 46] of Cohen and Suciu, and E. Hironaka, and the references therein). Hence, once the representation $\pi_1(\partial T) \to \mathbb{Z}_d$ associated with the covering is identified, one can try to recover all data about $\partial F$ from that of $\partial T$ (some of them can be done easily, some of them by rather hard work). Here we will consider only one such computation, which provides some numerical data still missing. This basically targets the rank of the 1-eigenspace $\dim H_1(\partial F)_1$ associated with the monodromy action.

**19.3.3. The rank of $H_1(\partial F)_1$ and** corank $A_G$.

By the above discussion we get $H_1(\partial F)_1 = H_1(\partial T)$. On the other hand, $H_1(\partial T)$ can be computed easily, via 18.1.9(b), since $\partial T$ is also a plumbed 3-manifold (cf. 18.1.9(c)). One can determine a possible plumbing graph for $\partial T$ as follows.

Start with the curve $C$, blow up the infinitely near points of its singularities (as in Chap. 8), and transform the combinatorics of the resulting curve configuration into a graph. In this way we get as a plumbing graph the graph $\Gamma_{\mathscr{C}}$ with some natural modifications: we have to keep the genus decorations, delete the arrowheads and the weight decorations of type $(m; n, \nu)$, and have to insert the Euler numbers. These can in turn be determined as follows: start with the intersection matrix of the components of $C$. This, by Bézout's theorem, is the $\Lambda \times \Lambda$ matrix $A_C$ with entries $(d_\lambda d_\xi)_{\lambda,\xi}$. This intersection matrix is modified by blow ups to get the intersection matrix $A_{\Gamma_{\mathscr{C}}}$ of the plumbing graph of $\partial T$. Hence, similarly as in 15.1.2, one has:

$$\dim H_1(\partial F)_1 = \dim H_1(\partial T) = 2g(\Gamma_{\mathscr{C}}) + c(\Gamma_{\mathscr{C}}) + \operatorname{corank} A_{\Gamma_{\mathscr{C}}}. \tag{19.3.4}$$

Using the notations of Chap. 8, we compute each summand. $g(\Gamma_{\mathscr{C}}) = \sum_\lambda g_\lambda$ is clear. Since each $\Gamma_j$ is a tree, one also gets

$$c(\Gamma_{\mathscr{C}}) = \sum_j (|I_j| - 1) - (|\Lambda| - 1).$$

Finally, by blowing up, the corank of an intersection matrix stays stable, hence $\operatorname{corank} A_{\Gamma_{\mathscr{C}}} = \operatorname{corank} A_C = \operatorname{corank}((d_\lambda d_\xi)_{\lambda,\xi})$, i.e.:

$$\operatorname{corank} A_{\Gamma_{\mathscr{C}}} = |\Lambda| - 1. \tag{19.3.5}$$

Therefore we get the identities (where $b_1(C)$ denotes the first Betti number of $C$):

$$\dim H_1(\partial F)_1 = 2g(\Gamma_{\mathscr{C}}) + \sum_j (|I_j| - 1) = b_1(C) + |\Lambda| - 1. \tag{19.3.6}$$

This has the following immediate consequence:

**Corollary 19.3.7.** $\operatorname{corank} A_G = |\Lambda| - 1$. *Hence, in general, $A_G$ is degenerate.*

*Proof.* $\operatorname{corank} A_G = \operatorname{corank} A_{\Gamma_{\mathscr{C}}}$ by (16.2.9) and (19.3.4). Then use (19.3.5). □

Now we start our list of applications.

## 19.4 The Characteristic Polynomial of $\partial F$

Using Theorem 16.2.8 and Corollary 19.3.7 we get:

**Corollary 19.4.1.** *If $f$ is homogeneous, then the characteristic polynomial of the Milnor monodromy acting on $H_1(\partial F)$ is*

$$\frac{(t-1)^{|\Lambda|-d} \cdot \prod_{w\in\mathscr{W}(\Gamma_{\mathscr{C}})} (t^{(m_w,d)} - 1)^{2g_w+\delta_w-2} \cdot (t^{\mathfrak{n}_w} - 1)}{\prod_{e\in\mathscr{E}_w(\Gamma_{\mathscr{C}})} (t^{\mathfrak{n}_e} - 1)}. \tag{19.4.2}$$

*For the integers $\mathfrak{n}_w$ and $\mathfrak{n}_e$ see 10.2, with the additional remark that all the second entries of the vertex-decorations are equal to $d$.*

Notice that this is a formula given entirely in terms of $\Gamma_{\mathscr{C}}$. If $c(G) = 0$, then via 16.1.3 it simplifies to:

$$(t-1)^{1+|\Lambda|-d} \cdot \prod_{w\in\mathscr{W}(\Gamma_{\mathscr{C}})} (t^{(m_w,d)} - 1)^{2g_w+\delta_w-2}. \tag{19.4.3}$$

Since the monodromy has finite order, the complex algebraic monodromy is determined completely by its characteristic polynomial. By Corollary 19.2.3, the complex monodromy is trivial if and only if $c(G) = c(\Gamma_{\mathscr{C}})$ and $g(G) = g(\Gamma_{\mathscr{C}})$, cf. also with 13.6.4.

**19.4.4.** Now, we provide a description of the characteristic polynomial in terms of the projective curve $C$.

Let $Div = \sum k_\lambda(\lambda) \in \mathbb{Z}[S^1]$ be the divisor of the characteristic polynomial.

By 19.3, the multiplicity $k_1$ of the eigenvalue $\lambda = 1$ is $b_1(C) + |\Lambda| - 1$.

On the other hand, for $\lambda \neq 1$, Sect. 14.2 and (19.1.1) provide the following identity in terms of the local horizontal/Milnor algebraic monodromies of the transversal types (i.e. in terms of the monodromy operators of the local singularities of $C$):

$$k_\lambda = \sum_j \sum_{\lambda^d=1} \#\{\text{Jordan block of } M'_{j,hor} \text{ with eigenvalue } \lambda\} \quad (\lambda \neq 1).$$

In order to see this, we apply 19.1 and the Wang homological exact sequence for $F'_j \to \partial_{2,j} \to S^1$ with 15.1.4.

Indeed, if $B$ is a 1-Jordan block of $M'_{j,hor}$ with eigenvalue $\lambda$, and $\lambda^d = 1$, then this creates a 1-block in $M'_{j,ver}$ with eigenvalue 1, and the corresponding 1-dimensional eigenspace survives in coker $(M'_{j,ver} - 1)$.

If $B$ is a 2-Jordan block of $M'_{j,hor}$ with eigenvalue $\lambda$, and $\lambda^d = 1$, then this creates a 2-block in $M'_{j,ver}$ with eigenvalue 1, and in coker $(M'_{j,ver} - 1)$ only a 1-dimensional space survives, and $M'_{j,hor}$ acts on it by multiplication by $\lambda$.

This and the last identity of 19.2.3 imply:

$$\sum_j \sum_{\lambda^d=1,\, \lambda\neq 1} \#\{\text{Jordan block of } M'_{j,hor} \text{ with eigenvalue } \lambda\}$$

$$= 2g(G) + c(G) - 2g(\Gamma_{\mathscr{C}}) - c(\Gamma_{\mathscr{C}}).$$

**Example 19.4.5.** Assume that $C$ is irreducible and it has a unique singular point $(C, p)$ which is locally equisingular to $\{(f' = u^2 + v^3)(u^3 + v^2) = 0\}$. For the embedded resolution graph of $(C, p)$ see 9.1.1. In order to get $\Gamma_{\mathscr{C}}$ one has to add one more vertex, which will have genus decoration $[g]$, where $2g = (d-1)(d-2) - 12$, and it will support $d$ arrows. The reader is invited to complete the picture of $\Gamma_{\mathscr{C}}$.

By (19.4.2), we get that the characteristic polynomial of the monodromy operator acting on $H_1(\partial F)$ is

$$\frac{(t-1)^{(d-1)(d-2)-10}}{t^{(d,2)} - 1} \cdot \left(\frac{t^{(d,10)} - 1}{t^{(d,5)} - 1}\right)^2.$$

In all the cases, the multiplicity of the eigenvalue 1 is $k_1 = (d-1)(d-2) - 11 = (d-1)(d-2) - \mu(f')$. If $d$ is odd, then there are no other eigenvalues, hence the complex monodromy on $H_1(\partial F, \mathbb{C})$ is trivial.

If $d$ is even then other eigenvalues appear too. If $5 \nmid d$ then only one appears, namely $-1$, otherwise their divisor is the divisor of $(t^5+1)^2/(t+1)$.

This result can be compared with 19.4.4: use that the algebraic monodromy of the local transversal singularity has characteristic polynomial $(t^5+1)^2(t-1)$, and exactly one 2-Jordan block, which has eigenvalue $-1$.

The graph $G^m$ for $d = 10$ is the following (for this $d$ all the Hirzebruch–Jung strings in the Main Algorithm are trivial):

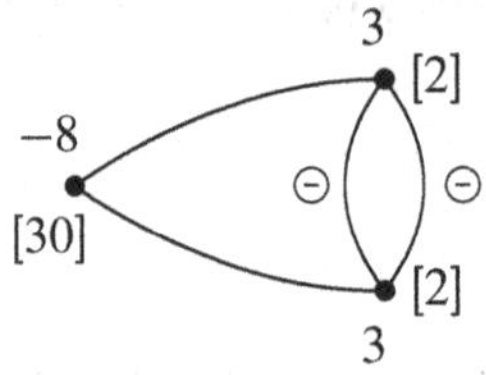


In this case $\dim H_1(\partial F) = 70$, $\dim H_1(\partial F)_1 = 61$, $A_G$ is non-degenerate with $|\det(A_G)| = 50$.

**Example 19.4.6.** The characteristic polynomial of the example 10.4.4 is $(t-1)^{d^2-3d+1}$, hence the complex monodromy on $H_1(\partial F, \mathbb{C})$ is trivial.

## 19.5 $M'_{j,hor}$, $M'_{j,ver}$, $M^{\Phi}_{j,hor}$, $M^{\Phi}_{j,ver}$, $M_{\Phi,hor}$ and $M_{\Phi,ver}$

In this section we show that the local horizontal/Milnor monodromy operators $\{M'_{j,hor}\}_j$ of the transversal types, together with the integers $d$ and $c(G)$ completely determine the $\mathbb{Z}^2$-representations generated by the commuting pairs $(M'_{j,hor}, M'_{j,ver})$, $(M^{\Phi}_{j,hor}, M^{\Phi}_{j,ver})$, respectively $(M_{\Phi,hor}, M_{\Phi,ver})$.

Indeed, the first two pairs agree by (19.1.1). Hence, by 19.1, it is enough to determine $M_{\Phi,hor}$. Moreover, for eigenvalues $\lambda \neq 1$, the following generalized eigenspaces, together with the corresponding horizontal actions, coincide:

$$H_1(F_\Phi)_{M_{\Phi,hor},\lambda\neq 1} = \oplus_j H_1(F_\Phi \cap T_j)_{M^{\Phi}_{j,hor},\lambda\neq 1} = \oplus_j H_1(F'_j)_{M'_{j,hor},\lambda\neq 1}. \quad (19.5.1)$$

Next, we determine the action of $M_{\Phi,hor}$ on the generalized 1-eigenspace $H_1(F_\Phi)_1$ of $M_{\Phi,hor}$.

The rank of $H_1(F_\Phi)_{M_{\Phi,hor},1}$ can be determined in several ways.

First, one can consider the rank of the total space $H_1(F_\Phi)$, which is $(d-1)^2$. Indeed, $F_\Phi$ is the Milnor number of an ICIS $(f,g)$, where $f$ is homogeneous of degree $d$ and $g$ is a generic linear form. Hence, their Milnor number is the Milnor number of a homogeneous plane curve singularity of degree $d$. Since we already determined the $(\lambda \neq 1)$-generalized eigenspaces, the rank of $H_1(F_\Phi)_1$ follows too.

There is another formula based on 13.4.11(a), which provides

$$\begin{aligned}\operatorname{rank} H_1(F_\Phi)_{M_{\Phi,hor},1} &= 2g(\Gamma_{\mathscr{C}}) + 2c(\Gamma_{\mathscr{C}}) + d - 1\\ &= d - |\Lambda| + b_1(C) + \sum_j (|I_j| - 1).\end{aligned}$$

In particular, since we have only one and two-dimensional Jordan blocks:

$$\#_1^1 M_{\Phi,hor} + 2\#_1^2 M_{\Phi,hor} = \operatorname{rank} H_1(F_\Phi)_{M_{\Phi,hor},1}.$$

On the other hand, by 19.1 and (19.5.1) we have

$$\#_1^2 M_{\Phi,ver} = \#_1^2 M_{\Phi,hor} + \sum_j \sum_{\lambda^d = 1 \neq \lambda} \#_\lambda^2 M'_{j,hor}.$$

Since by 19.2.3 the left hand side of this identity is $c(G)$, we get

$$\#_1^2 M_{\Phi,hor} = c(G) - \sum_j \sum_{\lambda^d = 1 \neq \lambda} \#_\lambda^2 M'_{j,hor}.$$

**Example 19.5.2.** Assume that we are in the situation of Example 19.4.5 with $d = 10$. Then both local operators $M'_{1,hor}$ and $M'_{1,ver}$ have exactly one 2-Jordan block, the first with eigenvalue $-1$, the other with 1. This is compatible with the fact that $c(G_{2,1}) = 1$.

Since the order of any eigenvalue of $M_{\Phi,hor}$ divides 10, we get that $M_{\Phi,ver}$ is unipotent. By Corollary 19.2.3, the number of 2-Jordan blocks of $M_{\Phi,ver}$ is $c(G) = 2$. Hence $M_{\Phi,hor}$ has two 2-Jordan blocks as well. By (19.5.1), there is only one with eigenvalue $\neq 1$, namely with eigenvalue $-1$. Hence the remaining block has eigenvalue 1.

## 19.6 When is $\partial F$ a Rational/Integral Homology Sphere?

By 19.4.4, or by the statements of 19.2, we get the following characterization:

**Corollary 19.6.1.** *$\partial F$ is a rational homology sphere if and only if $C$ is an irreducible, rational cuspidal curve (i.e. all its singularities are locally irreducible), and there is no eigenvalue $\lambda$ of the local singularities $(C, p_j)$ with $\lambda^d = 1$.*

This situation can indeed appear, see e.g. 19.2.1, or many examples of the present work.

We wish to emphasize that the classification of rational cuspidal projective curves is an open problem. (For a partial classification with $d$ small see e.g. [80], or [12] and the references therein.) It would be very interesting and important to continue the program of the present work: some restrictions on the structure of $\partial F$ might provide new data in the classification problem as well.

The next natural question is: *When is $\partial F$ an integral homology sphere?* The answer is simple: Never (provided that $d \geq 2$)! This follows from the following proposition:

**Proposition 19.6.2.** *Assume that $f$ is homogeneous of degree $d$. Then there exists an element $h \in H_1(\partial F, \mathbb{Z})$ whose order is either infinity or a multiple of $d$. If $H_1(\partial F, \mathbb{Z})$ is finite, one can choose such an $h$ fixed by the monodromy action.*

*Proof.* By 19.6 we may assume that $C$ is irreducible, rational and cuspidal. Consider the following diagram (for notations, see 19.3):

$$\begin{array}{ccccc} \pi_1(\partial F) & \xrightarrow{i} & \pi_1(\partial T) & \xrightarrow{p} & \mathbb{Z}_d \\ \downarrow{\scriptstyle q} & & \downarrow{\scriptstyle r} & & \\ H_1(\partial F, \mathbb{Z}) & \longrightarrow & H_1(\partial T, \mathbb{Z}) & = & \mathbb{Z}_{d^2} \end{array}$$

The first line, provided by the homotopy exact sequence of the covering, is exact. Moreover, $i$ is injective, and $q$, $r$ and $p$ are surjective. Let $\gamma \in \pi_1(\partial T)$ be the class of a small loop around $C$. Then $p(\gamma^d) = 0$, hence $\gamma^d \in \pi_1(\partial F)$.

On the other hand, $H_1(\partial T, \mathbb{Z}) = \mathbb{Z}_{d^2}$, and it is generated by $[\gamma]$. This follows from the fact (see 19.3.3 and 15.1.3) that $H_1(\partial T, \mathbb{Z}) = \operatorname{coker} A_C$, but in this case $A_C$ has only one entry, namely $d^2$.

Hence $r(\gamma^d)$ has order $d$ in $H_1(\partial T, \mathbb{Z})$. In particular, the order of $q(\gamma^d) \in H_1(\partial F, \mathbb{Z})$ is a multiple of $d$. Since the monodromy action in $\pi_1(\partial F)$ is conjugation by $\gamma$, the second part also follows. □

The above bound is sharp: there are examples when $H_1(\partial F, \mathbb{Z}) = \mathbb{Z}_d$, see e.g. 19.2.1. The above result also shows that the monodromy action over $\mathbb{Z}$ is trivial whenever $H_1(\partial F, \mathbb{Z}) = \mathbb{Z}_d$.

The fact that for $f$ homogeneous $\partial F$ cannot be an integral homology sphere was for the first time noticed by Siersma in [118], page 466, where he used a different argument.

## 19.7 Cases with $d$ Small

In this section we treat all possible examples with $d = 2$ and $d = 3$, and we present some examples with $d = 4$. The statements are direct applications of the Main Algorithm and the above discussions regarding the monodromy.

One of our goals in this "classification list" is to provide an idea about the variety of the possible 3-manifolds obtained in this way, and to check if the same 3-manifold can be realized from essentially two different situations.

**19.7.1.** (**d** $=$ **2**) There is only one case: $C$ is a union of two lines. A possible graph $G^m$ is $0 \bullet$ hence $\partial F \approx S^1 \times S^2$. The monodromy is trivial.

**19.7.2.** (**d** $=$ **3**) There are six cases:

(**3a**) $C$ is irreducible with a cusp. See 19.3.1, or 19.8.1(a) with $d = 3$ for the graph $G^m$ of $\partial F$. In this case $H_1(\partial F, \mathbb{Z}) = \mathbb{Z}_6 \oplus \mathbb{Z}_2$, hence the complex monodromy is trivial. The integral homology has the following peculiar form (specific to the homology of any Seifert 3-manifold):

$$H_1(\partial F, \mathbb{Z}) = \langle\, x_1, x_2, x_3, h \mid 2x_1 = 2x_2 = 2x_3 = h,\ x_1 + x_2 + x_3 = 0 \,\rangle.$$

$h$ has order 3 and it is preserved by the integral monodromy, while the elements $x_1, x_2, x_3$ are cyclically permuted (in order to prove this, use e.g. 19.3.1).

(**3b**) $C$ is irreducible with a node. A possible graph of $\partial F$ is

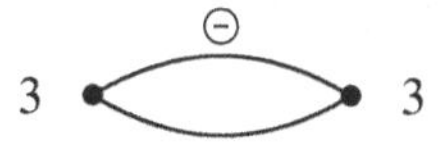


The rank of $H_1(\partial F)$ is 1, and the complex monodromy is trivial.

By calculus, one can verify that $G \sim -G$.

(**3c**) $C$ is the union of a line and an irreducible conic, intersecting each other transversely. A possible graph for $\partial F$ is $3 \bullet [1]$, and the complex monodromy is trivial.

(**3d**) $C$ is the union of an irreducible conic and one of its tangent lines. Then $\partial F \approx S^2 \times S^1$ and the monodromy is trivial.

(**3e**) $C$ is the union of three lines in general position; then $\partial F \approx S^1 \times S^1 \times S^1$ and the monodromy is trivial.

(**3f**) $C$ is the union of three lines intersecting each other in one point: then by *reduced* calculus we get for $\partial F$ the graph $G^m$:

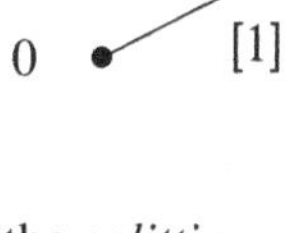


(the decoration * is irrelevant)

In fact, if we apply the *splitting operation* (not permitted by reduced calculus), we obtain that $\partial F \approx \#_4 S^2 \times S^1$.

The characteristic polynomial of the monodromy is $(t^3 - 1)(t - 1)$.

**19.7.3.** **(d = 4)** We will start with the cases when $C$ **is irreducible, rational and cuspidal** (i.e. all singularities are locally irreducible). There are four cases:

**(4a)** $C$ has three $A_2$ singular points. A possible equation for $f$ and $\Gamma_{\mathscr{C}}$ is given in 8.1.4. The boundary $\partial F$ is a rational homology sphere with graph:

$G$ :

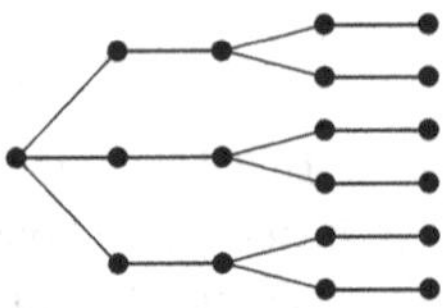

(with all the Euler numbers 2)

(In this case, neither $G$, nor $G$ with the opposite orientation can be represented by a negative definite graph.)

**(4b)** $C$ has two singular points with local equations $u^2 + v^3 = 0$ and $u^2 + v^5 = 0$. This case has an unexpected surprise in store: having two singular points we expect that the graph will have two nodes. This is indeed the case for the graph $G$, the immediate output of the Main Algorithm. Nevertheless, via reduced plumbing calculus the chain connecting the two nodes disappears by a final 0-chain absorption. We get that $\partial F$ is a Seifert manifold with graph:

$G^m$ :

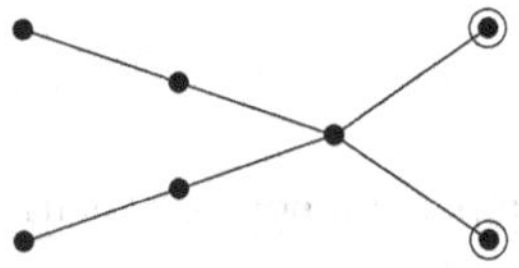

• are decorated by 2

⊙ are decorated by 5

The orbifold Euler number is $4/15 > 0$, and the order of $H_1(\partial F, \mathbb{Z})$ is 60.

**(4c)** $C$ has one singular point with local equation $u^3 + v^4 = 0$. The 3-manifold $\partial F$ is a rational homology sphere, its plumbing graph is given in 19.8.2(a) (with $d = 4$). It is worth mentioning that there are two projectively non-equivalent curves of degree four with this local data, namely $x^4 - x^3y + y^3z = 0$ and $x^4 - y^3z = 0$.

**(4d)** $C$ has one singular point with local equation $u^2 + v^7 = 0$. In this case, rather surprisingly, $\partial F$ is the lens space of type $L(28, 15)$ (that is, the graph is a string with decorations: $-2$, $-8$, $-2$).

For an example of **irreducible non-rational** $C$, see 10.4.4 or 19.4.6.

For **reducible** $C$ we give three more examples:

**(4a')** Let $C$ be the union of two conics intersecting each other transversely. The characteristic polynomial is $(t - 1)^4$, the complex monodromy is trivial, and the graph of $\partial F$ is:

$G^m$ :

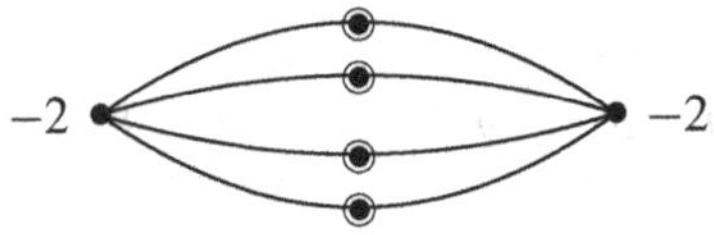


⊙ are decorated by $-4$

**(4b')** Let $C$ = be the union of a smooth irreducible conic and two different tangent lines. The characteristic polynomial is $(t^2+1)^2(t-1)^3$. The graph $G^m$ is $0 \bullet [3]$ .

**(4c')** Let $C$ be the union of two smooth conics intersecting tangentially in two singular points of type $A_3$ (e.g. the equation of $f$ is $x^2y^2+z^4$). Then the graph $G^m$ is $4 \bullet [3]$. The characteristic polynomial is $(t^2+1)^2(t-1)^2$.

**19.7.4.** It is also instructive to see the case when $C$ is given by $(x^3-y^2z)(y^3-x^2z)=0$. The curve $C$ has two components, and they intersect each other in 6 points. One of them, say $(C,p)$, has local equation $(u^3-v^2)(v^3-u^2)=0$, cf. 9.1.1, the others are nodes. Some of the cycles of $G$ are generated by these intersections, and one more by a 2-Jordan block of $(C,p)$: $c(\Gamma_{\mathscr{C}})=5$ and $c(G)=6$.

## 19.8 Rational Unicuspidal Curves with One Puiseux Pair

The present section is motivated by two facts.

First, by such curves we produce plumbed 3-manifolds with start-shaped graphs (cf. 19.2.2), some of which are irreducible 3-manifolds with Seifert fiber-structure, some of which are not irreducible. Seifert 3-manifolds play a special role in the world of 3-manifolds. Therefore, it is an important task to classify all Seifert manifolds realized as $\partial F$, where $F$ is a Milnor fiber as above. Even if we restrict ourselves to the homogeneous case, the problem looks surprisingly hard: although all unicuspidal rational curves with one Puiseux pair produce star-shaped graphs, in Sect. 19.7(4b) we have found a curve $C$ with two cusps, which produces a Seifert manifold as well. Since the classification of the cuspidal rational curves is not finished, the above question looks hard. On the other hand, notice that Seifert manifolds can also be produced by other types of germs as well, e.g. by some weighted homogeneous germs as in 10.4.5 or 21.1.6, but not all weighted homogeneous germs provide Seifert manifolds.

In this section we will list those star-shaped graphs which are produced by unicuspidal curves with one Puiseux pair.

It is also worth mentioning, that the normalization of $V_f$ (for any $f$ where $C$ is irreducible and rational) has a very simple resolution graph: only one vertex with genus zero and Euler number $-d$. This graph can be compared with those obtained for $\partial F$ listed below.

For the second motivation (of more analytic nature), see 19.8.5.

**19.8.1.** Assume that $C$ is rational and unicuspidal of degree $d \geq 3$, and the equisingularity type of its singularity is given by the local equation $u^a+v^b=0$. The possible triples $(d,a,b)$ are classified in [13].

In order to state this result, consider the Fibonacci numbers $\{\varphi_j\}_{j\geq 0}$ defined by $\varphi_0=0$, $\varphi_1=1$, and $\varphi_{j+2}=\varphi_{j+1}+\varphi_j$ for $j\geq 0$.

Then $(d,a,b)$ $(1<a<b)$ is realized by a curve with the above properties if and only if it appears in the following list:

(a) $(a,b) = (d-1,d)$;
(b) $(a,b) = (d/2, 2d-1)$, where $d$ is even;
(c) $(a,b) = (\varphi_{j-2}^2, \varphi_j^2)$ and $d = \varphi_{j-1}^2 + 1 = \varphi_{j-2}\varphi_j$, where $j$ is odd and $\geq 5$;
(d) $(a,b) = (\varphi_{j-2}, \varphi_{j+2})$ and $d = \varphi_j$, where $j$ is odd and $\geq 5$;
(e) $(a,b) = (\varphi_4, \varphi_8 + 1) = (3, 22)$ and $d = \varphi_6 = 8$;
(f) $(a,b) = (2\varphi_4, 2\varphi_8 + 1) = (6, 43)$ and $d = 2\varphi_6 = 16$.

(a) is realized e.g. by the curve $\{x^d = zy^{d-1}\}$, (b) by $\{zy - x^2)^{d/2} = xy^{d-1}$. The cases (c) and (d) appear in Kashiwara's list [52], while (e) and (f) were found by Orevkov in [98].

**19.8.2.** Now we list the graphs $G^m$ for $\partial F$. The computations are straightforward, perhaps except the cases (c) and (d); for these ones we provide more details.

- **Case (a)**

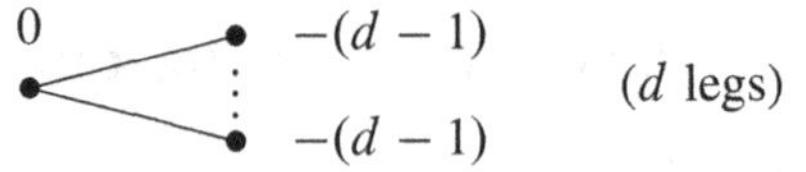


- **Case (b)**

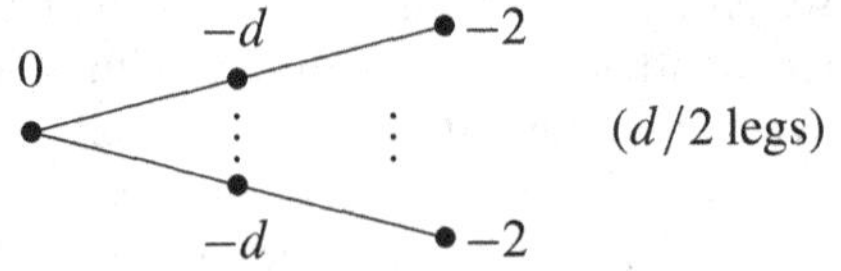


In the examples (a) and (b) the above "normal forms" are *not* negative definite, hence the graphs cannot be transformed by plumbing calculus into a negative definite graph (without changing the orientation).

- **Case (c)** Let $\Gamma$ be the minimal resolution graph of the Brieskorn isolated hypersurface singularity $\{u^a + v^b + w^d = 0\}$, where $(a,b,d) = (\varphi_{j-2}^2, \varphi_j^2, \varphi_{j-2}\varphi_j)$. This is a star-shaped graph, it can be determined via the procedure 5.3 or 5.3.3. Here we will indicate the steps of 5.3.3.

Some arithmetical properties of the Fibonacci numbers enter as important ingredients in the computations. Namely, we need the facts that $\gcd(\varphi_{j-2}, \varphi_j) = 1$, and for $j$ odd one also has:

$$\varphi_{j-2} \cdot \varphi_{j+2} = \varphi_j^2 + 1, \quad \varphi_{j-2} + \varphi_{j+2} = 3\varphi_j. \tag{19.8.3}$$

Using the notations from 5.3.3, we will write $(a_1, a_2, a_3) = (\varphi_{j-2}^2, \varphi_j^2, \varphi_{j-2}\varphi_j)$. Then $(d_1, d_2, d_3) = (\varphi_{j-2}, \varphi_j, \varphi_{j-2}\varphi_j)$, $(\alpha_1, \alpha_2, \alpha_3) = (\varphi_{j-2}, \varphi_j, 1)$, and $(\omega_1, \omega_2, \omega_3) = (1, 1, 0)$. Hence, the embedded resolution of the suspension germ is a star shaped graph, whose each leg has only one vertex. There are $\varphi_j$ legs each with Euler number $-\varphi_{j-2}$, and $\varphi_{j-2}$ legs each with Euler number $-\varphi_j$. The central vertex has Euler number $-3$ and genus decoration $g = (\varphi_j - 1)(\varphi_{j-2} - 1)/2$. The arrow of the function germ $w$ has multiplicity 1 and it is supported by the central vertex which has multiplicity $\varphi_{j-2}\varphi_j$.

Therefore, the graph $G^m$ is the following: it is a star shaped graph with all the legs having only one vertex. It has $\varphi_j$ legs each with Euler number $\varphi_{j-2}$, and $\varphi_{j-2}$ legs each with Euler number $\varphi_j$, as well as a leg with Euler number 0. The central vertex has genus decoration $g = (\varphi_j - 1)(\varphi_{j-2} - 1)/2$ and Euler number 3 (although this is not relevant anymore, because of the 0-leg).

This 3-manifold is not irreducible. The irreducible components can be seen if we apply the splitting operation to the unique 0-vertex. Then we get

$2g = (\varphi_j - 1)(\varphi_{j-2} - 1)$ copies of $\overset{0}{\bullet}$, $\varphi_{j-2}$ copies of $\overset{\varphi_j}{\bullet}$, and $\varphi_j$ copies of $\overset{\varphi_{j-2}}{\bullet}$.

• **Case (d)** In this case the graph $G^m$ of $\partial F$ is $\overset{d}{\bullet}$. This follows again from the very special arithmetical properties (19.8.3) of the Fibonacci numbers.

Similarly as in the previous case, we have first to determine the minimal resolution graph $\Gamma$ of the Brieskorn singularity with coefficients $(a, b, d)$. Using (19.8.3) it is easy to verify that these integers are pairwise relatively prime. Moreover, $\varphi_{j\pm 2}$ is not divisible by 3, and one also has the following congruences

$$\varphi_{j-2}\varphi_{j+2} \equiv 1 \ (mod\ \varphi_j) \quad \varphi_{j\pm 2}\varphi_j \equiv -3 \ (mod\ \varphi_{j\mp 2}). \tag{19.8.4}$$

Hence, by a computation, $\Gamma$ consists of a central vertex with three legs. The central vertex has decoration $-2$ (and genus 0), one leg consists of $\varphi_j - 1$ vertices each decorated by $-2$, the second has three vertices with decorations $-2, -2, -\lceil \varphi_{j\pm 2}/3 \rceil$ (where the first $-2$ vertex is connected to the central vertex). The third leg has two vertices with decorations $-3$ and $-\lceil \varphi_{j\mp 2}/3 \rceil$, where the $-3$ vertex is connected to the central vertex. (Note that among the integers $\varphi_{j-2}$ and $\varphi_{j+2}$ exactly one has the form $3k - 1$. The leg whose $\alpha$-invariant has the form $3k - 1$ will be the leg with two vertices, and the other leg will have three vertices.)

By the procedure 19.2 we have to change the orientation (hence the $(-2)$-string transforms into a 2-string) and add one more vertex to the end-curve of the 2-string, which will be decorated by 1. Hence this new extended string can be contracted completely. After this contraction the central curve becomes a 1-curve. This generates 3 more blow-downs and a 0-chain absorption.

As an interesting phenomenon: the graph of the normalization of $V_f$ is $\overset{-d}{\bullet}$. Hence the boundary of the Milnor fiber coincides with the link of the normalization with opposite orientation.

• **Case (e)**

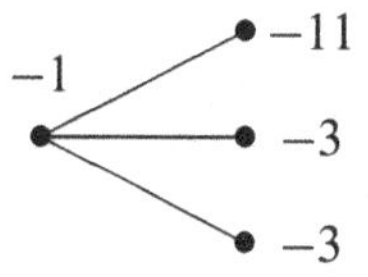

- **Case (f)**

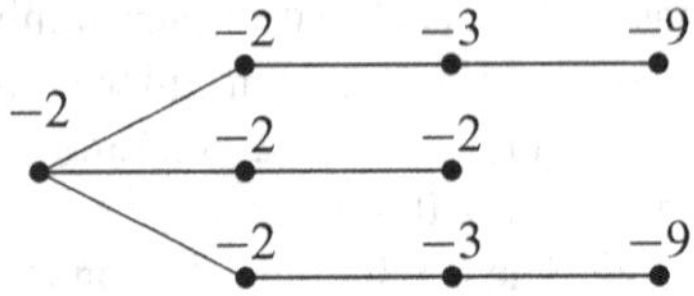


The normal forms in (e) and (f) are *negative definite*, a property which rather rarely happens for non-isolated singularities. In particular, in both cases, $\partial F$ is diffeomorphic (under a diffeomorphism which preserves orientation) with the link of a(n) (elliptic, not complete intersection) normal surface singularity.

**19.8.5.** Looking at the shapes and decorations of the above graphs, one can group them in four categories (a)–(b), (c), (d) and (e)–(f). This is compatible with other groupings based on certain analytic/combinatorial invariants of these curves, see [12, 13].

For example, if $\pi : X \to \mathbb{P}^2$ is the minimal embedded resolution of $C \subset \mathbb{P}^2$, and $\bar{C}$ denotes the strict transform of $C$, then $\bar{C}^2 = d^2 - ab = 3d - a - b - 1$. In the above six cases this value is the following: it is positive for (a) and (b), it is zero for (c), it is $-1$ for (d), and $-2$ for (e) and (f).

Perhaps, the most striking "coincidence" is that the cases (e)–(f) are the only cases when the logarithmic Kodaira dimension of $\mathbb{P}^2 \setminus C$ is 2 (for the first four cases it is $-\infty$), while from the point of view of the present discussion (e)–(f) are the only cases when $\partial F$ is an irreducible Seifert 3-manifold, representable by a negative definite graph (and which is not a lens space).

This suggests that the topology of $\partial F$ probably carries a great amount of analytic information about $f$.

## 19.9 The Weight Filtration of the Mixed Hodge Structure

**Proposition 19.9.1.** *If $f$ is homogeneous then the mixed Hodge structure of $H^*(\partial F)$ satisfies Conjecture 18.1.12.*

*Proof.* Let $d$ denote the degree of $f$, and let $C = \{f = 0\} \subset \mathbb{P}^2$ be the projective curve as above. Recall that $H_1(\mathbb{P}^2 \setminus C, \mathbb{Z}) = \mathbb{Z}^{|\Lambda|}/(d_1, \dots, d_{|\Lambda|})$. Therefore, the representation

$$\pi_1(\mathbb{P}^2 \setminus C, *) \xrightarrow{ab} H_1(\mathbb{P}^2 \setminus C, \mathbb{Z}) \xrightarrow{\sigma} \mathbb{Z}_d,$$

where $ab$ is the Hurewicz epimorphism and $\sigma([a_1, \dots, a_{|\Lambda|}]) = \sum a_\lambda$, is well-defined, it is onto, and it provides a cyclic covering $\pi : Y \to \mathbb{P}^2$, branched along $C$. Moreover, $F := \pi^{-1}(\mathbb{P}^2 \setminus C)$ is a smooth affine variety which can be identified with the open Milnor fiber of $f$, cf. 19.3.

The singularities of $Y$ are isolated and are situated above the singular points of $C$: if $\{f_j'(u,v) = 0\}$ is the local equation of a singular point of $C$, then its cyclic covering ($d$-suspension) $\{w^d = f_j'(u,v)\}$ is the local equation of the corresponding singular point in $Y$ above it. Let $r : \widetilde{Y} \to Y$ be the minimal good resolution of the singularities of $Y$ and set $\widetilde{\pi} := \pi \circ r : \widetilde{Y} \to \mathbb{P}^2$. Then $\widetilde{Y}$ is smooth and $F = \widetilde{Y} \setminus \widetilde{\pi}^{-1}(C)$ is the complement of the normal crossing curve configuration $\widetilde{\pi}^{-1}(C)$. Moreover, the dual graph associated with this curve configuration is exactly $-G$, where $G$ is the graph provided by the discussion 19.2 (which has the same $g$, $c$ and corank as the graph provided by the Main Algorithm). Therefore, for any sufficiently small tubular neighbourhood $\widetilde{T}$ of $\widetilde{\pi}^{-1}(C)$, we have $\partial F = -\partial \widetilde{T}$, cf. 19.3.

The point is that the natural mixed Hodge structure of $H^*(\partial \widetilde{T})$ is transported by this isomorphism into the mixed Hodge structure of $H^*(\partial F)$. This follows, for example, by analyzing the terms of the exact sequence (18.1.7). Namely, the MHS on the local vanishing cohomology of the Milnor fiber $F$ can be identified with Deligne's MHS on the affine smooth hypersurface $\widetilde{Y} \setminus \widetilde{\pi}^{-1}(C)$, see for example Example (3.12) of [122], or [123]. Similarly, the local (Steenbrink) mixed Hodge structures on $H_c^*(F)$ coincides with Deligne's MHS on $H^*(\widetilde{Y}, \widetilde{\pi}^{-1}(C))$. Since the maps $H_c^2(F) \to H^2(F)$ and $H^2(\widetilde{Y}, \widetilde{\pi}^{-1}(C)) \to H^2(\widetilde{Y} \setminus \widetilde{\pi}^{-1}(C))$ can also be identified, their cones also agree. This proves that the MHS of $H^*(\partial F)$ is the same as the MHS of $H^*(\partial \widetilde{T})$.

This combined with 18.1.9(b) (or with [36, (6.9)]) shows that this mixed Hodge structure satisfies Conjecture 18.1.10, hence its dual structure in cohomology satisfies Conjecture 18.1.12. □

Note that the above proof provides an alternative way to show that the plumbing graph constructed in 19.2 is indeed a possible graph for $\partial F$.

**Remark 19.9.2.** As we already mentioned, the oriented handle absorption modifies the integers $c(G^m)$, $g(G^m)$ and corank $A_{G^m}$, and this can happen even in the homogeneous case. For example, if $f = z(xy + z^2)$, (compare also with Sects. 19.7(3c) and 15.1.7):

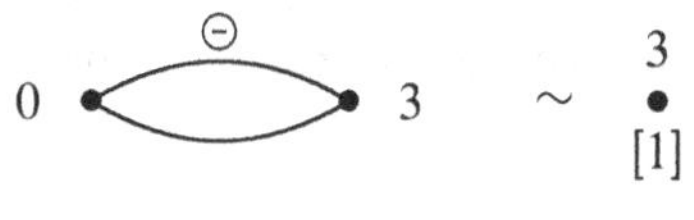


The integers $c(G^m)$, $2g(G^m)$ and corank $A_{G^m}$ read from the left hand side graph provide the ranks $Gr_\bullet^W H_1(\partial F)$, while the right hand side graph does not have this property. (In particular, the topological methods from 23.1 or 23.2 are perfectly suitable to determine the oriented 3-manifold $\partial F$, or even the characteristic polynomial of its algebraic Milnor monodromy, but they are not sufficiently fine to recover the weight filtration of the mixed Hodge structure of $H_1(\partial F)$.)

**Example 19.9.3. The weight filtration of the MHS of $H_1(\partial F)$ is not a topological invariant.**

Consider the homogeneous function $f = xy(xy + z^2)$ of degree $d = 4$. Its graph $G$ under *strictly reduced calculus*, that is, under reduced calculus excluding R5, cf. 4.2.7, can be transformed into

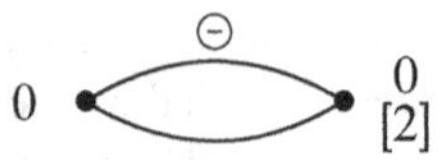


In particular, $c(G) = 1$, $2g(G) = 4$ and $\operatorname{corank} A_G = 2$, and these numbers equal the ranks of $Gr^W_\bullet H_1(\partial F)$.

Next, consider the homogeneous function $f = z(x^4 + y^4)$ of degree $d = 5$. For its graph $G$ see the left graph from 19.10.7 with $c(G) = 3$, $2g(G) = 0$ and $\operatorname{corank} A_G = 4$. In this case these are the ranks of $Gr^W_\bullet H_1(\partial F)$.

Note that using R3 and R5, both graphs can be transformed into $0 \bullet [3]$. Hence, in both cases, $\partial F$ is orientation preserving diffeomorphic with the product of $S^1$ with a surface of genus 3. In particular, the smooth type of $\partial F$ does not determine the weight filtration of $H_1(\partial F)$. (This phenomenon might happen with links of isolated singularities of higher dimension as well, see [126].)

Additionally, the two characteristic polynomials of the algebraic monodromies acting on $H_1(\partial F)$ associated with the above two examples are $(t^2+1)^2(t-1)^3$ and $(t-1)^7$ respectively. Hence, from $\partial F$ one cannot, in general, read the characteristic polynomial. (Note that the multiplicity of $f$ cannot be read either!)

**Example 19.9.4.** In fact, the Milnor fibers of the above two functions treated in 19.9.3 are also different. Indeed, let $S_g$ denote the oriented closed surface of genus $g$. Then for $f = xy(xy + z^2)$ the Milnor fiber is diffeomorphic to $[0,1] \times S^1 \times (S_1 \setminus 2 \text{ points})$ (cf. 23.1), while for $f = z(x^4 + y^4)$, the Milnor fiber is $[0,1] \times S^1 \times (S_0 \setminus 4 \text{ points})$ (cf. Chap. 21). Note that, though these spaces are not homeomorphic, in fact, they are homotopic.

**19.9.5.** The weight filtration on $H_1(\partial F)$ is compatible with the eigenspace decomposition of the algebraic monodromy action. Indeed, from previous computations and from 16.2.11(a) we obtain for the generalized eigenspaces the following decomposition.

**Proposition 19.9.6.**

$$\dim Gr^W_i(H_1(\partial F)_{\lambda=1}) = \begin{cases} \operatorname{corank} A_G = \operatorname{corank} A_{\Gamma_{\mathscr{C}}} = |\Lambda| - 1 & \textit{if } i = -2, \\ 2g(\Gamma_{\mathscr{C}}) & \textit{if } i = -1, \\ c(\Gamma_{\mathscr{C}}) & \textit{if } i = 0, \end{cases}$$

*and*

$$\dim Gr^W_i(H_1(\partial F)_{\lambda\neq 1}) = \begin{cases} 0 & \textit{if } i = -2, \\ 2g(G) - 2g(\Gamma_{\mathscr{C}}) & \textit{if } i = -1, \\ c(G) - c(\Gamma_{\mathscr{C}}) & \textit{if } i = 0. \end{cases}$$

**Remark 19.9.7. The case of weighted homogeneous germs.**

It is a natural task to generalize the above results valid for homogeneous singularities to weighted homogeneous non-isolated singularities. In fact, several examples of the present book are of this type, nevertheless, in their study we did not exploit their weighted homogeneous action.

The general strategy for the study of these germs, which exploits their $\mathbb{C}^*$-action, is rather straightforward: one has to consider the toric resolution associated with their Newton diagram. As an intermediate step, one has to consider a non-singular subdivision of the fan determined by the Newton diagram. This, usually is not unique, and depends on several choices. Then the identification of the curve $\mathscr{C}$ and of the decorations of $\Gamma_{\mathscr{C}}$ can be done via computations specific to toric resolutions.

We completed this program for several examples presented in this book, and using the Main Algorithm and plumbing calculus we verified that they provide the expected answer. Nevertheless, we were not satisfied with our computations: the general statement formulated *only* in terms of the Newton diagram and *independently of the choice of the subdivision* is still missing. Therefore, we decided not to include these results; they will be completed in the near future.

## 19.10 Line Arrangements

Assume that $C$ is a projective line arrangement. We will use the notations of 8.2. Recall (cf. 19.3.7) that

$$|\mathscr{A}(\Gamma_{\mathscr{C}})| = |\mathscr{A}(G)| = d, \quad \operatorname{corank} A_G = d-1, \text{ and } g(\Gamma_{\mathscr{C}}) = 0.$$

Moreover, since the covering data of $G$ over $\Gamma_{\mathscr{C}}$ is trivial, one has

$$c(G) = c(\Gamma_{\mathscr{C}}) = \sum_{j\in\Pi}(m_j-1) - (d-1),$$

and by 13.6.3

$$2g(G) = \sum_{j\in\Pi}(m_j-2)\big(\gcd(m_j,d)-1\big).$$

**19.10.1. Characteristic polynomial.** Since $\mathfrak{n}_w = \mathfrak{n}_e = 1$ for $w \in \mathscr{W}(\Gamma_{\mathscr{C}})$ and $e \in \mathscr{E}_{\mathscr{W}}(\Gamma_{\mathscr{C}})$, Corollary 19.4.1 via a computation transforms into

**Theorem 19.10.2.** *Let $\{L_\lambda\}_{\lambda\in\Lambda}$ be an arrangement with $d = |\Lambda|$ lines and with singular points $\Pi$. Then the characteristic polynomial of the monodromy acting on $H_1(\partial F)$ is*

$$(t-1)^{|\Pi|}\cdot\prod_{j\in\Pi}(t^{(m_j,d)}-1)^{m_j-2}.$$

**19.10.3. Jordan blocks of $M_{\Phi,hor}$ and $M_{\Phi,ver}$.** Since $M'_{j,hor}$ is semisimple, from 19.5 we get

$$\#^2_\lambda M_{\Phi,hor} = \#^2_\lambda M_{\Phi,ver} = 0 \text{ for } \lambda \neq 1, \text{ and}$$
$$\#^2_1 M_{\Phi,hor} = \#^2_1 M_{\Phi,ver} = c(\Gamma_{\mathscr{C}}).$$

**Example 19.10.4. The generic arrangement.** Consider the generic arrangement: all the intersection points of the $d$ lines are transversal. In this case there are $|\Pi| = d(d-1)/2$ intersection points, all with $m_j = 2$. Therefore, the characteristic polynomial is just $(t-1)^{|\Pi|}$.

A possible graph for $\partial F$ can be constructed as follows.

Consider the complete graph $\mathfrak{G}$ with $d$ vertices (i.e. any two different vertices are connected by an edge). Decorate all these vertices by $-1$. Put on each edge $e$ of $\mathfrak{G}$ a new vertex with decoration $-d$. In this way, the edge $e$ is "cut" into two edges; decorate one of them by $\ominus$ (and do this with all the edges of $\mathfrak{G}$).

Notice that any $(-1)$-vertex is adjacent to $d-1$ vertices (all decorated by $-d$). Hence, if $d \leq 3$, this graph is not minimal, but otherwise in this way we get the "normal form". The graph has $(d-1)(d-2)/2$ cycles and corank $A = d-1$.

**Example 19.10.5. The A$_3$ arrangement.** Consider the arrangement from 8.2.1. The characteristic polynomial is $(t^3-1)^4 \cdot (t-1)^7$. The graph can be deduced easily from $\Gamma_{\mathscr{C}}$, which is presented in 8.2.1. There are 4 vertices with $g = 1$, 6 cycles and corank $A_G = d-1 = 5$.

**Example 19.10.6. The pencil.** Assume that all the lines contain a fixed point. (For example, $f = x^d + y^d$.) Then the characteristic polynomial is $(t-1)(t^d-1)^{d-2}$, and $G^m$ is

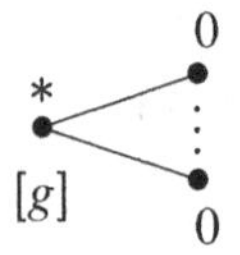


$d$ legs and $g = (d-1)(d-2)/2$

In fact, $\partial F \approx \#_{(d-1)^2} S^2 \times S^1$.

**Example 19.10.7.** Assume that $f = z(x^{d-1} + y^{d-1})$. Then the characteristic polynomial is $(t-1)^{2d-3}$, and $G^m$ is

$(d-1)$ 0-vertices

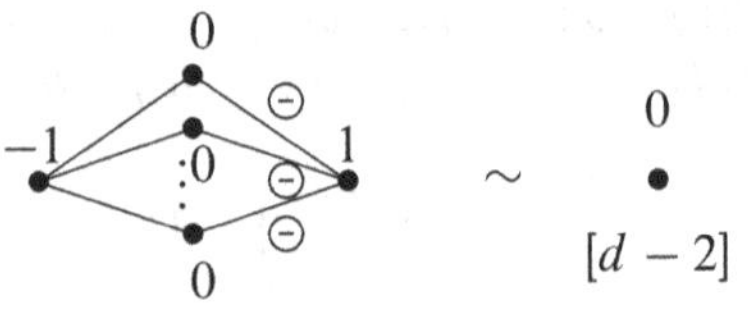


In particular, for any oriented surface $S$, the product $S \times S^1$ can be realized as $\partial F$.

**19.10.8. The torsion of the integral homology $H_1(\partial F, \mathbb{Z})$.** In general, even for homogeneous $f$, $H_1(\partial F, \mathbb{Z})$ might have nontrivial torsion. It would be important to characterize completely this torsion group in the case of arrangements. The interest in such a question is motivated by a conjecture which predicts that for arrangements $H_1(F, \mathbb{Z})$ has no torsion. Hence it is natural to attack this conjecture via the epimorphism $H_1(\partial F, \mathbb{Z}) \to H_1(F, \mathbb{Z})$.

**Example 19.10.9.** Assume that $f$ is the generic arrangement with $d = 4$. Then the torsion part of $H_1(\partial F, \mathbb{Z})$ is $\mathbb{Z}_4$.

Indeed, by 15.1.3, the torsion part of $H_1(\partial F, \mathbb{Z})$ is exactly the torsion part of $\operatorname{coker} A_G$. Since the intersection matrix $A$ is determined explicitly in 19.10.4, a computation provides the result.

In fact, for generic arrangements and larger $d$, the torsion part of $H_1(\partial F, \mathbb{Z})$ is even bigger. One can prove that $A \otimes \mathbb{Z}_d$ has corank $\geq (d^2 - 3d + 4)/2$, which is considerably larger than $\operatorname{corank} A = d - 1$.

# Chapter 20
# Cylinders of Plane Curve Singularities: $f = f'(x, y)$

## 20.1 Using the Main Algorithm: The Graph $G$

Assume that $f(x, y, z) = f'(x, y)$, as in 9.1. Assume that $f'$ has $\# = \#(f')$ local irreducible components, and let $\mu = \mu(f')$ be its Milnor number.

For $g(x, y, z) = z$ a graph $\Gamma_{\mathscr{C}}$ is determined in 9.1. In this section we determine the graph $G^m$ using the main steps of the collapsing algorithm. Nevertheless, we provide certain numerical invariants for the original $G$ as well.

Let

$$(m_w) \qquad\qquad (m'_w)$$
$$\bullet\rule{4cm}{0.4pt}\bullet$$
$$[g] \qquad\qquad [g']$$
$$w \qquad\qquad w'$$

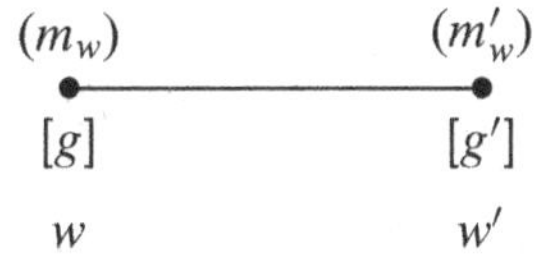

be an edge $e$ of the minimal embedded resolution graph $\Gamma' = \Gamma(\mathbb{C}^2, f')$ of $f'$. Then using the recipe of 9.1, the transformation step from 10.1.3, and the algorithm 10.2, this edge will be replaced in $G$ by $n_e = (m_{w_1}, m_{w_2})$ strings with the decorations:

$$\underset{[\tilde{g}_w]}{\bullet}\ \ \underset{0}{\bullet}\ \ominus\ \overset{\frac{m_w + m'_w}{n_e}}{\bullet}\ \ominus\ \underset{0}{\bullet}\ \ \underset{[\tilde{g}_{w'}]}{\bullet}$$

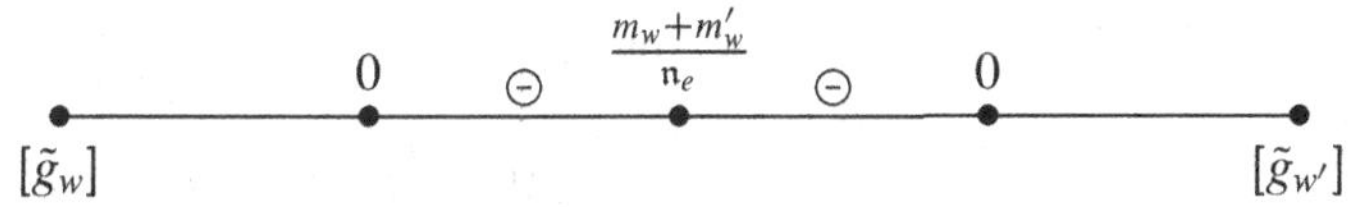

Compare also with 12.1.3.

Above $w$ there are $n_w$ vertices in $G$, where $n_w$ is the greatest common divisor of $m_w$ and all the multiplicities of the adjacent vertices of $w$ in $\Gamma'$. Their genus decoration $\tilde{g}_w$ is given by

$$n_w(2 - 2\tilde{g}_w) = (2 - \delta_w)m_w + \sum_{e \text{ adjacent to } w} n_e.$$

A. Némethi and Á. Szilárd, *Milnor Fiber Boundary of a Non-isolated Surface Singularity*, 201
Lecture Notes in Mathematics 2037, DOI 10.1007/978-3-642-23647-1_20,
© Springer-Verlag Berlin Heidelberg 2012

There is a similar picture for edges supporting arrowheads. Note that

$$|\mathscr{A}(G)| = \#,$$

the number of local irreducible components of $f'$.

In general, $G$ is not a tree, the number $c(G)$ of its cycles is given by

$$1 - c(G) = \sum_{w \in \mathscr{W}(\Gamma')} \mathfrak{n}_w - \sum_{e \in \mathscr{E}_w(\Gamma')} \mathfrak{n}_e. \tag{20.1.1}$$

After executing all the 0-chain absorptions we get

$G^{mod}$ :

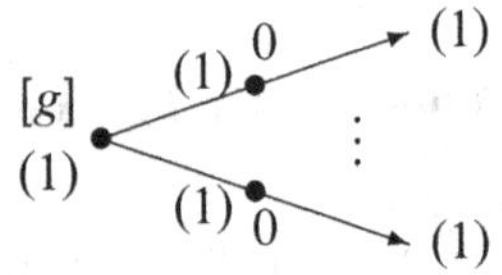


(# arrows)

Here, the Euler number of the central vertex is missing, since it is irrelevant. The genus $g$ is determined as a sum provided by the 0-chain absorption formula: $2g = 2c(G) + 2\sum_{w \in \mathscr{W}(\Gamma')} \mathfrak{n}_w \tilde{g}_w$, which via the above identities is equal to $2 - \# - \sum_w (2 - \delta_w) m_w$. Hence, by A'Campo's formula (5.2.8) for $\mu$, we get

$$g = (\mu + 1 - \#)/2.$$

**20.1.2.** Clearly, the fiber $F_\Phi$ of $\Phi = (f', z)$ is the same as the Milnor fiber $F'$ of $f'$. Moreover, the vertical monodromy of $F_\Phi$ is isotopic to the identity (this follows either from 17.2.6, or by the observation that $\Delta_\Phi = \{c = 0\}$, hence $\Phi$ is a trivial fibration over $D_\delta$). Therefore, from (13.6.6) we get

$$\mu = 2g(G) + 2c(G) + \# - 1. \tag{20.1.3}$$

From the above diagram of $G^{mod}$, and by 15.1.2 we get

$$\begin{aligned} \dim H_1(\partial F) &= 2g(G^{mod}) + c(G^{mod}) + \operatorname{corank} A_{G^{mod}} \\ &= (\mu + 1 - \#) + 0 + (\# - 1) = \mu. \end{aligned}$$

On the other hand, again by 15.1.2 applied for $G$, we get

$$\mu = 2g(G) + c(G) + \operatorname{corank} A_G. \tag{20.1.4}$$

Then, from (20.1.3), (20.1.4) and Corollary 15.1.5 we obtain

$$\begin{cases} \operatorname{corank} A_G = c(G) + \# - 1, \\ \operatorname{corank} (A, \mathfrak{I})_G = c(G) + \#. \end{cases} \tag{20.1.5}$$

This shows that the upper bounds from 15.2.1 can be realized.

In fact, $c(G)$ has an intrinsic meaning (hence, by (20.1.5), $\operatorname{corank}(A, \mathfrak{I})_G$ and $\operatorname{corank} A_G$ too): the expression (20.1.1) combined with (5.2.13) gives that

$$c(G) = \text{the number of 2-Jordan blocks of the monodromy of } f'. \tag{20.1.6}$$

**20.1.7.** If the arrowheads and multiplicities of $G^{mod}$ above are deleted, we get a graph which coincides with $\widehat{G}$ from Chap. 12.

It has $c(\widehat{G}) = 0$ and $\operatorname{corank} A_{\widehat{G}} = \# - 1$:

$\widehat{G}$ :

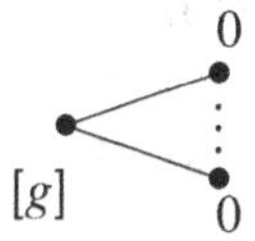


# legs and $g = (\mu + 1 - \#)/2$

By the "splitting operation" R6 of the plumbing calculus, we get another possible non-connected plumbing graph for $\partial F$: the disjoint union of $\mu$ vertices (without any edges), all of them decorated with Euler number and genus zero. In other words:

$$\partial F \approx \#_\mu S^2 \times S^1.$$

Although this presentation of $\partial F$ can be deduced by other methods too (see e.g. Sect. 20.2), the graph of $G$ associated with the open book decomposition of $g = z$, or any graph $G$ computed in this way associated with any germ $g$, is a novelty of the present method.

**20.1.8. The characteristic polynomial.** Since $M_{\Phi,ver} = id$, the characteristic polynomial $P_{H_1(\partial F)}$ of the Milnor monodromy acting on $H_1(\partial F)$ is the formula given in 13.4.11(a). It turns out that this expression coincides with the expression of the characteristic polynomial of the monodromy of the isolated plane curve singularity $f'$ provided by A'Campo's formula (5.2.8).

## 20.2 Comparing with a Different Geometric Construction

Let $F'$ be the Milnor fiber of $f'$. In the above situation it is easy to see (since $\Delta_\Phi = \{c = 0\}$) that $F \approx F' \times D$, where $D$ is a real 2-disc. In particular, we also have the following geometric description for $\partial F$:

$$\partial F = F' \times S^1 \cup_{\partial F' \times S^1} \partial F' \times D. \tag{20.2.1}$$

Using the Mayer–Vietoris exact sequence for this decomposition, one gets that

$$H_1(\partial F, \mathbb{Z}) = H_1(F', \mathbb{Z}). \tag{20.2.2}$$

Since the monodromy acts on this sequence, we also obtain that the monodromy on $H_1(\partial F, \mathbb{Z})$ is the same as the monodromy of the plane curve singularity $f'$ acting on $H_1(F', \mathbb{Z})$. In particular, in this way we get the Jordan block structure of the monodromy acting on $H_1(\partial F, \mathbb{Z})$ as well.

This shows that, in general, the *algebraic monodromy acting on $H_1(\partial F)$ is not finite*: take e.g. for $f'$ the germ from 9.1.1.

## 20.3 The Mixed Hodge Structure on $H_1(\partial F)$

The isomorphism (20.2.2) can also be proved as follows. Here we prefer to discuss the cohomological case, which is more traditional from the point of view of MHS.

Since $H^2_c(F) = H^0_c(F') = 0$, by (18.1.7) the inclusion induces an isomorphism $H^1(F) \to H^1(\partial F)$. Moreover, the inclusion of $F'$ into $F$ (cut out by $z = 0$) induces also an isomorphism $H^1(F) \to H^1(F')$. Being induced by inclusions, these isomorphisms preserve the mixed Hodge structures. Therefore, the mixed Hodge structures on $H^1(F')$ and $H^1(\partial F)$ coincide.

The mixed Hodge structure of $H^1(F')$ is Steenbrink's MHS defined on the vanishing cohomology of the plane curve singularity $f'$ [122, 124]. Its weight filtration is compatible with the generalized eigenspace decompositions. According to the general theory, $H^1(F')_{\lambda=1}$ is pure of weight 2, and it has rank $\# - 1$. On the other hand, $H^1(F')_{\lambda\neq 1}$, in general, has weights 0, 1 and 2, and the weight filtration is the monodromy weight of the algebraic monodromy. In particular, the ranks of $Gr^W_0 H^1(F')_{\lambda\neq 1}$ and $Gr^W_2 H^1(F')_{\lambda\neq 1}$ are equal, and they agree with the number of 2-Jordan blocks of the monodromy (recall that the monodromy has no 2-blocks with eigenvalue one). Hence, this rank is exactly $c(G)$ by (20.1.6). In particular, the rank of $Gr^W_0 H^1(F')$ is $c(G)$, while the rank of $Gr^W_2 H^1(F')$ is $c(G) + \# - 1 = \operatorname{corank} A_G$, cf. (20.1.5). Hence, by dimension computation, we get that the remaining $Gr^W_1 H^1(F')$ has rank $2g(G)$. This supports Conjecture 18.1.12:

**Corollary 20.3.1.** *If $f$ is a cylinder of an isolated plane curve singularity then the mixed Hodge structure of $H^*(\partial F)$ satisfies Conjecture 18.1.12.*

Note that the information about the ranks of $Gr^W_\bullet H^1(F')$ cannot be read from $\widehat{G}$: the graph $\widehat{G}$ contains information about $\#$ and $\mu$ only.

# Chapter 21
# Germs $f$ of Type $zf'(x,y)$

## 21.1 A Geometric Representation of $F$ and $\partial F$

In this section we assume that $f(x,y,z) = zf'(x,y)$, where $f'$ is an isolated plane curve singularity.

Similarly as in the case of cylinders (or, in the case of "composed singularities" [81]), consider the ICIS $\bar{\Phi}' : (\mathbb{C}^3, 0) \to (\mathbb{C}^2, 0)$ given by $\bar{\Phi}' = (f'(x,y), z)$. By similar notations as in Chap. 3, the Milnor fiber $F$ of $f$ is

$$F = B^3_\epsilon \cap \bar{\Phi}'^{-1}(\{cd = t\} \cap D^2_\eta),$$

where $0 < t \ll \eta \ll \epsilon$. For any $\eta > 0$ consider the disc $D_\eta := \{c : |c| \leq \eta\}$. Then, by isotopy, the above representation of $F$ can be transformed into

$$F = B^3_\epsilon \cap \bar{\Phi}'^{-1}(D_\eta \times \{t\} \setminus D^\circ_{\eta'} \times \{t\}),$$

for some $0 < \eta' \ll \eta$. From this the variable $z$ can be eliminated, that is, if $B^2_\epsilon$ is the $\epsilon$-ball in the $(x,y)$-plane, then

$$F = B^2_\epsilon \cap (f')^{-1}(D_\eta \setminus D^\circ_{\eta'}). \tag{21.1.1}$$

It can also be verified that the monodromy on $F$ is isotopic to the identity, since it is the rotation by $2\pi$ of the annulus $D_\eta \setminus D_{\eta'}$.

In particular, the homotopy type of $F$ is the same as the homotopy type of the complement of the link of $f'$ in $S^3_\epsilon$.

(21.1.1) provides a geometric picture for $\partial F$ as well, and proves that its monodromy action is trivial.

**21.1.2.** Projecting $D_\eta \setminus D_{\eta'}$ to $S^1 = \partial D_\eta$, and composing with $f'$ we get a map $\partial F \to S^1$ which is a locally trivial fibration. Hence,

$$\partial F \text{ fibers over } S^1.$$

A. Némethi and Á. Szilárd, *Milnor Fiber Boundary of a Non-isolated Surface Singularity*, Lecture Notes in Mathematics 2037, DOI 10.1007/978-3-642-23647-1_21,

**21.1.3.** From (21.1.1) one can read the following plumbing representation for $\partial F$. In order to understand the geometry behind the statement, let us analyze different constituent parts of $\partial F$. Notice that $B^2_\epsilon \cap (f')^{-1}(\partial D_\eta)$ can be identified with the complement of the link of $f'$ in $S^3$. Similarly, $B^2_\epsilon \cap (f')^{-1}(\partial D_{\eta'})$ is the same, but with opposite orientation. Moreover, they are glued together along their boundaries in a natural way. Therefore, the plumbing graph can be constructed as follows.

Take the (minimal) embedded resolution graph of $(\mathbb{C}^2, V_{f'})$. This has # arrows, where # is the number of irreducible components of $f'$. Keep all the Euler numbers and delete all the multiplicity decorations. Let the schematic form of the result be the next "box":

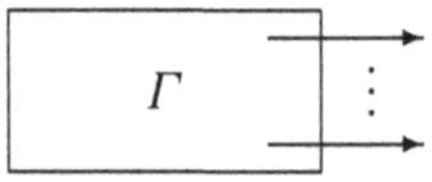


Then a possible plumbing graph for $\partial F$ is:

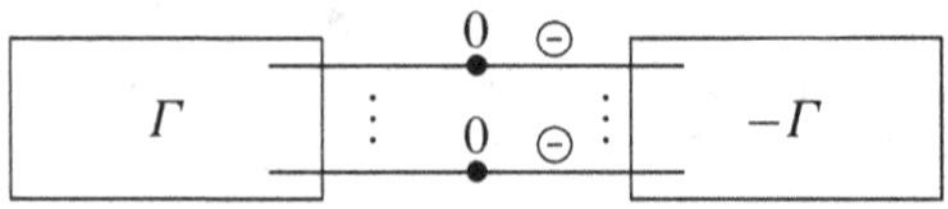


Here, in $-\Gamma$, we change the sign of all Euler numbers and edge-decorations, that is, we put $\ominus$ on all edges.

Notice that any 3-manifold $M$ obtained in this way is orientation preserving diffeomorphic with the manifold obtained by changing its orientation: $M \approx -M$.

The statement about the above shape of the graph can be tested using the examples listed in Sect. 9.2 as well.

**Example 21.1.4.** The first (and simplest) example, when $f' = x^{d-1} + y^{d-1}$, hence $f$ defines an arrangement, was already considered in 19.10.7. Its graph produced by the Main Algorithm is the left diagram of 19.10.7 supporting (together with the characteristic polynomial computation) the above statements.

**Example 21.1.5.** Assume that $f = z(x^2 + y^3)$ and take $g$ to be a generic linear form. The graph $\Gamma_{\mathcal{C}}$ is given in 9.2.1. Running the algorithm and reduced calculus, a possible $G^m$ is

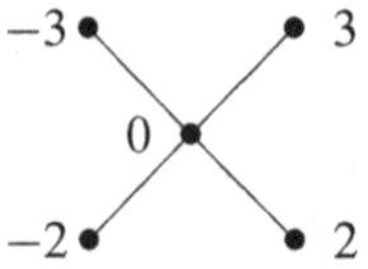


which is compatible with the predicted form from 21.1.3.

**Example 21.1.6.** More generally, if $f' = x^a + y^b$, then in the graph of 21.1.3, all the 0-vertices can be eliminated by 0-chain or handle absorption, hence we get a star-shaped graph with four legs and central vertex with genus $\gcd(a, b) - 1$. In particular, $\partial F$ has a Seifert structure.

**Remark 21.1.7.** Let us consider again the ICIS $\Phi = (f, g)$ as in the original construction.

(a) This family might also serve as a testing family for some of the characteristic polynomial formulae 13.4.11, regarding the vertical monodromies $M'_{j,ver}$ of the transversal types (which, in fact, are much harder to test).

Indeed, if $f = zf'$ and $g$ is a generic linear function, then $\Sigma_f$ has two components. $\Sigma_1 = \{x = y = 0\}$ with $d_1 = 1$, and $\Sigma_2 = \{z = f' = 0\}$. Let us concentrate on the first component $\Sigma_1$. The transversal type $T\Sigma_1$ is the same as the type of $f'$ (in two variables). Hence, the Milnor fiber $F'_1$ of the transversal type can be identified with the Milnor fiber of $f'$.

Since $\{f'e^{it} = \delta\} = \{f' = \delta e^{-it}\}$, we get that the vertical monodromy of $F'_1$ coincides with the inverse of the Milnor monodromy acting on the Milnor fiber of $f'$. Hence, the characteristic polynomial of $M'_{1,ver}$, determined by 13.4.11(c), should coincide with the characteristic polynomial of $f'$ provided by the classical A'Campo formula (5.2.8). The interested reader is invited to verify this on all the available graphs $\Gamma_{\mathscr{C}}$.

(b) Let us also test the invariants of the fiber of $\Phi$. Let us write $g$ as $z + g'(x, y)$, where $g'$ is a generic linear form with respect to $f'(x, y)$. Then, solving the system $zf'(x, y) = c$ and $z + g'(x, y) = d$, we get the fiber $F_\Phi$ of $\Phi$. Eliminating $z$ we get $(d - g'(x, y))f'(x, y) = c$, hence the fiber $F_\Phi$ is the same as the Milnor fiber of the plane curve singularity $g'f'$.

For example, in the case of $f' = x^2 + y^3$, whose graph $\Gamma_{\mathscr{C}}$ is given in 9.2.1, the first formula of (13.4.10) provides for $H_1(F_\Phi)$ the rank 5, which is the Milnor number of $y(x^2 + y^3)$.

It is interesting to identify and analyze the vertical monodromy of $F_\Phi$ via the local equation $(\eta e^{it} - g'(x, y))f'(x, y) = \delta$ in two variables with $0 < \delta \ll \eta \ll \epsilon$.

Note that the present strategy provides a method for the study of the monodromy of such a deformation for an arbitrary plane curve singularity pair $(f', g')$: only has to be computed the graph $\Gamma_{\mathscr{C}}$ for $(zf', z + g')$.

**21.1.8.** While computing the plumbing graph of $\partial F$ for this family $f = zf'$ via the Main Algorithm, the following amazing fact emerged: although the graph $\Gamma_{\mathscr{C}}$ shows no symmetry, after running the algorithm and calculus the output graph has the symmetry (up to orientation) predicted in 21.1.3. For example, looking at the starting graph 9.2.1, we realize absolutely no symmetry, nor even the hidden potential presence of symmetry. Indeed, the two parts of $\Gamma_{\mathscr{C}}$, which provide $\Gamma$ and $-\Gamma$ of 21.1.3 respectively, are rather different; they codify two different geometric situations. Nevertheless, after calculus we get the two symmetric parts represented by $\Gamma$ and $-\Gamma$.

To construct a resolution $r$ which produces $\Gamma$ (the embedded resolution of a plane curve singularity) is rather natural (see e.g. the case of cylinders), but to get a resolution $r$ which (or part of it) produces $-\Gamma$ in a natural way, is very tricky. But, in fact, this is what $\Gamma_{\mathscr{C}}$ does: a part of it produces $\Gamma$, another part of it produces $-\Gamma$. This anti-duality is still a mystery for the authors. This shows how hard it is to recognize and follow global geometric properties by manipulating (local) equations/resolutions.

# Chapter 22
# The $T_{*,*,*}$-Family

## 22.1 The Series $T_{a,\infty,\infty}$

We start our list of examples with the series associated with $T_{\infty,\infty,\infty}$, where $T_{\infty,\infty,\infty}$ denotes the germ $f = xyz$. This germ was already clarified either as an arrangement, see Sect. 19.7(3e), or by the algorithm of Chap. 21, hence $\partial F = S^1 \times S^1 \times S^1$.

Next we consider the series $T_{a,\infty,\infty}$ with one-parameter $a \geq 2$ given by $f = z^a + xyz$. For $g$ we take the generic linear function.

The case $a = 2$ can be rewritten as $z^2 = x^2y^2$ and its graph $\Gamma_{\mathscr{C}}$ is given in 9.3.2, Case 1. The case $a = 3$ is treated in Sect. 19.7(3c), the general $a > 3$ in 9.4. The Main Algorithm provides for the boundary of the Milnor fiber $\partial F$ the plumbing graph

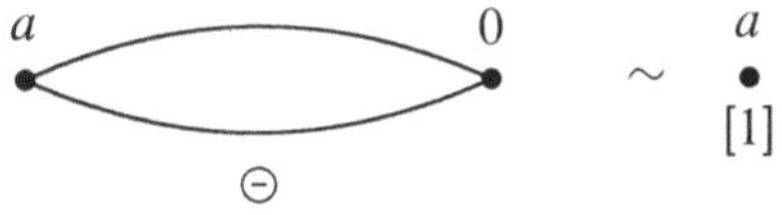


## 22.2 The Series $T_{a,2,\infty}$

Running the Main Algorithm and calculus for the graphs of 9.4.7 and 9.3.3, we get for $\partial F$ the plumbing graph:

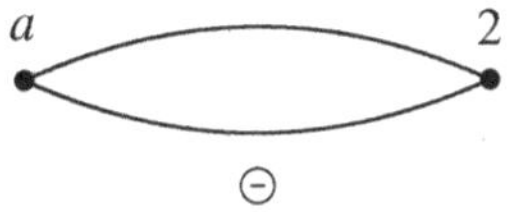


A. Némethi and Á. Szilárd, *Milnor Fiber Boundary of a Non-isolated Surface Singularity*, Lecture Notes in Mathematics 2037, DOI 10.1007/978-3-642-23647-1_22,

In fact, $\partial F$ for $T_{a,b,\infty}$ ($f = x^a + y^b + xyz$) is given by the plumbing graph:

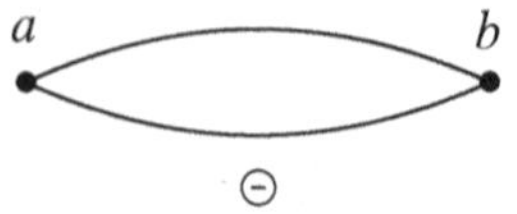


By a different geometric argument this was verified by the first author's student B. Sigurdsson in [120]. For $a = b = 3$ it follows from 10.4.4 or Sect. 19.7(3b).

# Chapter 23
# Germs $f$ of Type $\tilde{f}(x^a y^b, z)$: Suspensions

In this chapter we again provide an alternative way to identify the boundary of the Milnor fiber for a special class of germs.

## 23.1 $f$ of Type $\tilde{f}(xy, z)$

First, we sketch a geometric construction which provides $\partial F$ and its monodromy, provided that $f(x, y, z) = \tilde{f}(xy, z)$, where $\tilde{f}$ is an isolated plane curve singularity in variables $(u, z)$.

We will use the following notations: $\tilde{F} = \{\tilde{f} = \eta\}$ is the Milnor fiber and $\tilde{\mu}$ the Milnor number of $\tilde{f}$. Define $I$ as zero if $u$ is a component of $\tilde{f}$, otherwise $I$ denotes the intersection multiplicity $(\tilde{f}, u)_0$ at the origin. Set $\Delta := \tilde{F} \cap \{u = 0\}$. Then, clearly, $|\Delta| = I$.

Next, consider the Morse singularity $h : (\mathbb{C}^2, 0) \to (\mathbb{C}, 0)$, $h(x, y) = xy$. Then the Milnor fiber $F$ of $f$ can be reproduced via the projection $\bar{\Phi} : (\mathbb{C}^3, 0) \to (\mathbb{C}^2, 0)$ given by $(x, y, z) \mapsto (h(x, y), z) = (u, z)$. Indeed, $\bar{\Phi}$ maps $F$ onto $\tilde{F}$. The fibers of $\bar{\Phi}$ are as follows: above generic points of $\tilde{F}$ we have the Milnor fiber $F_h = S^1 \times [0, 1]$ of $h$, while over the special points $P \in \Delta$ we have the contractible central fiber $D \vee D$ of $h$ (here $D$ is the real 2-disc).

Hence, $\partial F$ decomposes into two parts, one of them is $(S^1 \sqcup S^1) \times \tilde{F}$, while the other is $\bar{\Phi}^{-1}(\partial \tilde{F})$, which is a locally trivial fiber bundle over $\partial \tilde{F}$ with fiber $F_h = S^1 \times [0, 1]$. The monodromy of this bundle is given by the variation map of $\bar{\Phi}$, hence it is the composition of $I$ Dehn twists (corresponding to the variation maps around the Morse points $P_i$). Therefore, if we define the "double of $\tilde{F}$" as $d\tilde{F} := \tilde{F} \sqcup_{\partial \tilde{F}} (-\tilde{F})$, then $\partial F$ is an $S^1$-bundle over $d\tilde{F}$: two trivial copies of $S^1 \times \tilde{F}$ are glued above $\partial \tilde{F}$ so that the Euler number of the resulting $S^1$-bundle is $I$.

A. Némethi and Á. Szilárd, *Milnor Fiber Boundary of a Non-isolated Surface Singularity*, Lecture Notes in Mathematics 2037, DOI 10.1007/978-3-642-23647-1_23,

Notice that the genus of $d\tilde{F}$ is exactly $\tilde{\mu}$, hence:

$$\partial F \textit{ is given by the plumbing graph} \quad \overset{I}{\underset{[\tilde{\mu}]}{\bullet}}$$

In particular, any $S^1$-bundle with arbitrary genus and non-negative Euler number can be realized as $\partial F$ (and any such bundle might have many realizations by rather different singularities).

It is interesting to note that even if $\tilde{f}$ has many Puiseux pairs (hence its embedded resolution graph has many rupture vertices), $\partial F$ is still Seifert –, in fact, it is an $S^1$-bundle.

**Example 23.1.1.** Assume that $f = z^a + xyz$ with $a \geq 2$ as in 22.1. Then $\tilde{F} = \{z^a + uz = \eta\}$ $(0 < \eta \ll 1)$ and $\Delta = \{u = 0,\ z^a = \eta\}$. Then $\tilde{F}$, via the projection $(u, z) \mapsto z$, is diffeomorphic with the annulus $A := \{x : \eta_1 \leq |z| \leq \eta_2\}$, where $\eta_1 < \sqrt[a]{\eta} < \eta_2$, and by this identification its special points from $\Delta$ are $\cup_{i=1}^a P_i = \{z^a = \eta\}$. Notice that $d\tilde{F}$ is a torus. Moreover, the monodromy is the rotation of $A$, hence it is isotopic to the identity. In particular, we get (a fact compatible with 22.1):

$$\partial F \textit{ is the torus bundle} \quad \overset{a}{\underset{[1]}{\bullet}} \quad \textit{with trivial monodromy.}$$

**23.1.2.** More generally, in the situation of 23.1, the monodromy on $\partial F$ is induced by the monodromy of $\tilde{f}$. Hence, if $P_{\partial F}$ (resp. $P_{\tilde{f}}$) denotes the characteristic polynomial of the algebraic monodromy acting on $H_1(\partial F)$ (resp. $H_1(\tilde{F})$), then

$$P_{\partial F}(t) = \left(P_{\tilde{f}}(t)\right)^2 \cdot (t-1)^{1-sign(I)} \tag{23.1.3}$$

where $sign(I)$ is 1 for $I > 0$ and zero for $I = 0$.

The above result can be applied for $f = xyz$, $f = z(z^2 + xy)$, $f = xy(z^2 + xy)$ and $f = z^a + x^k y^k$. These cases can be compared with Sect. 19.7(3e), (3c), (4b') or (4c'). Moreover, (23.1.3) can also be compared with the formula from Theorem 16.2.8 valid for the characteristic polynomial.

**23.1.4.** The above construction provides the structure of the Milnor fiber $F$ as well: one may extract from it key information, such as the Euler characteristic, zeta-function of the monodromy, etc.

## 23.2 $f$ of Type $\tilde{f}(x^a y^b, z)$

Assume now that $a$ and $b$ are two positive relative prime integers. Then, the discussion of 23.1 can be modified as follows.

Fix any isolated plane curve singularity $\tilde{f}(u, z)$, and replace $h$ by $u = h(x, y) := x^a y^b$ in order to get $f := \tilde{f} \circ \bar{\Phi}$. Then all the arguments of 23.1 remain valid with the following modification.

The difference is that in the case $h = xy$, $\bar{\Phi}$ restricted on $\partial F$ is a trivial fibration above any small neighbourhood of any point of $\Delta$. In the new situation of $h = x^a y^b$, above a small neighbourhood of a point $P$ of $\Delta$, $\bar{\Phi}|_{\partial F}$ is *not* a fibration; in fact, in $\partial F$ two special $S^1$-fibers appear above $P$. One of them has the same local neighbourhood (Seifert structure) as $\{x = 0\}$ in $S^3 = \{|x|^2 + |y|^2 = 1\}$ with $S^1$-fibers/orbits cut out by the family $x^a y^b$ = constant; the other has local behaviour as $\{y = 0\}$ in the same space. These two local Seifert neighbourhoods in $S^3$ are well-understood, they are guided by the continued fractions of $a/b$, respectively $b/a$.

Therefore, we get the following result:

**Theorem 23.2.1.** *Let $\tilde{f}(u, z)$ be an isolated plane curve singularity, and $a$ and $b$ two positive relative prime integers. Then $\partial F$ associated with $f = \tilde{f}(x^a y^b, z)$ is a Seifert 3-manifold whose minimal star-shaped plumbing graph can be constructed as follows:*

*(i) The central vertex has genus $\tilde{\mu}$, the Milnor number of $\tilde{f}$, while its Euler number is $I$, where $I$ is zero if $u|\tilde{f}$, otherwise $I = (\tilde{f}, u)_0$.*

*(ii) Let $a/b = [p_0, p_1, \ldots, p_s]$, respectively $b/a = [q_0, q_1, \ldots, q_t]$ be the Hirzebruch–Jung continued fraction expansions of $a/b$ and $b/a$ with $p_0, q_0 \geq 1$, $p_i \geq 2$ for $i \geq 1$ and $p_j \geq 2$ for $j \geq 1$. Then the graph has $2I$ legs (with all vertices having genus-decoration zero). $I$ strings have length $s$, the vertices are decorated by Euler numbers $p_1, \ldots, p_s$ such that the vertex decorated by $p_s$ is the one glued to the central vertex; while the other $I$ legs are strings decorated by Euler numbers $q_1, \ldots, q_t$, and the $q_t$-vertex is the one glued to the central vertex.*

*If $b = 1$ then the first set of $I$ legs is empty, if $a = 1$ then the second set of $I$ legs is empty. (If $a = b = 1$ then we recover 23.1, in which case the graph has no legs.)*

*(iii) The orbifold Euler number of the Seifert 3-manifold is $\frac{I}{ab} \geq 0$.*

*(iv) The characteristic polynomial of monodromy action on $H_1(\partial F)$ is*

$$P_{\partial F}(t) = \left(P_{\tilde{f}}(t)\right)^2 \cdot (t-1)^{1-sign(I)}.$$

*Proof.* We need to prove only (*iii*). For this use the fact that for any $a$ and $b$ one has:

$$\frac{[p_1, \ldots, p_{s-1}]}{[p_1, \ldots, p_s]} + \frac{[q_1, \ldots, q_{t-1}]}{[q_1, \ldots, q_t]} = 1 - \frac{1}{ab}.$$

This and (5.3.6) show that $e^{orb}(-\partial F) = -I/ab$. Then use (5.3.7). □

**Example 23.2.2.** If we take $\tilde{f} = z^n + u^d$, then $f = z^n + x^{da} y^{db}$, $\tilde{\mu} = (n-1)(d-1)$ and $I = n$. In this way we recover the main result of [76].

See also 10.4.5 for an explicit example determined via $\Gamma_{\mathscr{C}}$.

The reader can also verify that the graphs $\Gamma_{\mathscr{C}}$ from 9.3 produce compatible answers via the Main Algorithm. The next remark emphasizes certain advantages of the Main Algorithm.

**Remark 23.2.3.** Let us assume that $f = x^{2n}y^{2m} + z^2$ as in 9.3.2, Case 1. Set $d := \gcd(m,n)$. Since $\Gamma_{\mathscr{C}}$ is unicolored, the boundary $\partial F$ and its characteristic polynomial can be determined in two ways, either by 16.2.8, or by the above topological theorem 23.2.1. Both theorems provide for the characteristic polynomial

$$P_{\partial F}(t) = (P_{\tilde{f}}(t))^2 = \frac{(t^{2d}-1)^2(t-1)^2}{(t^2-1)^2}.$$

Nevertheless, the Main Algorithm also gives that $\#_1^2 M_{\Phi,ver} = 1$ provided that $g$ is the generic linear form, and also the following data about the graph $G$:

$$\operatorname{corank} A_G = c(G) = 1, \text{ and } 2g(G) = 4(d-1).$$

This data provides not only $\operatorname{rank} H_1(\partial F) = 2g(G) + c(G) + \operatorname{corank} A_G = 4d-2$, but also the ranks of its "weight" decomposition.

Note that in this case, the mixed Hodge structure on $H^1(\tilde{F})$ (where $\tilde{F}$ is the Milnor fiber of $\tilde{f}$ as above) has two weights, the 1-dimensional eigenspace $H^1(\tilde{F})_1$ has weight 2, while $H^1(\tilde{F})_{\lambda\neq 1}$ has weight 1.

On the other hand, the Conjecture (18.1.13) regarding the weight decomposition of $H^1(\partial F)$ predicts a similar decomposition compatible with the eigenspace decomposition of the monodromy: in case of eigenvalue 1 weights 0 and 2 both survive in rank one, while $H^1(\partial F)_{\lambda\neq 1}$ has weight 1 and rank $4d-4 = 2g(G)$.

Similar study can be done for all the examples of 9.3.

**Remark 23.2.4.** In Theorem 23.2.1, if the total number of legs is less than two and $\tilde{\mu} = 0$, then we get a lens space. In particular the graph (a posteriori) will have no central vertex.

For example, take $z^2 = xy^2$. Then $\tilde{f} = z^2 - u$, hence $I = 2$ and $\tilde{\mu} = 0$. Thus we have to glue to a vertex (with Euler number 2) two other vertices, both decorated by 2. Hence, $\partial F$ is the lens space $L(4,1)$, which can also be represented by a unique vertex decorated by $-4$, cf. 10.4.2.

# Part IV
# What Next?

# Chapter 24
# Peculiar Structures on $\partial F$: Topics for Future Research

In this chapter we list some topics that are closely related with the oriented 3-manifold $\partial F$, and are natural extensions of the present work. With this we plan to generate some research in this direction.

We omit definitions and do not strive for a comprehensive treatment of the subjects involved. We simply wish to arouse interest and generate further research by pointing out some new phenomena generated by the "new" manifold $\partial F$ exhibited.

## 24.1 Contact Structures

Recently, there is an intense activity in the theory of *contact structures of 3-manifolds*, see e.g. [100]. From the point of view of complex geometry, a central place is occupied by links of normal surface singularities. This targets the classification of their (tight) contact structures and the classification of the corresponding *Stein fillings*. In the case of normal/isolated complex surface singularities, the analytic structure of the singularity induces a *canonical* contact structure on the link. Moreover, all the *Milnor open book decompositions* (that is, open book decompositions associated with analytic map-germs) support (in the sense of Giroux, cf. [38]) exactly this canonical contact structure, for details, see [20]. In fact, [20] also proves that this canonical contact structure can be recovered from the topology of the link, and can be topologically identified among all the contact structures. Any resolution of the singularity (with perturbed analytic structure) appears as a natural Stein filling of the canonical contact structure. Furthermore, if the singularity is smoothable, then all the Milnor fibers (smoothings) appear as natural Stein fillings of this contact structure.

One may ask the validity of similar properties in the present situation, that is, for $\partial F$, where $F$ is the Milnor fiber associated with a non-isolated singularity $f : (\mathbb{C}^3, 0) \to (\mathbb{C}, 0)$.

A. Némethi and Á. Szilárd, *Milnor Fiber Boundary of a Non-isolated Surface Singularity*, Lecture Notes in Mathematics 2037, DOI 10.1007/978-3-642-23647-1_24,

Although, in this case, the link is not smooth, hence from the point of view of the theory of 3-manifolds it is not interesting, we can concentrate on the boundary of the Milnor fiber. As it was emphasized in several places in the body of this book, this class of 3-manifolds grows out from the class of negative definite plumbed 3-manifolds (although we do not know a precise characterization of it).

For any such $\partial F$, we can ask about the classification of its (tight) contact structures. Moreover, the Milnor fiber, as a Stein manifold with boundary, induces a contact structure on $\partial F$. A crucial question is to characterize and identify this structure among all the contact structures. Note that all the open book decompositions of $\partial F$ cut out by germs $g$ considered in this article (namely when the pair $(f, g)$ is an ICIS) support the very same contact structure induced by $F$, cf. [19]. Hence this structure too should have some universal property.

The point we wish to emphasize is that the present new geometric situation (considering non-isolated $f$ instead of isolated singularities) introduces a *new set of contact structures* together with a new set of *Stein fillings* realizable by singularity theory. Their classification is of major interest.

For example, consider a 3-manifold which can appear as $\partial F$ for some non-isolated $f$ as in the present work, and also as a singularity link (that is, it can be represented by a connected negative definite plumbing graph). In such a case, the contact structure induced by the Milnor fiber $F$ of $f$, let us call it *Milnor fiber contact structure*, and the canonical contact structure (as singularity link) are, in general, not contactomorphic.

The simplest proof of this statement is by comparison of the Chern classes of the corresponding structures in $H_1(\partial F, \mathbb{Z}) = H^2(\partial F, \mathbb{Z})$, a fact already noticed in [19]. The Chern class of the Milnor fiber contact structure is zero, since $F$ is parallelizable. On the other hand, the Chern class of the canonical contact structure is the class of the canonical cycle in $H_1(\partial F, \mathbb{Z})$ (that is, the restriction of the canonical class of a resolution to its boundary). In particular, the Chern class of the restriction of the canonical class to $\partial F$ is zero if and only if the singularity/link is numerically Gorenstein (see e.g. [85] for the terminology). Therefore, if $\partial F$ can be realized by a negative definite graph which is not numerically Gorenstein, then the two contact structures are different. This is happening in the case of all lens spaces which are not $A_n$-singularity links, and also in 19.8.2, cases (e)-(f).

## 24.2 Triple Product: Resonance Varieties

**24.2.1. Triple product.** For any oriented 3-manifold $Y$, the *cohomology ring* $H^*(Y, \mathbb{Z})$ of $Y$ carries rather subtle information. This can be reformulated in the *triple product*, induced by the cup product $\mu_Y \in \Lambda^3 H'$ given by $\mu_Y(a, b, c) = \langle a \cup b \cup c, [Y] \rangle$ for $a, b, c \in H^1(Y, \mathbb{Z})$ (and $H'$ is the dual of $H^1(Y, \mathbb{Z})$).

Sullivan in [129] proved that for any pair $(H, \mu)$ (where $H$ is a finitely generated free abelian group, and $\mu \in \Lambda^3 H'$), there exists a 3-manifold $Y$ with $H^1(Y) = H$ and $\mu_Y = \mu$. Moreover, for any singularity link, $\mu$ is trivial. In fact, by the proof of Sullivan, if $Y$ can be represented by a plumbing graph with non-degenerate intersection form, then the cup-product $H^1(Y) \otimes H^1(Y) \to H^2(Y)$ is trivial.

In our situation, however, $Y = \partial F$ might have non-trivial $\mu_Y$; see e.g. the example of $S^1 \times S^1 \times S^1$ realized by $f = xyz$. It would be very interesting to determine the triple product for all the 3-manifolds appearing as $\partial F$, and connect it with other singularity invariants. The same project can be formulated for the related numerical invariant introduced by T. Mark in [69].

For partial results see 16.2.11 and 16.2.20. For results regarding the ring structure for Seifert and graph 3-manifolds, see e.g. [1].

**24.2.2. Resonance varieties.** The $d$-th *resonance variety* of a space $X$ is the set $\mathcal{R}_d(X)$ of cohomology classes $\lambda \in H^1(X, \mathbb{C})$ for which there is a subspace $W \subset H^1(X, \mathbb{C})$, of dimension $d+1$, such that $\lambda \cup W = 0$. The resonance varieties of arrangements were introduced by Falk in [37], and since then they play a central role in several parts of mathematics.

Since the cup-product on $H^1(\partial F)$ might be non-trivial, and rather subtle, it would be of major interest to determine the resonance varieties of $\partial F$ and connect them with other singularity invariants.

## 24.3 Relations with the Homology of the Milnor Fiber

**24.3.1.** From the cohomology long exact sequence of the pair $(F, \partial F)$, one gets a monomorphism

$$H^1(F, \mathbb{Z}) \hookrightarrow H^1(\partial F, \mathbb{Z}),$$

which is compatible with the monodromy action. In particular, we get an upper bound for the first Betti number of $F$: $\operatorname{rank} H_1(F) \leq \operatorname{rank} H^1(\partial F)$. In fact, the characteristic polynomial of the monodromy acting on $H^1(F)$ divides the characteristic polynomial of $H^1(\partial F)$, which is expressed combinatorially in 19.10.1.

This is important, since, in general, the behaviour of $\operatorname{rank} H_1(F)$ can be rather involved, mysterious. Even in the case of arrangements (that is, in the "simplest homogeneous case"), it is not known whether $\operatorname{rank} H_1(F)$ can be deduced from the combinatorics of the arrangement or not; see for example [17, 25, 64] and the references therein. Note that by our algorithm, $\partial F$ is deduced from the combinatorics of the arrangement, hence we get a combinatorial upper bound for $H_1(F)$ as well.

In fact, one can determine an even better bound. Notice that $a \cup b \cup c = 0$ for any $a, b, c \in H^1(F)$. Therefore, $H^1(F)$ should be injected in such a subspace of $H^1(\partial F)$ on which the restriction of the triple product vanishes.

For homogeneous singularities, 19.9 combined with known facts regarding the mixed Hodge structure of $H^1(F)$ might produce even stronger restrictions. Recall that in the homogeneous case, $H^1(F)_{\lambda \neq 1}$ is pure of weight 1, while $H^1(F)_{\lambda = 1}$ is pure of weight 2.

**24.3.2.** From a different point of view, more in the spirit of 24.1, the Milnor fiber $F$ is a Stein filling of $\partial F$. One of the most intriguing questions is if this Milnor fiber can be characterized universally from $\partial F$ (eventually under some restrictions on $\partial F$).

The Milnor fiber also might help in the classification of the possible boundaries as well. For example, we have the following statements:

**Proposition 24.3.3.** *Let $f : (\mathbb{C}^3, 0) \to (\mathbb{C}, 0)$ be a hypersurface singularity with Milnor fiber $F$.*

*(a) If $F$ is a rational ball (that is, $\widetilde{H}^*(F, \mathbb{Q}) = 0$) then $f$ is smooth.*
*(b) If the boundary $\partial F$ of the Milnor fiber is $S^3$ then $f$ is smooth.*

*Proof.* (a) follows from A'Campo's theorem [3], which says that for $f$ non-smooth the Lefschetz number of the monodromy is zero. Part (b) follows from (a) and a theorem of Eliashberg, which says that the *only* Stein filling of $S^3$ is the ball [34]. □

This result can be compared with the celebrated theorem of Mumford which states that if the link of a normal surface singularity is $S^3$ then the germ is smooth, and also with the famous conjecture of Lê Dũng Tráng and M. Oka, which predicts that if the *link* of the hypersurface germ $f$ with 1-dimensional singular locus is homeomorphic to $S^3$, then $V_f$ is an equisingular family of irreducible plane curves.

## 24.4 Open Problems

Some general questions/open problems, or natural tasks for further study:

**24.4.1.** Determine/characterize all oriented plumbed 3-manifolds which might appear as $\partial F$ for some non-isolated hypersurface singularity $f$. (Recall that the classification of those normal surface singularity links which might appear as hypersurface singularity or complete intersection links is also open.)

**24.4.2.** Classify all lens spaces realized as $\partial F$. Classify all Seifert manifolds realized as $\partial F$.

**24.4.3.** Find a $\partial F$ (with $f$ a non-isolated singularity) which is an integral homology sphere (cf. also Question 3.21 of [119]).

Note that although we have the criteria from Remark 2.3.5, the structure of monodromy operators is so rigid, that simultaneous realizations of those unimodularity properties of the operators is seriously obstructed. This open problem might lead to some compatibility conditions connecting these operators too.

**24.4.4.** Consider germs of map $h : (\mathbb{C}^2, 0) \to (\mathbb{C}^3, 0)$, and let $V_f$ be its image. Connect the invariants of the present book with the invariants of $h$ as they are treated, for example, in the series of articles of D. Mond, see e.g. [78].

**24.4.5.** Similarly, compare the analytic invariants and the more algebraic study of non-isolated surface singularities, as it is presented for example in [50, 101, 128], with the invariants of the present book.

**24.4.6.** Determine the Jordan block-structure of the algebraic monodromy acting on $H_1(\partial F)$.

**24.4.7.** Develop the mixed Hodge structure of $H^1(\partial F)$. Prove Conjecture 18.1.10 from Chap. 18. (For the case of homogeneous germs and cylinders, see 19.9 and 20.3.)

**24.4.8.** Determine the graph $G^\circ$ for which $c(G^\circ)$ is minimal among all $c(G^m)$, where $G^m \sim G$. Has $G^\circ$ got any intrinsic significance? Has $c(G^\circ)$ got any intrinsic significance? (For example, is $c(G^\circ)$ independent of $g$? In what situations is this minimum realized by $\widehat{G}$?)

**24.4.9.** Determine completely the ring structure of $H^*(\partial F)$ (that is, the triple product on $H^1(\partial F)$), and the resonance varieties, cf. Chap. 24.

**24.4.10.** Can $H^1(F)$ be determined from $\partial F$ (at least in particular cases, say, for arrangements)? Or, from $G$ of $\Gamma_{\mathscr{C}}$ used here? Understand the monomorphism from 24.3 better. (Recall, that it is a famous conjecture for arrangements, that rank $H^1(F)$ is determined combinatorially.)

**24.4.11.** In the case of arrangements, can the combinatorics of the arrangement be recovered from $\partial F$? (We conjecture that yes.)

**24.4.12.** Classify/characterize all the "Milnor fiber contact structures" induced on 3-manifolds realized as $\partial F$, cf. 24.1. Are there natural families for which one can classify all the Stein fillings of the Milnor fiber (or all the) contact structures? Find examples when the Milnor fiber contact structure (besides the Milnor fiber) has other Stein fillings as well.

**24.4.13.** Compute the Seiberg–Witten invariants, or more generally, the Heegaard Floer homologies (or generalized versions of the lattice homologies) for those 3-manifolds $\partial F$ which are rational homology spheres. How are they related to the signature of $F$? Is there any analogue of the Seiberg–Witten Invariant Conjecture of the first author and Nicolaescu, cf. [87, 89]?

**24.4.14.** Clarify the open question 11.7.1 (i.e. determine the resolution of the real analytic variety $\{(u, v, w) \in (\mathbb{C}^3, 0) \,:\, u^m v^{m'} w^n = |w|^k\}^+$).

**24.4.15.** Establish the relationship between the present work (which determines the boundary $\partial F$) and [92], which determines the links $K(f, g)_k$ of the Iomdin series $f + g^k$ ($k \gg 0$). In what sense is $\partial F$ the "limit" of $\{K(f, g)_k\}_k$?

**24.4.16.** We conjecture that there exists some kind of rigidity property restricting the pairs $(\partial_1 F, \partial_2 F)$ (cf. 3.3) which form together a possible $\partial F$. That is, we expect that the normalization has some effect on the possible types of transversal singularities, and vice versa.

**24.4.17.** Determine $\partial F$ for any quasi-ordinary singularity $f$ in terms of the characteristic pairs of $f$. For some related homological results see [39, 55].

**24.4.18.** Determine the variation operator $VAR^{III}$ of Siersma [118, 119] in terms of $\Gamma_{\mathscr{C}}$ in the spirit of the present work.

**24.4.19.** Find a nice formula for the torsion of $H_1(\partial F, \mathbb{Z})$, at least for arrangements.

**24.4.20.** Determine $\partial F$ for weighted homogeneous or Newton non-nondegenerate singularities in terms of their Newton diagram. For the isolated singularity case, see Oka's algorithm [96]. See also 19.9.7.

**24.4.21.** Develop the analytic aspects and study the analytic invariants related to $\partial F$; moreover, analyze also its relations with deformation theory (for this last subject see for example the thesis of D. van Straten or T. de Jong and their series of articles [50, 51, 128]).

**24.4.22.** Are the techniques and results of the present book applicable in the context of the equisingularity problems of non-isolated germs? (For such problems, consult for example the article of F. de Bobadilla [11] containing several key conjectures as well).

Questions related mostly to the technicalities used in the proofs and results:

**24.4.23.** Develop the "calculus" of decorated graphs such as $\Gamma_{\mathscr{C}}$, cf. 6.2.5.

**24.4.24.** Analyze the possible relations connecting the decorations of $\Gamma_{\mathscr{C}}$. Provide an independent proof of the fact that the expressions 13.4.11(c) are independent of $g$.

**24.4.25.** Find closed formulae for corank $A$ and corank $(A, \mathfrak{I})$. In what situations can we expect the validity of the relation $\operatorname{corank}(A, \mathfrak{I})_{\widehat{G}} = c(\widehat{G}) + |\mathscr{A}(G)|$ (or similar formula for $G^\circ$ instead of $\widehat{G}$)?

**24.4.26.** Is it true that $\#_1^2 M_{\Phi,ver} = c(\widehat{G})$? Or for $G^\circ$ instead of $\widehat{G}$?

**24.4.27.** Is the technical Lemma 17.1.11 true in general?

**24.4.28.** Discuss the case of *all* eigenvalues of the vertical monodromy $M_{\Phi,ver}$.

# List of Examples




A. Némethi and Á. Szilárd, *Milnor Fiber Boundary of a Non-isolated Surface Singularity*, Lecture Notes in Mathematics 2037, DOI 10.1007/978-3-642-23647-1,

# List of Notations

| Symbol | Description | Appears in |
|---|---|---|
| $(a, b)$ | $\gcd(a, b)$ | |
| $\chi$ | Euler characteristic of a space | |
| $[k_1, k_2, \ldots, k_s]$ | Continued fraction | 4.3.8 |
| $\ominus, +, -$ | Edge decorations of a plumbing graph | 4.1.1 |
| $\#_\lambda^k$ | Number of Jordan blocks of size $k$ with eigenvalue $\lambda$ | 13.2.1 |
| $\#, \#(f)$ | Number of local irreducible components of plane curve singularity | 2.1.5, Sect. 20.1 |
| $\#T\Sigma_j$ | Number of local irreducible components of $T\Sigma_j$ | Sect. 2.2 |
| $A$ | Intersection matrix | 4.1.7 |
| $(A, \mathfrak{I})$ | Block matrix | 4.1.7 |
| $(\alpha_\ell, \omega_\ell)_\ell$ | Seifert invariants | 5.3.5 |
| $\mathscr{A}(\Gamma), \mathscr{A}$ | The set of arrowhead vertices of a graph $\Gamma$ | 4.1.3 |
| arg | $\arg(g) = g/\lvert g\rvert$ | 3.2 |
| $\arg_*$ | Induced in homology by arg | 4.4.5 |
| $B_\epsilon$ | Closed ball of radius $\epsilon$ in $\mathbb{C}^n$ | Sect. 2.1 |
| $c(\Gamma)$ | The number of independent cycles of a graph $\Gamma$ | 4.2.3 |
| $C_\Phi$ | Critical locus of an ICIS $\Phi$ | 3.1.2 |
| $\mathscr{C}$ | The "special curve arrangement" in **D** | 6.1.2, 7.1.9 |
| $\mathscr{C}_{\Sigma_j}, \mathscr{C}_{\Sigma_i}$ | Collections of certain irreducible components of $\mathscr{C}$ | 7.4.2 |

A. Némethi and Á. Szilárd, *Milnor Fiber Boundary of a Non-isolated Surface Singularity*, Lecture Notes in Mathematics 2037, DOI 10.1007/978-3-642-23647-1,

| | | |
|---|---|---|
| $C=\cup_{\lambda\in\Lambda}C_\lambda$ | Projective curve | 8.1 |
| $(C_{j,i},p_j)_{i\in I_j}$ | Local analytic irreducible components of $(C,p_j)$ | 8.1 |
| $c$ | Identification map $c:\cup_j I_j\to\Lambda$ | 8.1 |
| $d_\lambda$ | Degree of $C_\lambda$, $\lambda\in\Lambda$ | 8.1 |
| $Div(H;M)$ | The divisor of a characteristic polynomial | Sect. 13.3 |
| $D_r$ | Complex disk of radius $r$ | Sect. 2.1 |
| $D^2_r$, $D^2_\eta$ | bidisc | Sect. 2.1, Sect. 3.1 |
| $\mathbf{D}$ | Total transform of $V_{fg}$ in an embedded resolution | 6.1.2 |
| $\mathbf{D}_c$, $\mathbf{D}_d$, $\mathbf{D}_0$ | Collections of certain components of $\mathbf{D}$ | 6.1.2 |
| $\delta(f)$ | Serre-invariant (or delta-invariant) of $f$ | 2.1.5 |
| $\partial$ | Homological operator | 14.1 |
| $\delta'_j$ | Serre-invariant of $T\Sigma_j$ | Sect. 2.2 |
| $d_j$ | Covering degree of $g\vert_{\Sigma_j}$ | 3.1.8 |
| $d(e)$ | Invariant of a cutting edge | 7.5.1 |
| $\Delta_\Phi$, $\Delta_j$ | Discriminant of an ICIS $\Phi$; its irreducible components | 3.1.2 |
| $\mathbf{d}$ | Covering data, *see* $(\mathbf{n},\mathbf{d})$ too | 5.1.2 |
| $\delta_w$ | The number of vertices adjacent to a vertex $w$ | 5.2.7 |
| $\partial F$, $\partial F_{\epsilon,\delta}$ | Boundary of the Milnor fiber | 1.1.1, Sect. 2.1 |
| $\partial_1 F$, $\partial_2 F$ | Decomposition subsets of $\partial F$ | 1.1.1, Sect. 2.3 |
| $\partial_{2,j}F$ | Connected components of $\partial_2 F$ | Sect. 2.3 |
| $\overline{\partial_{2,j}F}$ | Canonical closure of $\partial_{2,j}F$ | Remark 10.3.8 |
| $\partial'\Phi^{-1}(D_\delta)$ | Subset of $\partial F_{\epsilon,\delta}$ | 3.1.11, 3.3.1 |
| $\epsilon_e$ | Edge decoration | 4.1.4 |
| $e_w$ | Euler number decoration of $E_w$ | 4.1.4, 4.3.2 |
| $e^{orb}$ | Orbifold Euler number | 5.3.5 |
| $\mathscr{E}(\Gamma)$, $\mathscr{E}$ | The set of edges of a graph $\Gamma$ | 4.1.1 |
| $\mathscr{E}_{cut}$ | The set of cutting edges of $\Gamma$ | Sect. 7.2 |
| $\mathscr{E}_{\mathscr{W}}(\Gamma)$ | The set of edges connecting non-arrowhead vertices | 13.6.1 |
| $F$, $F_{\epsilon,\delta}$ | Milnor fiber of $f$ | Sect. 2.1 |
| $F_\Phi$, $F_{c,d}$ | Milnor fiber of an ICIS $\Phi$ | 3.1.6 |
| $F'_j$ | Milnor fiber of transversal type singularity $T\Sigma_j$ | Sect. 2.2 |
| $\widetilde{F}_\Phi$ | Lifted Milnor fiber | Chap. 6 |
| $\widetilde{F}_v$, $\widetilde{F}_w$ | Subsets of $\widetilde{F}_\Phi$ near curves $\mathscr{C}_v$ or $\mathscr{C}_w$ | Chap. 6 |

| | | |
|---|---|---|
| $\widetilde{F}_{\mathscr{W}}$ | Collection of subsets $\widetilde{F}_w$ | 17.1.8 |
| $\Phi$ | Good representative of an ICIS | 6.1.2 |
| $\widetilde{\Phi}$ | A lift of $\Phi$ | 6.1.2 |
| $\varphi_j$ | Fibonacci numbers | 19.8.1 |
| $g_w$ | Genus of $E_w$ | 4.3.2 |
| $d_\lambda$ | Degree of $C_\lambda$, $\lambda \in \Lambda$ | 8.1 |
| $g(\Gamma)$ | The sum of all genera in a graph $\Gamma$ | 4.2.3 |
| $\Gamma_1 + \Gamma_2$ | Disjoint union of graphs | 4.1.2 |
| $\Gamma_{\mathscr{C}}$ | The dual graph of $\mathscr{C}$ | 6.1.2, Sect. 6.2 |
| $\Gamma^1_{\mathscr{C}}$, $\Gamma^2_{\mathscr{C}}$ | "Complementary" subgraphs of $\Gamma_{\mathscr{C}}$ | Sect. 7.2 |
| $\Gamma^2_{\mathscr{C},j}$ | Connected components of $\Gamma^2_{\mathscr{C}}$ | 7.4.2 |
| $\Gamma^2_{\mathscr{C},j}/\sim$ | A "base graph" derived from $\Gamma^2_{\mathscr{C},j}$ | 7.4.10 |
| $\Gamma_{van}$ | A subgraph of $\Gamma_{\mathscr{C}}$ having vanishing 2-edges only | 12.1.5 |
| $\Gamma(X)$ | Resolution graph of a normal surface singularity $X$ | 4.3.3, Sect. 5.2 |
| $\Gamma(X, f)$ | Embedded resolution graph of $X$ and a germ $f$ | 4.3.2 Sect. 4.3 |
| $\widehat{\Gamma_{\mathscr{C}}}$ | An undecorated graph obtained from $\Gamma_{\mathscr{C}}$ | 12.2 |
| $\Gamma(f', g')$ | Shorthand for $\Gamma(\mathbb{C}^2, f'g')$ | Sect. 9.3 |
| $G, G_1, G_2$ | Graphs derived from $\Gamma_{\mathscr{C}}$, $\Gamma^1_{\mathscr{C}}$, $\Gamma^2_{\mathscr{C}}$ using the Main Algorithm | 10.2.9 |
| $G_{2,j}$ | Components of the graph $G_2$; related to $\Gamma^2_{\mathscr{C},j}$ | 10.3.4 |
| $\overline{G_{2,j}}$ | Canonical closure of $G_{2,j}$ | Remark 10.3.8 |
| $G^m$, $G^m_1$, $G^m_{2,j}$ | Graphs obtained from $G$, $G_1$, $G_2$ using plumbing calculus | 10.2.9 |
| $G(K_\ell)$ | Subgraphs of $\Gamma^2_{\mathscr{C},j}$ | 7.4.6 |
| $G^1_{\mathscr{C}}$ | An embedded resolution graph derived from $\Gamma^1_{\mathscr{C}}$ | 7.3.2 |
| $G(T\Sigma_j)$ | Embedded resolution graph of $T\Sigma_j$ | 7.4.10 |
| $G(X, f)$ | Universal cyclic covering graph of $\Gamma(X, f)$ | Sect. 5.2 |
| $\widehat{G}$ | Output graph of Collapsing Main Algorithm | 12.2 |
| $\widehat{G_1}$, $\widehat{G_2}$ | Graphs associated with $\partial_1 F$ and $\partial_2 F$; outputs of Collapsing Main Algorithm | 12.3.1 |
| $\mathscr{G}(\Gamma, (\mathbf{n}, \mathbf{d}))$ | Equivalence classes of covering graphs of $\Gamma$ associated with covering data $(\mathbf{n}, \mathbf{d})$ | 5.1.4 |
| $\Gamma_1 \sim \Gamma_2$ | Equivalent graphs | 4.2.6 |
| $-\Gamma$ | $\Gamma$ with reversed orientation (Euler and edge decorations reversed) | 4.2.9 |

| | | |
|---|---|---|
| $\lvert Gr \rvert$ | Topological realization of the graph $Gr$ | 5.2.7(3) |
| $\gamma_w$, $\gamma_a$ | Oriented meridian | 4.1.8, 4.1.10 |
| $Gr_m^W H$ | $W_m H / W_{m-1} H$ associated with a weight filtration | Sect. 18.1 |
| $H_{M,\lambda}$ | Generalized $\lambda$-eigenspace | Sect. 13.3 |
| $\mathfrak{I}$ | Incidence matrix | 4.1.7 |
| $I$ | Identity matrix | |
| $K$, $K_X$ | The link of a surface singularity $X$ | Sect. 2.1 |
| $K_f \subset K_X$ | The link $V_f \cap K_X \subset K_X$ of a germ $f$ defined on $X$ | 4.4.3 |
| $K^{norm}$ | Link of the normalization $V_f^{norm}$ | 3.4 |
| $L$, $L_j$ | Singular part of the link $K$; its components | Sect. 2.2 |
| $\Lambda(m; n, \nu)$ | A special divisor associated with $(m; n, \nu)$ | Sect. 13.3 |
| $\cup_{\lambda \in \Lambda} L_\lambda$ | Line arrangement | 8.2 |
| $M(\Gamma)$, $M$ | Oriented plumbed 3-manifold associated with $\Gamma$ | Chap. 4 |
| $-M$ | The oriented manifold $M$ with reversed orientation | 1.2 |
| $M_1 \# M_2$ | Oriented connected sum of $M_1$ and $M_2$ | 4.1.2 |
| $m_{geom,\Phi}$ | Geometric monodromy representation of an ICIS $\Phi$ | 3.1.6 |
| $M_\Phi$ | Algebraic monodromy representation of an ICIS $\Phi$ | 3.1.6 |
| $M_q$, $M$ | A monodromy operator | Remarks 2.3.5, 5.2.7, Chap. 13 |
| $m'_{j,hor}$, $m'_{j,ver}$ | Horizontal, vertical geometric monodromies of $T\Sigma_j$ | Sect. 2.2 |
| $M'_{j,hor}$, $M'_{j,ver}$ | Horizontal, vertical algebraic monodromies of $T\Sigma_j$ | Sect. 13.1 |
| $m_{\Phi,hor}$, $m_{\Phi,ver}$ | Horizontal, vertical geometric monodromies of $(\Phi, \Delta_1)$ | 3.1.9 |
| $M_{\Phi,hor}$, $M_{\Phi,ver}$ | Algebraic monodromies induced by $m_{\Phi,hor}$ and $m_{\Phi,ver}$ | Sect. 13.1 |
| $m^\Phi_{j,hor}$, $m^\Phi_{j,ver}$ | Horizontal, vertical geometric monodromies of $\Phi$ near $\Sigma_j$ | 3.3 |
| $M^\Phi_{j,hor}$, $M^\Phi_{j,ver}$ | Algebraic monodromies induced by $m^\Phi_{j,hor}$ and $m^\Phi_{j,ver}$ | Sect. 13.1 |
| $\mu(f)$, $\mu$ | Milnor number of $f$ | 2.1.4 |

| | | |
|---|---|---|
| $\mu'_j$ | Milnor number of $T\Sigma_j$ | Sect. 2.2 |
| $(m_v)$, $(m)$ | Multiplicities of a plumbing graph | 4.1.4 |
| $\{m_v\}_{v\in\mathscr{V}}$ | Multiplicity system | 4.1.4, Sect. 6.2 |
| $(m_w; n_w, \nu_w)$, $(m; n, \nu)$ | Multiplicity weights of the graph $\Gamma_{\mathscr{C}}$ | Chap. 6 |
| $\{m_j\}_{j\in\Pi}$ | Cardinality of $I_j = \{L_\lambda : L_\lambda \ni p_j\}$ for an arrangement | 8.2 |
| $n$ | Normalization | 3.4 |
| $\mathbf{n}$, $(\mathbf{n}, \mathbf{d})$ | Covering data | 5.1.2 |
| $\nu(e)$ | Invariant of a cutting edge | 7.5.1 |
| $\nu$ | The covering data for $\Gamma^2_{\mathscr{C},j}/\sim$ | 7.4.10 |
| $P(\Gamma)$, $P$ | Plumbed 4-manifold associated with $\Gamma$ | Chap. 4 |
| $P_M(t)$ | Characteristic polynomial of $M$ | 5.2.7, Sect. 13.3 |
| $P^\#$, $P^\#_j$ | Certain characteristic polynomials | 14.2.9 |
| $\mathscr{P}$, $\mathscr{P}(t_{hor}, t_{ver})$ | Subsets of $\mathbb{C}^* \times \mathbb{C}^*$ | Chap. 13 |
| $\{p_j\}_{j\in\Pi}$ | Singular points of $C = \cup_{\lambda\in\Lambda} C_\lambda$ | 8.1, 8.2 |
| $R0$, $R1$, …, $R7$ | Operations of plumbing graphs | 4.2 |
| $S_\epsilon$, $S^{2n-1}_\epsilon$ | $(2n-1)$-dimensional sphere of radius $\epsilon$ | Sect. 2.1 |
| $S^k$ | $k$-dimensional sphere | Sect. 2.1 |
| $Sl_q$ | Transversal slice | Sect. 2.2, 7.4.3 |
| $\mathscr{S}_k$ | A real analytic surface germs | 10.2.11 |
| $\widetilde{\mathscr{S}}_k$ | Strict transform of $\mathscr{S}_k$ | 11.4 |
| $\mathscr{S}^{norm}_k$ | Normalization of $\widetilde{\mathscr{S}}_k$ | 11.6.1 |
| $\overline{\mathscr{S}_k}$ | "Resolution" of $\mathscr{S}_k$ | 11.2, 11.7 |
| $St$, $St(\Sigma)$ | Strict transform in a resolution | 4.3.1, 3.4 |
| $St_e$ | Strict transform associated with a cutting edge | 7.5.7 |
| $Str$, $Str(a, b; c \mid i, j; k)$ | A (graph) string | 4.3.10 |
| $Str^{\ominus}(a, b; c \mid i, j; k)$ | The same string with $\ominus$ edge-decorations | 4.3.10 |
| $\Sigma$, $\Sigma_f$; $\Sigma_j$ | Singular locus; its irreducible components | 2.1, 2.1.4 |
| $T(E_v)$, $T(\mathscr{C})$ | Tubular neighborhoods of certain curves | 5.2 |
| $T^\circ$ | Interior of $T$ | 2.1 |
| $T\Sigma_j$ | Transversal type singularity | 2.2 |
| twist$(h; A)$ | Twist | 17.1.2 |

| | | |
|---|---|---|
| $T_{a,*,*}$ | Family of certain singularities | 9.4 |
| $\mathscr{V}(\Gamma)$, $\mathscr{V}$ | The set of all vertices of a graph $\Gamma$ | 4.1.1 |
| $\mathscr{V}_v$ | Set of vertices adjacent to the vertex $v$ | 5.2.2 |
| $V_f$ | The zero locus of a germ $f$ | 2.1 |
| $V_f^{norm}$ | The normalization of $V_f$ | 3.4 |
| $\mathscr{W}(\Gamma)$, $\mathscr{W}$ | The set of non-arrowhead vertices of a graph $\Gamma$ | 4.1.3 |
| $\mathsf{W}_{\eta,M}$ | Wedge neighbourhood | Chap. 3, Sect. 7.1 |
| $W_\bullet$ | Weight filtration of a mixed Hodge structure | 18.1 |
| $X_{f,N}$ | Cyclic covering | 4.3.5 |

# References


1. Aaslepp, K., Drawe, M., Hayat-Legrand, C., Sczesny, Ch.A., Zieschang, H.: On the cohomology of Seifert and graph manifolds. Topology. Appl. **127**, 3–32 (2003)
2. A'Campo, N.: Sur la monodromie des singularités isolées d'hypersurfaces complexes. Inventiones math. **20**, 147–169 (1973)
3. A'Campo, N.: Le nombre de Lefschetz d'une monodromie. Indag. Math. **35**, 113–118 (1973)
4. A'Campo, N.: La fonction zêta d'une monodromie. Comment. Math. Helv. **50**, 233–248 (1975)
5. Arnold, V.I., Gusein-Zade, S.M., Varchenko, A.N.: Singularities of Differentiable maps, Vol. 1 and 2, Monographs Math. **82–83**. Birkhäuser, Boston (1988)
6. Artal–Bartolo, E.: Forme de Seifert des singularités de surface. C. R. Acad. Sci. Paris **t. 313**, Série I, 689–692 (1991)
7. Ban, C., McEwan, L.J., Némethi, A.: The embedded resolution of $f(x,y)+z^2 : (\mathbb{C}^3,0) \to (\mathbb{C},0)$. Studia Sc. Math. Hungarica **38**, 51–71 (2001)
8. Barth, W., Peters, C., Van de Ven, A.: Compact complex surfaces, Ergebnisse der Mathematik und ihrer Grenzgebiete, 3. Folge, Band **4**, A Series of Modern Surveys in Mathematics. Springer, New York (1984)
9. Blank, S., Laudenbach, F.: Isotopie des formes fermées en dimension 3. Invent. Math. **54**, 103–177 (1979)
10. de Bobadilla, F.J.: Answers to some equisingularity questions. Invent. Math. **161**, 657–675 (2005)
11. de Bobadilla, F.J.: Topological equisingularity of function germs with 1-dimensional critical set. arXiv:math/0603508
12. de Bobadilla, J.F., Luengo, I., Melle-Hernández, A., Némethi, A.: On rational cuspidal projective plane curves. Proc. London Math. Soc. **92**(3), 99–138 (2006)
13. de Bobadilla, J.F., Luengo, I., Melle-Hernández, A., Némethi, A.: Classification of rational unicuspidal projective curves whose singularities have one Puiseux pair, Proceedings of Sao Carlos Workshop 2004, Real and Complex Singularities, Series *Trends in Mathematics*, Birkhäuser, 31–46 (2007)
14. Bochnak, J., Coste, M., Roy, M-F.: Real algebraic geometry. Ergebnisse der Mathematik und ihrer Grenzgebiete, pp. 36(3). Springer, Berlin (1998)
15. Brieskorn, E.: Die Monodromie der isolierten Singularitäten von Hyperflächen. Man. Math. **2**, 103–161 (1970)
16. Brieskorn, E., Knörrer, H.: Plane Algebraic Curves, Birkhäuser, Boston (1986)
17. Budur, N., Dimca, A., Saito, M.: First milnor cohomologyof hyperplane arrangements, topology of algebraic varieties and singularities. Contemp. Mathematics **538**, 279–292 (2011)



A. Némethi and Á. Szilárd, *Milnor Fiber Boundary of a Non-isolated Surface Singularity*, Lecture Notes in Mathematics 2037, DOI 10.1007/978-3-642-23647-1,

18. Burghelea, D., Verona, A.: Local analytic properties of analytic sets. Manuscr. Math. **7**, 55–66 (1972)
19. Caubel, C.: Contact Structures and Non-isolated Singularities, Singularity Theory, pp. 475–485. World Scientific Publishing, Hackensack, NJ, (2007)
20. Caubel, C., Némethi, A., Popescu–Pampu, P.: Milnor open books and Milnor fillable contact 3-manifolds. Topology **45**(3), 673–689 (2006)
21. Cisneros-Molina, J.L., Seade, J., Snoussi, J.: Refinements of Milnor's fibration theorem for complex singularities. Adv. Math. **222**(3), 937–970 (2009)
22. Clemens, H.: Picard–Lefschetz theorem for families of non-singular algebraic varieties acquiring ordinary singularities. Trans. Am. Math. Soc. **136**, 93–108 (1969)
23. Cohen, D.C., Suciu, A.I.: Boundary manifolds of projective hypersurfaces. Adv. Math. **206**(2), 538–566 (2006)
24. Cohen, D.C., Suciu, A.I.: The boundary manifold of complex line arrangements. Geomet. Topology Monogr. **13**, 105–146 (2008)
25. Cohen, D.C., Suciu, A.I.: On Milnor fibrations of arrangements. J. London Math. Soc. (2) **51** (1), 105–119 (1995)
26. Deligne, P.: Theorie de Hodge, II and III. Publ. Math. IHES **40**, 5–58 (1971) and **44**, 5–77 (1974)
27. Dimca, A.: Sheaves in Topology, Universitext, Springer (2004)
28. Du Bois, Ph., Michel, F.: The integral Seifert form does not determine the topology of plane curve germs. J. Algebr. Geomet. **3**, 1–38 (1994)
29. Durfee, A.H.: The signature of smoothings of complex surface singularities. Math. Ann. **232**(1), 85–98 (1978)
30. Durfee, A.H.: Mixed Hodge structures on punctured neighborhoods. Duke Math. J. **50**, 1017–1040 (1983)
31. Durfee, A.H, Hain, R.M.: Mixed Hodge Structures on the Homotopy of Links. Math. Ann. **280**, 69–83 (1988)
32. Durfee, A.H., Saito, M.: Mixed Hodge structures on intersection cohomology of links. Composition Math. **76** (1–2), 49–67 (1990)
33. Eisenbud, D., Neumann, W.: Three-dimensional link theory and invariants of plane curve singularities. Ann. Math. Studies, Princeton University Press, Princeton **110** (1985)
34. Eliashberg, Y.: Filling by holomorphic discs and its applications. London Math. Soc. Lecture Notes Series **151**, 45–67 (1991)
35. Elzein, F.: Mixed Hodge structures. Trans. AMS **275**, 71–106 (1983)
36. El Zein, F., Némethi, A.: On the Weight Filtration of the Homology of Algebraic Varieties: The Generalized Leray Cycles. Ann. Scuola Norm. Sup. Pisa, Cl. Sci. **I**(5), 869–903 (2002)
37. Falk, M.: Arrangements and cohomology. Ann. Combin. **1**, 135–157 (1997)
38. Giroux, E.: Géométrie de contact: de la dimension trois vers les dimensions supérieures, Proceedings of the ICM, Vol II, (Beijing 2002), 405–414, Higher Ed. Press, Beijing (2002)
39. González Pérez, P.D., McEwan, L.J., Némethi, A.: The zeta-function of a quasi-ordinary singularity II, Topics in Algebraic and Noncommutative Geometry. Contemporary Math. **324**, 109–122 (2003)
40. Grauert, H.: Über Modifikationen und exzeptionelle analytische. Mengen. Math. Ann. **146**, 331–368 (1962)
41. Griffiths, Ph.A.: Periods of integrals on algebraic manifolds, summary of main results and discussion of open problems. Bull. Am. Math. Soc. **76**, 228–296 (1970)
42. Griffiths, Ph., Harris, J.: Principles of Algebraic Geometry, Reprint of the 1978 original. Wiley Classics Library. Wiley, New York (1994)
43. Hamm, H.: Lokale topologische Eigenschaften komplexer Raüme. Math. Ann. **191**, 235–252 (1971)
44. Harer, J., Kas, A., Kirby, R.: Handlebody decomposition of complex surfaces. Mem. Am. Math. Soc. **62**(350) (1986)
45. Hartshorne, R.: Algebraic Geometry. Graduate Texts in Mathematics, vol. 52. Springer, New york (1977)

46. Hironaka, E.: Boundary manifolds of line arrangements. Math. Ann. **319**, 17–32 (2001)
47. Hirzebruch, F.: Über vierdimensionale Riemannsche Flaĉhen mehrdeutiger analytischer Functionen von zwei complexen Veränderlichen. Math. Ann. **126**, 1–22 (1953)
48. Hirzebruch, F., Neumann, W.D., Koh, S.S.: Differentiable manifolds and quadratic forms. Math. Lectures Notes, vol. 4. Dekker, New York (1972)
49. Iomdin, I.N.: Complex surfaces with a one-dimensional set of singularities (Russian), Sibirsk. Mat. Ž., **15**, 1061–1082 (1974)
50. Jong, T.de: Non-isolated hypersurface singularities, Thesis, Univ. Nijmegen (1988)
51. de Jong, T., van Straten, D.: Deformation theory of sandwiched singularities. Duke Math. J. **95**, 451–522 (1998)
52. Kashiwara, H.: Fonctions rationelles de type (0, 1) sur le plan projectif complexe. Osaka J. Math. **24**, 521–577 (1987)
53. Kato, M., Matsumoto, Y.: On the connectivity of the Milnor fiber of a holomorphic function at a critical point, Manifolds Tokyo 1973, University of Tokyo Press, 131–136 (1975)
54. Kaup, L., Kaup, B.: Holomorphic Functions of Several Variables, An Introduction to the Fundamental Theory. Walter de Gruyter, Berlin, New York, (1983)
55. Kennedy, G., McEwan, L.J.: Monodromy of plane curves and quasi-ordinary surfaces. J. Singul. **1**, 146–168 (2010)
56. Kulikov, V.S.: Mixed Hodge Structures and Singularities, Cambridge Tracts in Mathematics, pp. 132. Cambridge University Press, Cambridge (1998)
57. Laufer, H.B.: Normal two-dimensional singularities. Ann. Math. Studies, pp. 71. Princeton University Press, Princeton (1971)
58. Laufer, H.B.: On $\mu$ for surface singularities. Proc. Symp. Pure Math. **30**, 45–49 (1977)
59. Landman, A.: On Picard–Lefschetz transformation for algebraic manifolds acquiring general singularities. Trans. Am. Math. Math. **181**, 89–126 (1973)
60. Lê Dũng Tráng: Calcul du nombre de cycles évanouissants d'une hypersurface complexe. Ann. Inst. Fourier (Grenoble) **23**(4), 261–270 (1973)
61. Lê Dũng Tráng: Some remarks on relative monodromy. In: Holm, P. (ed.) Real and Complex Singularities, Oslo 1976, pp. 397–403. Sijthoff & Noordhoff, Alphen a/d Rijn (1977)
62. Lê Dũng Tráng: Ensembles analytiques complexes avec lieu singulier de dimension un (d'après I. N. Iomdine), Seminar on Singularities (Paris, 1976/1977), pp. 87–95, Publ. Math. Univ. Paris VII, 7, Univ. Paris VII, Paris (1980)
63. Lê Dũng Tráng, Ramanujam, C.P.: The invariance of Milnor's number implies the invariance of the topological type. Am. J. Math. **98**, 67–78 (1976)
64. Libgober, A.: On combinatorial invariance of the cohomology of Milnor fiber of arrangements and Catalan equation over function field. arXiv: 1011.0191
65. Lipman, J.: Introduction to resolution of singularities. Proc. Symp. Pure Math. **29**, 187–230 (1975)
66. Lojasiewicz, S.: Triangulations of semi-analytic sets. Annali Scu. Norm. Sup. Pisa **18**, 449–474 (1964)
67. Looijenga, E.J.N.: Isolated Singular Points on Complete Intersections, London Math. Soc. Lecture Note Series 77, Cambridge University Press, Cambridge (1984)
68. Looijenga, E., Wahl, J.: Quadratic functions and smoothing surface singularities. Topology **25**(3), 261–291 (1986)
69. Mark, T.E.: Triple prducts and cohomological invariants for closed 3-manifolds. Michigan Math. J. **56**(2), 265–281 (2008)
70. Massey, D.: The Lê variaties I. Invent. Math. **99**, 357–376 (1990)
71. Massey, D.: The Lê variaties II. Invent. Math. **104**, 113–148 (1991)
72. Massey, D.: Lê Cycles and Hypersurface Singularities, Lecture Notes in Mathematics, vol. 615. Springer, New york (1995)
73. Michel, F., Pichon, A.: On the boundary of the Milnor fibre of nonisolated singularities. Int. Math. Res. Not. **43**, 2305–2311 (2003)
74. Michel, F., Pichon, A.: Erratum: "On the boundary of the Milnor fibre of nonisolated singularities", [Int. Math. Res. Not. 2003, no. 43, 2305-2311;] Int. Math. Res. Not. **6**, 309–310 (2004)

75. Michel, F., Pichon, A.: Carrousel in family and non-isloated hypersurface singularities in $\mathbb{C}^3$, arXiv:1011.6503v1, 30 November (2010)
76. Michel, F., Pichon, A., Weber, C.: The boundary of the Milnor fiber of Hirzebruch surface singularities. Singularity theory, World Scientific Publication 745–760 (2007)
77. Milnor, J.: Singular Points of Complex Hhypersurfaces, Ann. Math. Studies, vol. **61**. Princeton University Press, Princeton (1968)
78. Mond, D.: Vanishing Cycles for Analytic Maps, Singularity theory and its applications, Part I (Coventry, 1988/1989), 221–234. Lecture Notes in Math. **1462**, Springer, Berlin (1991)
79. Mumford, D.: The topology of normal surface singularities of an algebraic surface and a criterion of simplicity. IHES Publ. Math. **9**, 5–22 (1961)
80. Namba, M.: Geometry of projective algebraic curves. Monographs and Textbooks in Pure and Applied Math, pp. 88. Marcel Dekker, New York (1984)
81. Némethi, A.: The Milnor fiber and the zeta function of the singularities of type $P(f, g)$. Compositio Math. **79**, 63–97 (1991)
82. Némethi, A.: The equivariant signature of hypersurface singularities and eta-invariant. Topology **34**(2), 243–259 (1995)
83. Némethi, A.: The eta–invariant of variation structures. I. Topology. Appl. **67**, 95–111 (1995)
84. Némethi, A.: The signature of $f(x, y) + z^n$, Proceedings of Real and Complex Singularities, Liverpool (England), August 1996; London Math. Soc. Lecture Note Series **263**, 131–149 (1999)
85. Némethi, A.: Five lectures on normal surface singularities; lectures delivered at the Summer School in Low dimensional topology, Budapest, Hungary 1998; Bolyai Soc. Math. Studies **8**, 269–351 (1999)
86. Némethi, A.: Resolution graphs of some surface singularities, I. (cyclic coverings). Contem. Math. **266**, 89–128 (2000)
87. Némethi, A.: Invariants of normal surface singularities. Contemp. Math. **354**, 161–208 (2004)
88. Némethi, A.: Lattice cohomology of normal surface singularities. Publ. Res. Inst. Math. Sci. **44**(2), 507–543 (2008)
89. Némethi, A., Nicolaescu, L.I.: Seiberg-Witten invariants and surface singularities. Geomet. Topology **6**, 269–328 (2002)
90. Némethi, A., Popescu–Pampu, P.: On the Milnor fibers of cyclic quotient singularities. Proc. London Math. Soc. **101**(2), 497–553 (2010)
91. Némethi, A., Steenbrink, J.: On the monodromy of curve singularities. Math. Zeitschrift **223**, 587–593 (1996)
92. Némethi, A., Szilárd, Á.: Resolution graphs of some surface singularities, II. (Generalized Iomdin series). Contemporary Math. **266**, 129–164 (2000)
93. Némethi, A., Szilárd, Á.: The boundary of the Milnor fibre of a non-isolated hypersurface surface singularity. arXiv:0909.0354, 2 September (2009)
94. Neumann, W.D.: A calculus for plumbing applied to the topology of complex surface singularities and degenerating complex curves. Trans. Am. Math. Soc. **268**(2), 299–344 (1981)
95. Neumann, W.D.: Splicing algebraic links, advanced studies in pure math. Complex Analytic Singularities **8**, 349–361 (1986)
96. Oka, M.: On the Resolution of Hypersurface Singularities, Complex Analytic Singularities, Advanced Study in Pure Math. vol. 8, pp. 437–460. North-Holland, Amsterdam-New York-Oxford (1986)
97. Ono, I., Watanabe, K.: On the singularity of $z^p + y^q + x^{pq} = 0$. Sci. Rep. Tokyo Kyoika Daigaku Sect. A. **12**, 123–128 (1974)
98. Orevkov, S.Yu.: On rational cuspidal curves, I. Sharp estimate for degree via multiplicity. Math. Ann. **324**, 657–673 (2002)
99. Orlik, P., Wagreich, Ph.: Isolated singularities of algebraic surfaces with $\mathbb{C}^*$ action. Ann. Math. **93**(2), 205–228 (1971)
100. Ozbagci, B., Stipsicz, A.: Surgery on Contact 3-manifolds and Stein Surfaces, vol. 13. Bolyai Society Mathematical Studies, Springer and János Bolyai Mathematical Society (2004)

101. Pellikaan, G.R.: Hypersurface singularities and resolutions of Jacobi modules. PhD Thesis, University of Utrecht (1985)
102. Peters, C.A.M., Steenbrink, J.H.M.: Mized Hodge Structures, Ergebnisse der Mathematik und ihrer Grenzgebiete, vol. 52, p. 3. Folge, Springer (2008)
103. Pichon, A.: Variétés de Waldhausen et fibrations sur le cercle. C. R. Acad. Sci. Paris Sér. I Math. **324**(6), 655–658 (1997)
104. Pichon, A.: Singularities of complex surfaces with equations $z^k - f(x, y) = 0$. Internat. Math. Res. Notices **5**, 241–246 (1997)
105. Pichon, A.: Three-dimensional manifolds which are the boundary of a normal singularity $z^k - f(x, y)$. Math. Z. **231**(4), 625–654 (1999)
106. Pichon, A., Seade, J.: Real singularities and open-book decompositions of the 3-sphere. Ann. Fac. Sci. Toulouse Math. (6), **12**(2), 245–265 (2003)
107. Pichon, A., Seade, J.: Fibred multilinks and singularities $f\overline{g}$. Math. Ann. **342**(3), 487–514 (2008)
108. Randell, R.: On the topology of non-isolated singularities, Proceedings 1977 Georgia Topology Conference, 445–473
109. Riemenschneider, O.: Deformationen von Quotientensingularitäten (nach zyklischen Gruppen). Math. Ann. **209**, 211–248 (1974)
110. Riemenschneider, O.: Zweidimensionale Quotientensingularitäten: Gleichungen und Syzygien. Arch. Math. **37**, 406–417 (1981)
111. Saito, M.: On Steenbrink's conjecture. Math. Ann. **289**(4), 703–716 (1991)
112. Scharf, A.: Faserungen von Graphenmannigfaltigkeiten, Dissertation, Bonn, 1973; summarized in Math. Ann. **215**, 35–45 (1975)
113. Seade, J.A.: A cobordism invariant for surface singularities. Singularities, Part 2 (Arcata, Calif., 1981), Proc. Sympos. Pure Math. **40**, 479–484; Am. Math. Soc., Providence, R.I., (1983)
114. Seade, J.A.: On the topology of isolated singularities in analytic spaces. Progress in Mathematics, vol. 241. Birkhuser Verlag, Basel (2006)
115. Sebastiani, M., Thom, R.: Un résultat sur la monodromie. Invent. Math. **13**, 90–96 (1971)
116. Serre, J.P.: Groupes algébriques et corps de classes. Hermann, Paris (1959)
117. Siersma, D.: The monodromy of series of hypersurface singularities. Comment. Math. Helv. **65**, 181–197 (1990)
118. Siersma, D.: Variation mappings of singularities with 1-dimensional critical locus. Topology **30**, 445–469 (1991)
119. Siersma, D.: The vanishing topology of non isolated singularities, New developments in singularity theory (Cambridge 2000), pp. 447–472. Kluwer, Dordrecht (2001)
120. Sigurdsson, B.: The Milnor fiber of $f(x, y) + zg(x, y)$, manuscript in preparation
121. Steenbrink, J.: Limits of Hodge Structures. Inv. Math. **31**, 229–257 (1976)
122. Steenbrink, J.H.M.: Mixed Hodge structures of the vanishing cohomology, Nordic Summer School/NAVF, Symp. in Math., Oslo, August 5–25, 525–563 (1976)
123. Steenbrink, J.H.M.: Intersection form for quasi-homogeneous singularities. Compositio Math. **34**(2), 211–223 (1977)
124. Steenbrink, J.H.M.: Mixed Hodge structures associated with isolated singularities. Proc. of Symp. in Pure Math. **40**, Part 2, 513–536 (1983)
125. Steenbrink, J.H.M.: The spectrum of hypersurface singularities, Actes du Colloque de Theorie de Hodge (Luminy, 1987), Astérisque **179–180**, 163–184 (1989)
126. Steenbrink, J.H.M., Stevens, J.: Topological invariance of the weight filtration. Indagationes Math. **46**, 63–76 (1984)
127. Stipsicz, A., Szabó, Z., Wahl, J.: Rational blowdowns and smoothings of surface singularities. J. Topol. **1**(2), 477–517 (2008)
128. Straten, D. van: Weakly normal surface singularities and their improvements. Thesis Univ. Leiden (1987)
129. Sullivan, D.: On the intersection ring of compact three manifolds. Topology **14**, 275–277 (1975)

130. Teissier, B.: Cycles évamescents, section planes, et conditions de Whitney. Astérisque **7–8**, 285–362 (1973)
131. Teissier, B.: The hunting of invariants in the geometry of discriminants, Real and complex singularities (Proc. Ninth Nordic Summer School/NAVF Sympos. Math., Oslo, 1976), 565–678; Sijthoff and Noordhoff, Alphen aan den Rijn (1977)
132. Tibăr, M.: The vanishing neighbourhood of non-isolated singularities. Israel J. Math. **157**, 309–322 (2007)
133. Wagreich, Ph.: The structure of quasihomogeneous singularities. Proc. Symp. Pure Math. **40**(2), 593–611 (1983)
134. Wahl, J.: Smoothings of normal surface singularities. Topology **20**(3), 219–246 (1981)
135. Waldhausen, F.: On irreducible 3-manifolds that are sufficiently large. Ann. Math. **87**, 56–88 (1968)
136. Wall, C.T.C.: Singular points of plane curves. London Math. Soc. Student Texts, vol. 63. Cambridge University Press, Cambridge (2004)

# Index

# *LECTURE NOTES IN MATHEMATICS*

Springer

Edited by J.-M. Morel, B. Teissier; P.K. Maini

**Editorial Policy** (for the publication of monographs)

1. Lecture Notes aim to report new developments in all areas of mathematics and their applications - quickly, informally and at a high level. Mathematical texts analysing new developments in modelling and numerical simulation are welcome.
   Monograph manuscripts should be reasonably self-contained and rounded off. Thus they may, and often will, present not only results of the author but also related work by other people. They may be based on specialised lecture courses. Furthermore, the manuscripts should provide sufficient motivation, examples and applications. This clearly distinguishes Lecture Notes from journal articles or technical reports which normally are very concise. Articles intended for a journal but too long to be accepted by most journals, usually do not have this "lecture notes" character. For similar reasons it is unusual for doctoral theses to be accepted for the Lecture Notes series, though habilitation theses may be appropriate.

2. Manuscripts should be submitted either online at www.editorialmanager.com/lnm to Springer's mathematics editorial in Heidelberg, or to one of the series editors. In general, manuscripts will be sent out to 2 external referees for evaluation. If a decision cannot yet be reached on the basis of the first 2 reports, further referees may be contacted: The author will be informed of this. A final decision to publish can be made only on the basis of the complete manuscript, however a refereeing process leading to a preliminary decision can be based on a pre-final or incomplete manuscript. The strict minimum amount of material that will be considered should include a detailed outline describing the planned contents of each chapter, a bibliography and several sample chapters.
   Authors should be aware that incomplete or insufficiently close to final manuscripts almost always result in longer refereeing times and nevertheless unclear referees' recommendations, making further refereeing of a final draft necessary.
   Authors should also be aware that parallel submission of their manuscript to another publisher while under consideration for LNM will in general lead to immediate rejection.

3. Manuscripts should in general be submitted in English. Final manuscripts should contain at least 100 pages of mathematical text and should always include
   - a table of contents;
   - an informative introduction, with adequate motivation and perhaps some historical remarks: it should be accessible to a reader not intimately familiar with the topic treated;
   - a subject index: as a rule this is genuinely helpful for the reader.

   For evaluation purposes, manuscripts may be submitted in print or electronic form (print form is still preferred by most referees), in the latter case preferably as pdf- or zipped psfiles. Lecture Notes volumes are, as a rule, printed digitally from the authors' files. To ensure best results, authors are asked to use the LaTeX2e style files available from Springer's web-server at:

   ftp://ftp.springer.de/pub/tex/latex/svmonot1/ (for monographs) and
   ftp://ftp.springer.de/pub/tex/latex/svmultt1/ (for summer schools/tutorials).

Additional technical instructions, if necessary, are available on request from lnm@springer.com.

4. Careful preparation of the manuscripts will help keep production time short besides ensuring satisfactory appearance of the finished book in print and online. After acceptance of the manuscript authors will be asked to prepare the final LaTeX source files and also the corresponding dvi-, pdf- or zipped ps-file. The LaTeX source files are essential for producing the full-text online version of the book (see http://www.springerlink.com/openurl.asp?genre=journal&issn=0075-8434 for the existing online volumes of LNM). The actual production of a Lecture Notes volume takes approximately 12 weeks.

5. Authors receive a total of 50 free copies of their volume, but no royalties. They are entitled to a discount of 33.3 % on the price of Springer books purchased for their personal use, if ordering directly from Springer.

6. Commitment to publish is made by letter of intent rather than by signing a formal contract. Springer-Verlag secures the copyright for each volume. Authors are free to reuse material contained in their LNM volumes in later publications: a brief written (or e-mail) request for formal permission is sufficient.

**Addresses:**

Professor J.-M. Morel, CMLA,
École Normale Supérieure de Cachan,
61 Avenue du Président Wilson, 94235 Cachan Cedex, France
E-mail: morel@cmla.ens-cachan.fr

Professor B. Teissier, Institut Mathématique de Jussieu,
UMR 7586 du CNRS, Équipe "Géométrie et Dynamique",
175 rue du Chevaleret
75013 Paris, France
E-mail: teissier@math.jussieu.fr

*For the "Mathematical Biosciences Subseries" of LNM:*

Professor P. K. Maini, Center for Mathematical Biology,
Mathematical Institute, 24-29 St Giles,
Oxford OX1 3LP, UK
E-mail : maini@maths.ox.ac.uk

Springer, Mathematics Editorial, Tiergartenstr. 17,
69121 Heidelberg, Germany,
Tel.: +49 (6221) 4876-8259

Fax: +49 (6221) 4876-8259
E-mail: lnm@springer.com